CURRENT ISSUES IN CLIMATE RESEARCH

Proceedings of the information symposium under the EEC programme on climatology, Sophia Antipolis, France, 2 – 5 October 1984

Commission of the European Communities
Directorate-General for Science, Research and Development

Publication arrangements

P. P. ROTONDÓ

Commission of the European Communities
Directorate-General Information Market and Innovation

Commission of the European Communities

Current Issues in Climate Research

Proceedings of the EC Climatology Programme Symposium,
Sophia Antipolis, France, 2-5 October 1984

Edited by

A. GHAZI

and

R. FANTECHI

Commission of the European Communities,
Directorate-General for Science, Research and Development, Brussels, Belgium

SPRINGER-SCIENCE+BUSINESS MEDIA, B.V.

Library of Congress Cataloging in Publication Data
Main entry under title:

Current issues in climate research.

    Includes index.
    1. Europe–Climate–Congresses.   2.   Paleoclimatology–
Europe–Congresses.   3. Climatology–Mathematical models–Congresses.
4. Climatic changes–Congresses.   I.   Ghazi, A., 1940-   II.
Fantechi, Roberto.   III.   Title: Climatology Research Programme at
Sophia Antipolis, France, 2–5 October 1984.
QC989.A1C87   1985   551.6   85-25746

ISBN 978-94-010-8925-8   ISBN 978-94-009-5494-6 (eBook)
DOI 10.1007/ 978-94-009-5494-6

Publication arrangements by
Commission of the European Communities
Directorate-General Information Market and Innovation, Luxembourg

EUR 10225
© 1986 Springer Science+Business Media Dordrecht
Originally published by D. Reidel Publishing Company in 1986
Softcover reprint of the hardcover 1st edition 1986

LEGAL NOTICE
Neither the Commission of the European Communities nor any person acting on behalf of the
Commission is responsible for the use which might be made of the following information.

FOREWORD

The Climatology Research Programme of the Commission of the European Communities started in 1980 after a few years of preparation which followed the concern caused by the European drought of 1976. It was mainly a concern about European land and water resources, which then as never before appeared threatened by climate vagaries. It was mainly an economic concern which led the Commmission to propose, and the Council to adopt, a five-year pilot research programme which in the meantime has proved to be both useful and successful. The best specialists in many interrelated fields were brought together for the first time to join into a unique cooperation effort in the area of European climatological research.

The programme also enjoyed the efficient and competent cooperation of an Advisory Committee for Programme Management (ACPM) which, being essentially composed of climatologists, proved to be a scientific body of high quality, whose work has been one of the factors for the success of the programme.

The Advisory Committee shared with the Services of the Commission the responsibility of preparing a Review Symposium, thought to be the best means of collecting and presenting the results of the research performed during 1980-1984.

This volume contains the proceedings of that Review Symposium of the EC Climatology Research Programme, organized by the Directorate General of Science, Research and Development jointly with the Centre d'Etudes et de Recherches Géodynamiques et Astronomiques (CERGA) at Sophia Antipolis, France, 2-5 October 1984.

The proceedings themselves are introduced by the rapporteurs' reports, covering the four research areas of the Programme : (1) the Reconstruction of Past Climates; (2) Climate Modelling; (3) Anthropogenic Climate Perturbations; (4) Climate impacts.

The rapporteurs had the difficult task of synthesizing the essence of presentations and of reviewing the summaries of research. Their reports are therefore a guideline for the development of the future activities of EC Climatological research.

With the assistance of the ACPM a selection of speakers had to be made out of more than 70 European scientists belonging to the programme. Since no parallel sessions were held, this stringent choice of speakers Ywas necessary in order to respect the time limit and to maintain a balance between the different research areas. Unfortunately, this implied that several competent scientists did not get the opportunity to present their results verbally. Their valuable contributions to the programme, however, were reviewed by the rapporteurs on the basis of the summaries of all the contract work, included in the volume "Final Summary Reports of the Climatology Programme", EUR 9920, 1985, published by the CEC.

The symposium programme also included presentations by invited representatives from the World Meteorological Organization (WMO), the US Dept. of Energy-CO_2 Research Division and the Climate Programme of Switzerland.

We are grateful to all the speakers, participants, session chairmen, rapporteurs and especially to those colleagues who prepared their research summaries for review in order to make the symposium so successful.

In general, the volume contains the highlights of research of the EC Climatology Programme and together with the rapporteurs' reports, provides a sound basis for the evaluation of results obtained during the first phase of the programme.

Ph. Bourdeau,
Director

TABLE OF CONTENTS

CHAPTER I : RECONSTITUTION OF
PAST CLIMATE

CHAPTER II : CLIMATE MODELLING

CHAPTER V : GENERAL TOPICS

O P E N I N G   A D D R E S S

J.P. ROZELOT
Director of Cerga

OPENING ADDRESS

J.P. ROZELOT

Director of Cerga

Chairman of the Organization Committee for the EEC Climatology Symposium

Ladies and gentlemen,

First of all, on behalf of the local organizing committee, the members of
the advisory committee on the management of the climatology programme and
on behalf of Directorate General XII of the Commission of the European
Communities, I would like to welcome you here to Sophia-Antipolis for this
first symposium on the first phase of the European climatology programme.

This symposium is almost unique in bringing together a large scientific
community that includes atmospheric and oceanographic physicists, geochemists
and glaciologists, palaeontologists, meteorologists, astrophysicists, fluid
mechanics specialists, and so on.

I believe that climate forms a unifying theme within Europe not just for the
obvious geographic reasons - the geographic scale of climate is such that it
transcends frontiers - but also because our ten countries possess a research
potential quite adequate to achieve progress in this field which, in the
final analysis, amounts to understanding the changes in the earth's climate
which have taken place in the past, and understanding the mechanisms of today,
in order to have a better conception of the climate of tomorrow.

There are imperative reasons for being able to understand our climate, make
models of it and be able to predict it. These are scientific, art for art's
sake, socio-economic and political in nature. In this respect, we need only
think of carbon dioxide, the increased concentration of which in the atmosphere is likely to lead in less than a century to major climatic changes, or
of the drought in the Sahel, or optimum soil management or even of biomass
resources, dependent as they are on climatic conditions. Many different
scenarios are possible - it will be our job, as scientists, to supply abso-
lutely objective responses.

It is a fact that the climate, defined as the entirety of average values of
variables that cover the full range of both meteorology and the physical
environment and that have been studied in an adequately broad spatial and
temporal context, has undergone significant variations within the historical
period. Man has adapted to his climate but we now see that, by his activities,
he is slowly beginning to modify it. At a time in which we are barely beginning to know what happened in the past, i.e. several thousands or millions
of years ago, and in which we are just beginning the very difficult tasks of
structuring models of what we already know, we are confronted by a new situation - namely understanding how human activity will upset climatic zones
which have up to now remained stable, and thus induce other climatic changes

with effects that are difficult to predict. The current and acute problems facing us are those of desertification, the advance or retreat of glaciers and ice caps, deforestation, the progressive destruction of soils etc.

This symposium has been organized around the various themes I have just mentioned, these being :

- the reconstruction of former climates
- climatic modelling
- human disruption of climate
- impact of climatic changes.

These themes are based on a whole package of scientific considerations that are, in turn, now well established as a result of the following factors - although the list is by no means complete :

- The progress made over recent years in dating land and sea deposits have made it possible to achieve a better understanding of what happened during deglaciation and for example, to evolve new theories.

- More and more powerful computers have been developed (for example CRAY 1) and have thus made it possible to carry out increasingly complete numerical simulations.

- Ever increasing quantities of data have been made available by space technology so that virtually complete and continuous monitoring of the Earth's atmosphere is now possible. Although a few days will not be sufficient to examine all these problems, I am convinced that the discussion will be fruitful. We have not reserved a great deal of time for questions in order to allow the largest possible number of people to present the results of their contract. Nevertheless, there will be time for questions, since the discussion they generate sheds light on the subject. It is also our intention that each session be analysed and summarized by a rapporteur so that a general report can be made at the end of the symposium to serve as a guideline for our representatives in Brussels in drawing up the next climatology programme.

The European Community cannot rest on the laurels of the first phase initiated in 1981. Indeed, a European campaign of oceanographic studies (TOPEX, POSSEI-DON, etc.) has been launched by the European Space Agency, a worldwide climate observation experiment is underway, and various international programmes under a variety of auspices are taking shape. So it is also my earnest hope that Europe, by its own efforts, will lead the world in the field.

I will, if you will permit me, make a small suggestion at this juncture : I would like our work to result in scientific approaches that are original and promising - mouth - wateringly so, in fact.

One of these could be the study of how changes in the sun affect the upper layers of the Earth's atmosphere. It is very striking that, apart from a rash of studies correlating sunspot activity with climate - or even with the production of burgundy - that we now know have to be approached with

circumspection, little or no physics has been done on time scales between
100 and 10 000 years although these are precisely the periods of interest
to palaeoclimatologists in determining the evolution of the sun. We know
nothing of the history of the sun which, to put it midly, must have had some
influence on the history of our climate. The theory that the sun was weaker
some millions of years ago, perhaps by 30% compared to the present day, should
encourage us to pose further questions. What was it doing 18 000 years ago,
during the glacial period?

One other approach could be satellite climatology which has been rather
ignored in this symposium. Satellites have, however, been engaged in the
synoptic collection of surface data over a period of several years that
is now comparable with the characteristic evolutionary periods of oceans :
the "climatic signal" can be read - it just has to be extracted from data
now scattered over most of the globe. I am also thinking of the significance
of work monitoring the thickness of sea ice and, especially, of ice caps since
these are very sensitive climatic indicators, as has been particularly well
indicated by the dramatically different conditions of the various glacials.

There are no limitations on our attempts to imagine the climatology of
tomorrow, but our main task will be the organization, around several major
and incontestable concepts, of our future programme.

RAPPORTEUR'S REPORTS

- Palaeoclimate in the E.C. climate programme
- Climate modelling
- Anthropogenic climate perturbations
- Impact of climate and climate variations on European resources

PALAEOCLIMATE IN THE E.C. CLIMATE PROGRAMME

J.DUPLESSY[1] and W.A. WATTS[2]

The study of marine and terrestrial sedimentary records reveals the climatic evolution of the last million years as a series of cyclic alternations of glacial and non-glacial climates. Climatic conditions similar to those of today are of brief duration in comparison with the much longer and somewhat complex glacial portions of each cycle. As little as 10,000 years in each cycle of approximately 125,000 years is occupied by temperate climates. The Holocene has already occupied some 10,000 years. Some data suggest that climatic changes may be rather abrupt, so it is of special interest to focus on the detailed events during change from one climatic condition to another and to generate precise time-scales so that rates of change can be established accurately. Ice-core data record with precision the abruptness of transition from one environment to another, both in phases of transition and during the last glaciation. Changes in ocean surface temperature and CO_2 variation may be detected in ice-cores. Some of the cyclic climatic variations observed appear to be correlated with orbital variations which are responsible for subtle but significant changes in radiative boundary conditions. The extent to which orbital variation provides a comprehensive explanation for Quaternary climatic evolution still requires further mathematical investigation which would also consider apparent short-term periodicities on 2500 and 9000 year frequencies. Climatic modelling is intimately associated with the study of Quaternary climates. At one level models of today's climate reveal inadequacies in understanding the physical processes which govern climate, at another general models of former climates reveal anomalies between the inferred climate and the data from fossils and sediments which point to areas of misinterpretation or error requiring further study.

Aspects of many of the topics referred to in the introductory paragraph were subjects for study during the first stage of the E.C. Climate Programme. The last 130,000 years on land in Europe and in the eastern North Atlantic and Mediterranean attracted particular interest because it permits the study of the last temperate period, the Eemian, and the processes and rates by which it was succeeded by glacial conditions. On land in southern Europe forest disappeared progressively and was replaced by steppe in a cooling less humid climate. Marine records and pollen data tfrom the same deep-sea core for the period show the correlation between changes in forest composition on land and isotopic composition of the eastern North Atlantic. They allow the inference that ice formation was beginning at high latitudes in North America while forested phases still occurred in Europe, a lag in the European response also suggested by modelling experiments. The glacial vegetation of southern and Mediterranean Europe is a subject of continuing investigation. The largely herbaceous vegetation with much grass, herbs and Artemisia (sage) points to a steppe-like environment, but a modern analogue does not suggest itself and the nature of the climate of which this peculiar

1   Centre des Faibles Radioactivités, CNRS-CEA, Gif-sur-Yvette, France

2   Trinity College, Dublin, Ireland.

vegetation is a reflection merits continuing study. In general, the utility of pollen surface-sampling as a data base for the transfer of pollen data into climatic data is confirmed. The detailed events and timing of deglaciation in northwest Europe and the adjoining oceans is still of interest. The timing of events is not established with critical accuracy and the geographic distribution and expression of response to the Younger Dryas cooling episode is not yet fully understood. The "signal" from ocean cores off southwest Europe appears to be stronger than on land. Studies of sapropels in marine cores from the eastern Mediterranean off Crete have contributed importantly to the new understanding that a stronger expression of the African monsoon, expressed as much greater Nile floods than at present and the northward extension of Sahel vegetation belts, was a characteristic of the early Holocene. In addition to data obtained by laboratory techniques from sediments, valuable progress in the reconstruction of past meteorological conditions over the last centuries has been made from documentary sources while studies of recent variability in upwelling intensity in equatorial areas may provide an explanation for some aspects of climatic change.

For the future several points might be noted which emerge form the present research programme. There is a high degree of complementarity between ice-core, marine-core and long land-core studies. Oxygen-isotope and pollen stratigraphies provide a unifying framework within which data can be correlated to answer many important questions. For many purposes more exact dating is required, by radiocarbon for correlation of relatively recent sediments, by oxygen-isotope stratigraphy for older marine cores. It would now be possible, on the basis of existing skills and experience, to develop a large research programme on the dynamics of the last glacial/interglacial cycle involving co-operation by modellers. Work should be planned on a continental scale with specific international co-operation between Community scientists. Finally, the development of a data bank for marine information has been very useful both for people describing and for people modelling climatic conditions. The establishment of a similar data bank for pollen data would considerably advance terrestrial studies.

CLIMATE MODELLING

A. Gilchrist[1], K.H. Hasselmann[2]

INTRODUCTION

Climatologists attempt to synthesise their knowledge and understanding of basic physical processes that are important in determining
climate by means of models which can simulate climate or aspects of it.
The models come in many forms : from simple formulations involving little
calculation to large computer programmes which aim to recreate the global
climate from the fundamental equations of physics and require minimal
initial information about the state of the components of the climate
system.   The range of interests within the modelling community as regards
both the models used and their applications is consequently very wide.

In establishing the European climate programme, it was
recongized that research in climate had to be pursued on a broad front.
Thus, for example, it is not scientifically acceptable to seek to investigate the effects of enhanced carbon dioxide concentrations on climate in
isolation, as even a cursory examination of the problem demonstrates that
it involves virtually all the problems which arise in understanding the
unperturbed climate.   Consequently, it must be tackled within the context
of broad climate programme.   The contracts which were awarded therefore
cover a wide spectrum of modelling activity and went to a selection of the
foremost modelling groups in Europe.

REVIEW

The first, outstanding challenge that faces climate models is to reproduce
the existing global climate, including its variability, realistically.
This is a prerequisite to establishing their credibility in potential
applications ; for example, simulating past climates and predicting the
future climate.   Considerable progress has been made towards this aim,
particularly through the use of atmospheric global circulation models
(GCMs).   They are in essence similar to numerical weather prediction
models used operationally by national meteorological services, and their
development and refinement have benefited immensely from the ongoing
intensive efforts to improve weather forecasts.   Being essentially
atmospheric, relevant properties of other components of the climate system
- the oceans, land- and sea-ice, and land-surface conditions particularly
- are prescribed or, possibly, allowed, at most, to respond to atmospheric
conditions in a simple way.   General circulation models of this kind are
capable of simulating, with considerable realism, the main features that
are responsible for the observed global climate and its regional variations. · However, it must be recongized that because the oceans,
cryosphere and bio-chemosphere - all of which have a pronounced influence
on our climate - are not modelled realistically in such numerical
atmospheric circulation simulations, they cannot be regarded as climate

---

1   Meteorological office, Bracknell, UK

2   Max Planck Institut für Meteorologie, Hamburg, FRG

simulations as such. They describe only the adjustment of one component
of the climatic system - the atmosphere - for prescribed states of the
other components, and cannot address the question of the structure of the
overall equilibrium (or evolving non-equilibrium) climate state.

ATMOSPHERIC MODELS

        Atmospheric GCMs are used in several of the investigations
reported in this volume. Some outstanding climatic anomalies of recent
years have been attributed to unusual conditions at the earth's surface,
which are capable of reacting back on the atmospheric circulation. For
example, there is evidence that the early drying out of the land surface
was a factor in prolonging the 1976 summer drought in parts of northern
Europe, and that the exceptional warmth of equatorial Pacific Ocean
surface waters in 1982-83 produced widespread climatic anomalies. In the
paper by Slingo (p.139) GCMs are used to explore the sensitivity of the
simulated European climate to minor variations in the formulation of the
model or in conditions at the earth's surface. It is shown that European
rainfall is dependent on the land wetness, and that the simulated rainfall
distribution changes substantially when the models' parametrization of
cloud/radiation interaction changes. The paper by Palmer (*) is
concerned with the impact of the warm Pacific sea surface temperatures on
the atmospheric circulation. It provides evidence that they probably
influenced conditions over Europe.

        Davies(*) used a rather simple form of GCM to investigate
aspects of the large-scale global circulation. The results indicate a
clear sensitivity of the flow to the presence of momentum barriers, to
variations in sea-ice cover and to the introduction of a simple mixed
layer representation of the upper ocean, interacting with the atmosphere.
If confirmed independently by other models, they point to important
factors that must be taken into account in attempts to predict European
climate variations on time-scales of weeks to a season. The work of
Opsteegh(*) is concerned with a similar time-scale and uses an even
simpler model to assess the extent to which the influence of a surface
anomaly on the atmosphere (e.g. the study by Palmer (*) might be
masked by the atmosphere's inherent variability. For practical forecasting this is a crucial issue. The results presented indicate only a
small direct influence of the anomaly on the extra-tropical circulation,
in apparent contradiction of the conclusion from more complex models.
Clarification of the reason for the differences is obviously desirable.

        Although atmospheric GCMs have been remarkably successful in
reproducing certain aspects of the global circulation, their simulation,
when examined in greater detail, show many deficiencies, particularly when
attention is focused on specific areas, such as Europe. The deficiencies
may be largely attributed to our lack of scientific understanding of many
basic physical processes, especially concerning the interaction of the
atmosphere with the earth's surface whose enormous variability in
roughness, wetness, colour and elevation, gives rise to formidable
difficulties. The scope for improving model formulations is great, and
investigations on how they can be achieved are an important area of
climate research which figures prominently in the World Climate Research
Programme approved internationally and coordinated by WMO and ICSU. Thus,
Slingo (p. 139) draws attention to deficiencies concerning the strength of
mid-latitude westerlies which are characteristic of GCMs in general. An

improvement in perfromance in this respect is the essential preliminary to using the model more exetensively to study the causes of year-to-year variability in European winters. Also, and perhaps as part of the 'westerlies' problem, it is known that models have difficulty in handling the phenomenon of 'blocking' whose presence or absence in particular years profoundly influences European winters. The study by Speranza (p.132) is a significant contribution to understanding the mechanisms which initiate 'blocking', and could lead on to an improvement in the treatment of orography in models.

An aspect of models which can clearly be made more realistic, is their treatment of conditions at a land surface. Renner (p. 219) describes a soil model for incorporation within GCMs. This is an important contribution to experiments on European climate, such as those of Slingo (p.139), where the known sensitivity to surface conditions indicates that the latter have to be simulated accurately.

OCEAN MODELS

In contrast with the generally healthy situation which prevails concerning our ability to model the general circulation of the atmosphere, knowledge and understanding of the other components of the climate system are still very deficient. For example, whereas data to define the global atmospheric state are available for the purposes of weather forecasting more or less continuously, data for the oceans, being as yet dependent for the most part on observations from specialized ships, are fragmentary, and in some parts of the ocean, non-existent. A better understanding of the oceans is dependent on obtaining a better coverage of observation (possibly from satellites), so that such important climatological quantities as the transport of heat from low to high latitudes within the oceans can be evaluated more accurately.

Two of the investigations reported in this volume are concerned with obtaining data that are essential for improving understanding of the global oceanic circulation. Jarrige et al (*) are concerned with observations for the tropical Atlantic, which because of the paucity of islands is less well observed than the equivalent Pacific area in which the strong El Nino sea surface temperature variations are known to occur. For Europe, the Atlantic variations are probably the more important, even though the geographical extent of the anomalies is likely to be less. This work is providing a valuable source of data for the prediction of European climatic variations. Gascard's (p.207) area of interest is also the Atlantic but is focused on the northmost part of it, close to the marginal sea-ice zone. The phenomenon of deep ocean convection, which carries cold surface waters deep into the ocean interior, is very significant for the whole oceaninc circulation. In the northern hemisphere, its occurrence is confined to locations near Labrador and Greenland, where, in winter, there is strong cooling of the ocean surface. Part of the work has been undertaken in the context of the MIZEX international experiment, whose object is to study ice-atmosphere interaction in the marginal ice zone. This, too, is accepted as a crucial area of research for global climate.

More theoretical investigations on the oceanic circulation and its interaction with the atmosphere are represented in the papers by Nihoul (p.163), Gill and Killworth (p.195), and Hasselmann (p.172).

Nihoul's study seeks a means of overcoming the problem encountered by
ocean modellers when they try to represent the 'forcing' of the ocean
circulation by the atmosphere. Motions in the oceans are slower than in
the lighter medium, and the time-scales for the life-span of eddies is
measured in terms of months, rather than, as for the atmosphere, in days.
Nihoul has therefore recast the basic equations to use values of the
variables averaged over a month and "synoptic Reynolds stress" derived
from meterological analyses. If this simplification is demonstrated to be
generally acceptable, it would have a substantial impact on methods for
modelling the oceans. The work reported by Gill and Killworth covers a
variety of topics using a hierarchy of ocean models with the object of
exploring some of the fundamental issues which need to be clarified and
understood to improve ocean modelling. The mechanisms which drive the
mid-latitude gyres, and the eddies and instabilities within them, have to
be understood before they can be successfully modelled ; high latitude
convection (considered from a less theoretical point of view by Gascard
(p. 207) and the influence of bottom topography are essential ingredients
in driving the observed ocean circulation : the sea surface interaction
is obviously a highly significant quantity in simulating climate, and its
variation due to the passage of mid-latitutde storms or in the tropical
oceans where 'El Nino type' events are observed, are matters that require
theoretical interpretation. These topics being investigated by the
Cambridge group, are, in an international sense, among the most active at
the present time. The investigations reported here by Hasselmann (p.172)
deal with global ocean circulation models, and their application to carbon
cycle modelling, and with statistical and inverse modelling techniques.
The global circulation model developed at Hamburg uses simplified
dynamics, appropriate for large scale, quasi-geostropic flow, enabling
economic computations of the mean ocean circulation and the time dependent
response of the ocean to external forcing. The model has been used to
construct a global carbon cycle model, including the storage and transport
of carbon in the ocean, carbon transformations through oceanic biota and
interaction with the terrestrial biosphere (modelled by Lieth's group).
Statistical techniques have been developed to study the response of the
atmosphere to imposed sea-surface temperature anomalies. As already noted
by Opsteegh(*) these may be obscured by the low frequency variability,
which occurs in GCMs equally as in the real atmosphere. A method of
isolating the atmospheric 'signal' has been proposed and tested on some
GCM results. Such an investigation leads naturally to attempts to predict
short-term variations of climate using sea-surface temperature, but also
atmospheric data are represented by mean sea-level pressure values ; the
results for seasonal forecasts showed some skill in the winter and spring
seasons. Similar methods have enabled sea-ice anomalies in the Arctic and
Antarctic to be estimated for up to a season ahead. Inverse modelling
techniques, implying the use of series of observations, combined with
models, to optimise the value of parameters used in the models, have been
used successfully in establishing rates of advection for sea surface
temperature anomalies in the Pacific. These results and the methods being
developed at Hamburg seem likely to influence the evolution of practical
forecasting techniques, applicable to periods of a season to a year ahead.

PALAEOCLIMATIC MODELS

When we turn to understanding climate changes over very long
periods of time, generally including glacials and interglacials, the
modelling required is of a different kind. The global climatic state has

to be represented by only a small number of variables and the aim of the models is usually to test the reasonableness of specific hypotheses. For example, Berger's investigations (p.16) are on the effects of changes in the earth's orbit. They influence the solar energy reaching specific parts of the earh's surface, with periods of around 40, 23, and 19 Kyr, and so change the climate. A single-column, 'radiative-convective' model interacting with oceanic mixed-layer and sea-ice models is being used to test whether this effect, allied to internal feedbacks caused, for example, by the increase of albedo due to snow and ice, can explain the observed palaeoclimatic record. Nicolis' work(*) is concerned with the more general theory of long period climate changes, and the extent to which they might be produced by changes internal to the climate system. Energy-balance thermodynamic models are used, and the possibility of multiple equilibrium, such as might produce an alternation between cold and relatively mild climatic states is studied.

DISCUSSION

The preceding brief review demonstrates that the spectrum of modelling topics included in the EC Climatology Programme is very wide indeed. Even so, it should be pointed out that some very significant areas of research, for example, modelling the behaviour of glaciers and sea-ice, or stratospheric investigations involving radiative, chemical and dynamical interactions are not represented.

Nevertheless, it emerges rather clearly that in addition to the specific results of the research which is supported, one of the most useful functions an EC programme can perfom is to bring modellers into contact and encourage a flow of information between them, so that they become aware of work going on elsewhere in Europe and have the opportunity to compare results. It is now widely recognized that in this area of research, where it is extremely difficult to investigate mechanisms by normal scientific techniques because the climate system itself can neither be experimented with nor reproduced adequately in the laboratory, scientists can learn a great deal by close intercomparison of sets of results derived independently. In this way sensitivity can be more quickly established, and crucial differences between model results and observations isolated and understood.

There is already good evidence that modellers throughout Europe have been brought together to their general benefit through the EC Climatology Programme. For example, climate modelling groups in France, F.R.G. and the U.K. are now in regular contact ; and co-operative programmes have been set up between groups in France and Belgium. These contacts need to be further strengthened and, indeed, in any succeeding programme, it may well be acceptable and desirable to make the co-operative nature of research a precondition for EC support.

However, it is not only among modellers that information exchange needs to be encouraged. Because of the very general basis of most climate models, they may be applied to many different problems. They can provide a means for testing the adequacy of current scientific understanding, pin-point areas of ignorance, and, when fully developed, have the essential requirements for predicting the future climate. Modelling the climate system is therefore an important central unifying activity within a scientifically respectable climate programme.

Among the uses to which climate models may be put are : first, in conjunction with palaeoclimatic evidence, to present a coherent picture of past global climatic states ; second, to provide data on the posible impact of climate changes due to man-made increases in the concentration of carbon dioxide ; and third, to provide a framework for the assimilation and interpretation of the large quantities of data which will be provided by earth observing, climate oriented satellites in the next years. As regards palaeoclimatology, the contacts that have been established in the EC programme have already proved fruitful. Particularly through the joint meeting of the "Reconstruction of Past Climates" and "Climate Modelling" contact groups at the PRaM meeting (Paleaoclimatic Research and Models, 1983, ed. A. Ghazi, Reidel) palaeoclimatologists have become more aware of the possibility of interpreting their observations by means of models ; equally through this interaction, modellers have learned about past climatic states, the simulation of which provides additional tests of the applicability of their models. Concerning the prediction of climate with a high-CO_2 atmospheric concentration, the work by Mitchell (p. 228), using a GCM, has provided climatic data for climate impact evaluation. Although the model is not yet sufficiently proven for its results to be taken as 'predictions' rather than as one of many possible sets of estimates, it is useful, and indeed probably essential for the development of a soundly based methodology of impact evaluation, that typical results of this kind be available at an early stage for those studying the impacts of climatic change on society. Again, an interchange of information between the two groups of scientists under conditions which allow misinterpretations of each other's data and motivations to be easily corrected, is a laudable objective, and one that can be achieved through a continuing EC Climatology Programme. The application of climate modelling to satellite data, finally, is an important field which is still in its infancy. It should be considered as a possible candidative objective for a future EC Climatology Programme, particularly in view of the European commitment to launch a satellite, ERS-1, towards the end of this decade, which is strongly directed towards climate applications.

(*)   Final Summary Reports of the Climatology Programme, EUR 9920, CEC, Brussels, 1985.

ANTHROPOGENIC CLIMATE PERTURBATIONS

A. Berger[1], C.D. Schönwiese[2]

1. INTRODUCTION

Twelve papers were presented to summarize the research sponsored by the Commission in order to study the anthropogenic climate perturbations. The results of one additional project belonging partly to the section on Reconstruction of Past Climates and Climate Modelling is also included in this reviewers report as it deals with modelling the radiative impacts of human activities in general.

This session is divided into 3 sub-sections dealing respectively with the climatic impacts of CO_2, with man-made aerosols and with urban climate. The chapter on CO_2 contains the largest number of papers (9); 3 papers are related to man-made aerosols and 1 to urban climatology.

Table 1 gives the list of subjects which have been treated, the principal investigators and the number of the Contract.

In order to better understand what are the expected anthropogenic impacts on climate, the following points require more investigation :

(i) the atmospheric concentrations of man-made versus natural polluants and their evolution in time.

(ii) the geochemical cycles (*)

(1) U.C.L., Institut d'Astronomie et de Géophysique Georges Lemaître, 2, Chemin du Cyclotron, 1348 Louvain-la-Neuve, Belgique.

(2) Institut für Meteorologie und Geophysik, Johann Wolgang Goethe-Universität, Feldbergstrasse 47, 6000 Frankfurt a.M.1, Deutschland.

(*) as strongly recommended by Department of Energy, Carbon Dioxide Climate Program, Cfr F.A. Koomanoff's discussion at the CE symposium Oct. 1984.

Table 1. Problems related to Anthropogenic Climate Perturbations as analysed during the CEC Climatology Research Programme 1981-1985.

(*) Summaries of these projects are published in Final Summary Reports of the Climatology Programme, EUR-Report 9920, CEC, Brussels, 1985.

Sub-sections		Principal investigators	Contract n° CLI
1. CO_2	1.1 Long-term and Short-term variations	Lambert	003-F
		Raynaud	013-F
	1.2 CO_2 cycle fresh water system	Degens	041-D (*)
	ocean surface	Merlivat	053-F (*)
	budget	Reiter	066-D (*)
	modelling	Kolhmaier	043-D (*)
	1.3 CO_2 Signal in Stratosphere	Labitzke	051-D (*)
	1.4 Modelling – radiative impacts of CO_2 and other trace gases	Crutzen	045-D (*)
		Berger (::)	026-B
	– impact of CO_2 increase on European Climate	Mitchell	030-U.K
2. Man-Made Aerosols	Saharan dust	Prodi	068-I (*)
	Radiative impacts	Hinzpeter	044-D
3. Urban Climate		Colacino-Dalu	019-I (*)

(::) belongs partly to Reconstruction of Past Climates and Climate Modelling

(iii) the aerosol optical properties

(iv) time dependent models which include the interactions between the relevant parts of the climate system and the polluants cycles; they must allow to estimate the transient response of the climate system to any perturbation (3).

(v) detailed climatic, CO_2 ... analogues as they can be found in the Earth historical and geological past.

(vi) detection of the climatic variation signal against the background noise.

Table 1 clearly shows that the most important components of the international effort conducted for a better understanding of the anthropogenic impacts on climate are covered by European laboratories. Among them, most original significant results were obtained for (1) past CO_2 atmospheric concentration as revealed by ice cores, (2) CO_2 cycle, (3) detection of a CO_2-induced climatic signal, and (4) CO_2 GCM experiment where SST were increased to take into account the feedback of the CO_2 increase on the sea temperature. 2-D time dependent climate models including man made polluants cycles are also under construction. A high priority must be given to such research projects to continue.

2.   CO2 AND CLIMATE

Short term variations of the CO_2 atmospheric concentrations at Amsterdam Island (37°S, 77°E) have been investigated by means of a non-dispersive IR analyser (Lambert). They are related to the transport of South African air masses and front passages, where 222 Rn is used as a tracer. The daily cycle is entirely attributed to the photosynthesis of the island vegetation. The trend shows a weak increase of 0.8 ppm between 1981 and 1982 and a larger one (1.6 ppm) between 1982 and 1983.

Air bubbles trapped in glacial ice have allowed the reconstruction of longer term variations of the worldwide atmospheric CO_2 concentration, the relative accuracy being now less than 3% (Raynaud). The estimates from Dome C ice core (Antarctica) indicate a pre-industrial value of 258 ppmv with significant natural fluctuations of the order of 10 ppmv or more between 1500 and 1850 AD. At the last glacial maximum the atmospheric CO_2 concentration was as low as 200 ppmv. It then increased before or simultaneously to the warming associated to the end of the last ice age before reaching, in 2 steps (?), the Holocene mean value of 270 ppmv.

Local modern measurements of atmospheric CO_2 must now be extended to South Africa, Europe and oceanic upwelling areas. In order to dissociate causes and effects in long term climatic variations, refined detailed analysis of CO_2 ice cores must be continued at the ice age/Holocene transition and extended over the whole last glacial/interglacial cycle with a priority given to the 6/5e and 5e/5d transitions.

One more topic concerns the global carbon cycle and its particular components. For the latter, measurements of hydrochemical parameters (carbon, nutrients, etc.) in the major rivers of the world are performed for several years (Degens, international cooperation). The flux of organic carbon in rivers is assessed as well as its deposition in coastal areas. The total global carbon flux in rivers is estimated to be in the order of 10^9 tons of carbon (the total anthropogenic flux from fossil fuel energy consumption is of the order of 5×10^9 tons C).

Furthermore, experiments (under controlled conditions) are performed in order to study the exchange of CO_2 and other gases through the ocean-atmosphere boundary (Merlivat). From gas injections, time-dependent gas distribution functions are measured by means of spectrometric and chromatographic methods. It is concluded that the ocean has absorbed at least 45% of total CO_2 injected to the atmospheredue to fossil fuel (previous assessments are 30-60%) burning.

At three stations in the German Alps (Garmisch, 0.7 km, and neighboured mountain stations 1.8, 2.7 km height) the atmospheric CO_2 concentration is measured along with meteorological parameters (Reiter). The daily and annual CO_2 variations are studied in relation to the underlying surface conditions of the biosphere (vegetation type etc.).

The model aspect (Kohlmaier) is taken into account by the development of a global carbon cycle box model of the atmosphere-ocean-biosphere system, where the biosphere is subdivided into several boxes (vegetation types, geographical latitude) and several ocean layers are discerned (diffusion effects involved). This model shall be applied to future energy scenarios and to global temperature variations (past and future) in order to assess the magnitude of the carbon fluxes and of the variations in the atmospheric CO_2 concentration.

Further investigations of both more detailed measurements and improvements of carbon cycle models are planned. It is recommended to compare instruments, methods and results and to improve the available box models of particular CO_2 cycle components in order to integrate them in an appropriate improved global carbon cycle model and to couple them to climate models.

Quite another problem is the "detection" of the CO_2 signal in observational data. This is very important to realize the actuality of the CO_2 problem. Stratospheric radiosonde data available since 1962 are analysed in respect to temperature variations (30 hPa level emphasized (4)). When the annual cycle and the QBO (quasi-biennial oscillation) are removed from these data, two trends ($0°$-$60°N$) can be seen : a cooling until c.1973 and a warming later on. Although the volcanic signal can be clearly identified in some cases, the "detection" of the CO_2 signal in these data remains uncertain.

The extension of this stratospheric data base (daily instead of monthly data, higher levels) is planned. It is recommended to apply statistical multiple regression techniques in this forcing study and to compare the results with similar investigations going on for the tropospheric data.

───────────────────────────

(4) See also Stratosphere Report (1984) of the European Communities R&D Programme Environment (Ed. A. Ghazi).

Two 1-D steady state Radiative convective photochemical models show a global temperature increase of 1.9°C for a CO_2 doubling and of 3°C when the combined effects of other anthropogenic trace gases are taken into account. The difference in their lapse rate parameterization and in their methods to compute the equilibrium temperature profile could explain why the maximum warming occurs in the lower troposphere of Bruehl-Crutzen model but not in the Tricot-Berger one.

A 4xCO_2 and increased sea surface temperatures experiement simulated with a 5-layer GCM (Mitchell), shows an increase in surface temperature throughout the year, which was most marked in winter, particularly in Northern Europe. Precipitation decreased in Southern Europe over the whole year, but increased in Northern Europe in winter. Although this study indicated a drier surface in summer over the EC area, the results should be treated with great caution in view of the assumptions and simplifications made in the model.

1-D and 2-D time dependent models with improved hydrological cycle should be built and intercomparison of R-C as well as GCM models encouraged. Investigations towards a realistic parameterization of the interactions between ocean-atmosphere and sea-ice must continue and meso-scale models must be developed for a better interpretation of the GCM results at the scale of Europe. A careful analysis of changes in the frequencies of extreme values of observed/model variables should continue.

3.   MAN-MADE AND NATURAL AEROSOLS

Sahara dust transported over Italy (its optical and size characteristics) is measured by means of multi-wavelength photometers at mountain stations and a climatology of the Angström parameters was performed (Prodi). The importance of precipitation scavenging and wet removal was confirmed.

Man-made aerosol particles play also an important role in global climate changes, as indicated by a 2-D model which simulated the steady state concentration distributions of SO_2 and aerosols in the troposphere (Hinzpeter). Anthropogenic emissions of SO_2 and particle enhance the particle concentration at nearly all latitudes and heights in the northern hemisphere, causing an increase of local planetary albedo even if 20% of the particles are assumed to be soot. If all feed-back mechanisms are neglected, this net flux reduction of solar radiation at the ground (0.7 to 1.1 Wm^{-2}) compensates more or less the net flux increase (1.2 Wm^{-2}) due to CO_2 doubling.

Aerosol circulation must be included in usual 2-D R-C models to properly take into account all feedbacks and the radiative treatment of atmospheric man-made particles.

4. URBAN CLIMATE

The urban climate (with special reference to the Roma area) is studied by means of micro-meteorological measurements in the atmospheric boundary layer and of statistical analysis of the climatic time series which include also the long-term records (e.g. Roma temperature since 1782 AD). The urban heat island effect can be satisfactorily simulated by means of a relatively simple primitive equations meso-scale circulation model.

It may be possible that more complicated problems, like precipitation and cloud patterns within an urban area, can be approximately simulated by means of appropriate simple meso-scale models. It is recommended, however to better use sophisticated meso-scale circulation models (in an international cooperation) for the simulation of the urban climate island effect. Statistical studies of particular local (regional) data set are useful for the verification of the circulation models.

5. CONCLUSIONS

The report of the EC symposium session on "anthropogenic climate perturbations" covers only part of the overall problem as it is studied in Europe. A conference, which would present the highlights of the results related to this topic and obtained from all the national climate research activities in CEC countries is advisable.

The following list of questions and goals may characterize the complexity and importance of the problem.

1. Which human activities do we have to consider in the order of priority? (Fossil fuel and energy use, agriculture practices, water use, deforestation etc.).

2. What are their observed effects? (Atmospheric CO_2 increase, corresponding increase of other trace gases and aerosols, variations in land surface processes and related urban climate, etc.).

3. Try to understand the related processes by means of data collection, data analysis and modelling. This includes also paleoclimatological techniques (proxy data). (Carbon cycle, meso- and macro-scale climate simulations, etc.).

4. Try to understand the role of anthropogenic climate perturbations among all climate forcing mechanisms (Signal to noise ratio, natural CO_2 and climate variations, combined effects in time-dependent way, etc.).

5. Discussion of future extrapolations, based on all these ideas and on possible scenarios.

It is evident that the investigation of anthropogenic perturbations of the climate requires very different techniques. The knowledge of human perturbation on climate is a prerequisite for a better understanding of the impacts of climate variations on man but needs climate data and climate models to quantify the phenomena. Therefore this session deals with a privileged subject matter which is situated at a crossing between reconstruction of past climates, climate modelling and impact studies.

Impact of climate and climate variations on European resources
C. Schuurmans *

1. ## Why

    In establishing the content of the Climatology Research Programme
    of the Commission of the European Communities, it was decided that
    the impact of climate and climatic variations on society should be
    an integrated part of the research activities. Although it was
    recognized that this part is quite different from the study of
    climate itself (reconstruction of past climate, climate modelling
    and the sensitivity of climate to anthropogenic influences), in
    order to cover the field of climatology as completely as possible,
    impact studies had to be included. Additional considerations were:
    1. the need for such studies in view of future practical
    applications and 2. the possible advantage of having such studies
    performed in close collaboration and contact with the study of
    climate itself.

    In reviewing the proposals it turned out however that it was not at
    all easy to select projects, acceptable for funding under this
    item. This was due to various factors like, lack of clear
    definition of climatic impacts, uncertainty about the value of the
    research on a European scale, getting involved into research in
    other disciplines (e.g. hydrology) etc.

2. ## Status

    With these difficulties in mind, one must say that the outcome of
    the first phase of the programme, · as far as impact studies is
    concerned, is certainly promising. The programme included 13
    different projects. Impacts on at least 4 different sectors of
    European society have been studied. These 4 sectors are :
    agriculture, hydrology, energy and to a smaller extent activities
    at sea.

    Such sectors or economic activities are commonly called resources,
    but in some sense of course climate itself may also be called a
    resource. Although its products can be sold less easily as those
    of the more conventional resources, like agriculture, etc.

    In the foregoing, impact of climate and impact of climate
    variability or change have been taken together. In most impact
    studies, however, one has to do with the latter. More specifically
    society is mostly concerned with the impact of deviations of the
    so-called climate and furthermore with the question of what happens
    if certain changes of our climatic regime are to be foreseen.

*   Royal Netherlands Meteorological Institute, De Bilt, The Netherlands.

In this respect the impact studies of our programme are indeed of both kinds : studies involving an analysis and description of climatic events which are known to have caused certain societal impact (like e.g. the potato blight disaster in Europe in 1845/46) and studies trying to answer the question of what will happen to European activities when the foreseen climate change due to the increase of atmospheric CO2 is going to materialize.

3. Results

To be sure, no definite answers have been obtained in neither of the two types of studies (although many useful order of magnitude estimates became available. The reasons are manifold (lack of observational data, uncertainty or ambiguity of "predicted" climate change, etc.) but the most important one is that the field in itself is in a pioneering stage. There is only a beginning of agreement on the methodology to be used.

A useful and effective methodology which has been applied in a number of studies described in this section is one in which a model is built for a certain activity or product (like crops e.g.) which can be investigated upon its sensitivity to climate variables (like temperature, precipitation etc.). The outcome of such sensitivity studies then gives an answer as to the likely impact of climate variability, given some estimates or predictions of climate variability or change (also called scenario's). Such models concerned may be of a statistical or more or less physical nature (also called mechanistic). In any case however, climatic variables are a few input parameters among a large number of other variables of quite different nature.

The first phase of the EC-climatology research programme has been successful in collecting, building and testing of impact models, some of them being ready for application, given suitable long-range weather and climate predictions. These are crop yield models, hay drying models and general ecosystem models. Evidently, the development of such models for agricultural or more general vegetational purposes has received much more attention, not only in Europe but also elsewhere, than for other applications. Water, especially drinking water and energy resources have been much less studied in their relation to climate variables.

4. Omissions

A field which is notoriously lacking in this programme is human health and human well being (including recreation). Of course some other important sectors are also missing like : fisheries, transportation and trade in general. Development of climate-oriented models for each of such activities seems to be a requirement for the future.

Having these models available however, the execution of sensitivity studies, in finding out what climate variability might to do the activity or product concerned, is not a trivial thing. Already in the development stage of the model one has to make sure that the climate variables used are readily available or in case they have to be produced by a climate model, that such simulations are feasible. Here exactly is the meeting ground between scientist of the disciplines concerned and the climatologist, or if it can be one person, he/she should know as much as possible of both fields.

Further detailed suggestions for future research in this field, are given in the individual reports.

5.   CO2-effects

Two studies in the programme explicitly aimed at an assessment of the impact of the increase of CO_2 in the atmosphere. Due to conflicting climate scenarios, especially regarding moisture conditions over the European territory only order of magnitude indications of some socio-economic impacts could be obtained. As such, these figures are alarming enough to take the CO2-problem seriously. However, to solve the problem, improvement of predictions of the relevant climate variables on a European scale, undoubtedly must have a higher priority, than improving impact studies.

6.   Conclusion

The results of the climatic impact studies as available now cannot be considered as operational tools for decision making on a European scale. However, apart from results which possibly can be applied nationally, the programme as a whole has shown that impact studies successfully can be incorporated in a climatology research programme. Further collaboration, as established in the EC-programme, is therefore to be recommended.

CHAPTER I : RECONSTITUTION OF
PAST CLIMATE

- The impact on Europe of large-scale climatic changes:
  the onset of glaciation and the last deglaciation

- Climatic history from ice core studies in Greenland
  - Data correction procedures

- Air-Sea interaction during the last 100 years and time scales
  of climatic fluctuations

- Modelling the astronomical theory of paleoclimates in the time
  and frequency domain - An example of the relationship between long
  term and short term climate changes

- Climate reconstruction using historical sources

- Stages of climatic change from full glacial to holocene in northwest
  Spain, southern France and Italy: a comparison of the Atlantic
  coast and the Mediterranean basin

- Paleoclimatic reconstruction in Belgium and in Greece based on
  Quaternary lithostratigraphic sequences

THE IMPACT ON EUROPE OF LARGE-SCALE CLIMATIC CHANGES :

THE ONSET OF GLACIATION AND THE LAST DEGLACIATION

J.C. Duplessy, L. Labeyrie,
Centre des Faibles Radioactivités, Laboratoire mixte CNRS-CEA,
Parc du CNRS, 91190 Gif sur Yvette (France),

J. Moyes, J.L. Turon, J. Duprat, C. Pujol,
Département de Géologie et Océanographie,I.G.B.A.,
Université de Bordeaux 1,Avenue des Facultés, 33405 Talence (France),

J.L. de Beaulieu, J. Clerc, M. Couteaux, A. Pons, M. Reille,
Laboratoire de Botanique Historique et Palynologie, Faculté des Sciences
et Techniques de St Jérôme, rue Henri Poincaré, 13397 Marseille cédex 13

M. Van Campo,
Laboratoire de Palynologie, U.S.T.L., Place Eugène Bataillon,
34060 Montpellier cédex (France),

G. Jalut,
Service de Botanique et Biogéographie, Université Paul Sabatier,
39 Allées Jules Guesde, 31400 Toulouse (France),

R. Sabatier,
Unité de Biométrie,E.N.S.A.M., Place Viala, 34060 Montpellier cédex
(France)

Abstract

When the Earth's climate changes from interglacial to glacial
conditions, temperature and precipitation patterns over the European
continent are profoundly modified : at the height of the last ice
age, surface water temperatures in the neighbouring Atlantic Ocean
were more than 10°C lower than those of the present day. A cold
dry climate had led to the development of a polar desert in northern
Europe and to steppe conditions further to the south.

The use of marine isotopic stratigraphy has made it possible to show
that temperatures remained high in the Northeast Atlantic and over
Europe at the same time that ice sheets were developing on the north
of the American continent about 115 000 years ago. The development
of cold climatic conditions over the European continent, and doubtless
that of the Scandinavian ice sheet, were only possible once cold
waters had invaded the Northern Atlantic and reached the coast of
Europe.

Deglaciation began over the whole of the northern hemisphere soon after full glaciation 18 000 years ago and a major thaw began about 15 500 years ago at a time of low temperatures all over the northern hemisphere. The melting of the ice was interrupted, between 13 000 years B P and 10 000 years B P, by a period of little overall change in which temperature oscillation of more than $10^{o}C$ occurred within a few centuries.

INTRODUCTION

The astronomical theory of paleoclimates makes it possible to establish the main trends in the evolution of the Earth's climate during the last few millions of years (1), and can also serve as reference standard making it now possible to reconstruct the evolution of thermal and hydrological conditions at a regional level. The differences observed between one region and another comprise in themselves indications of the nature and mechanisms of the great changes that have taken place in the planet's climate as well as supplying data for comparison with simulations obtained independently using general circulation models of the atmosphere (2). It is for this reason that we have attempted to reconstruct the evolution of the climate of Europe during two major transitional periods - the onset of glaciation approximately 11 500 years ago and the last deglaciation which, over the period between 15 500 years ago and 9 000 years B P (Before Present), resulted in the present interglacial conditions.

Europe is one of the zones most affected by shifts in climate : the interglacial periods show considerable similarity to present day conditions. During the glacial period, problably as a response to the disappearance of the Gulf Stream (3) and the invasion of the North Atlantic by polar waters which were between 10° and $14^{o}C$ colder than they are now (4), regions now famous for their agricultural wealth were transformed into a polar desert stretching from the Netherlands to the USSR (5, 6). Further south, there was a generalized fall in temperature and increase in aridity (7). The extent of these variations is such as to make one hopeful that the paleoclimatic indicators will be sufficiently precise to permit a detailed reconstruction of the transition between these extreme states and an understanding of the mechanisms involved.

2. COMMENCEMENT OF THE ICE AGE

Approximately 11 500 years ago, after a period of some 12 000 years in which the climate was similar to that of the present day and the seas were about 6 metres higher, the Earth's climate entered a new, long glacial phase.

2.1 Material and methods used in the reconstruction

The traditional methods of reconstructing former climates use geological deposits either on the surface of the continents or from the seabed in very deep water.

Isolated remains such as former beaches and coral reefs have allowed the use of radiochemical methods to date the last interglacial period at about 12 000 years B P (8), but do not make it possible to establish a time series showing the evolution of the environment over a long period.

It is therefore necessary to resort to cores drilled either on land, in bogs and lakebeds, or in the deep marine sediments of an oceanic environment. Unfortunately, these cores cannot be directly dated by radiochemical methods in a way which would permit the precise calculation of the length of the transitional period leading to th establishment of glacial conditions.

In the cores of continental sediments, the presence of pollens makes it possible to reconstitute the evolution of the local plant cover (9,10) which is very sensitive to temperature and humidity conditions. Ocean cores supply a wider range of information : to found within a single sediment sample, and deposited simultaneously, are pollens transported by rivers and winds from the land and the shells of foraminifera - microscopic animals that lived near the surface of the water. By analysing the foraminiferal fauna it is possible to establish with a precision of $+ 1.5^{\circ}$ C the summer and winter temperatures of the surface water of the ocean (11,12). At the same time, the $^{18}O/^{16}O$ ratio of these shells provides the best stratigraphic tool presently available (13, 14) since the variations in this ratio reflect variations in the volume of continental ice sheets which are low in ^{18}O. The variations in the $^{18}O/^{16}O$ ratio between cores coincide with the mixing time in the ocean water. This is of the order of 1500 years for the ocean basin such as the north east Atlantic.

The commencement of the ice age is defined in terms of marine isotopic curves as the transition between the isotopic stages 5e and 5d (Figure 1). The change in the quantity of foraminiferal ^{18}O observable during this transition is approximately two thirds the maximum change in the $^{18}O/^{16}O$ ratio during the whole of a climatic cycle. Isotopic stratigraphy therefore constitutes a very precise tool for revealing either synchronism or any time lag between climatic changes in different parts of the globe.

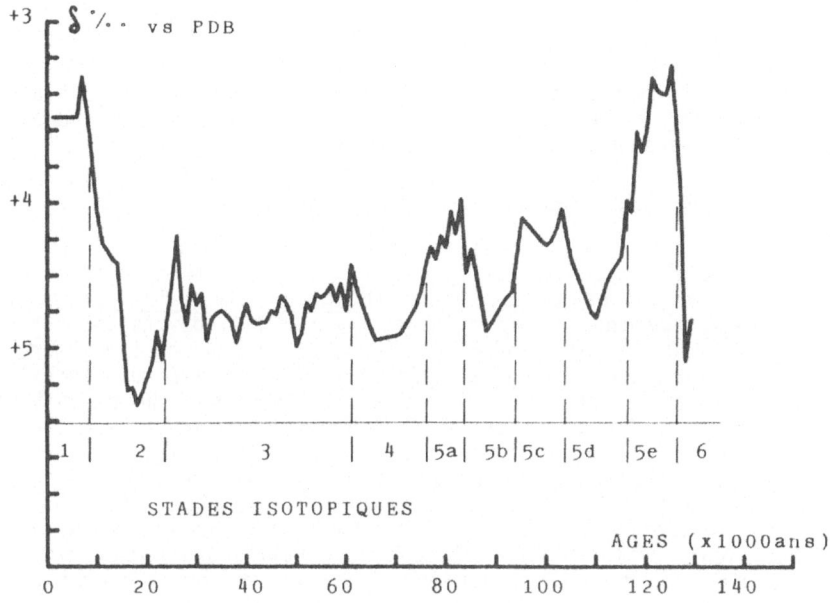

Figure 1

Variations in the $18_O/16_O$ ratio of benthic foraminifera in the
core labelled MD 73/025 drilled in the southern Indian Ocean.
The high 18_O levels (high values of δ) indicate the development
of enormous ice sheets on the continents (50 million cubic
kilometers) 18 000 years

In consequence, we have adopted a dual strategy to reconstitute the conditions under which the climate of the North Atlantic and Europe moved from interglacial to glacial conditions. On the one hand, we looked for a detailed continental pollen sequence showing the evolution of the vegetation. This was found at Les Echets in the Lyons region (10, 15) and it was possible to establish a very close correlation between this and the now classic sequence of La Grande Pile in the Vosges (9) which covers the same period. On the other hand, using ocean cores, we followed the development of the continental ice sheets by monitoring variations in the $^{18}O/^{16}O$ ratio of foraminifera. Simultaneously, we reconstructed the history of changes in surface water temperatures in the Atlantic Ocean, as revealed by changes in the foraminiferal fauna, and the evolution of European vegetation using the evidence of pollens.

2.2 The pollen sequence 'G' at Les Echets'

The last interglacial period - the Eemian interglacial - can be recognized at the base of the sequence by the development of a forest with successive domination by birch (Betula), elm (Ulmus), oak (Quercus), hornbeam (Carpinus) and fir (Abies). A simplified version of the complete diagram (10, 15) is plotted on Figure 2. The end of the interglacial has proved to be a complex affair. Starting from a temperate forest of fir in which a large number of oaks were still present, the proportion of firs declines in favour first of common spruce (Picea) and then of pines (Pinus) (15). After a new resurgence of fir, there is an evolution towards a boreal forest of pines and birch, the precursor of a cold-climate herbaceous vegetation heralding the first phase of glaciation after the Eemian interglacial. The accuracy of this complex evolution has been lent considerable credence by a detailed sedimentological study and by analysis of a large number of levels corresponding to the climatic transition. Observations similar to those obtained at Les Echets can also be made from diagram X of La Grande Pile (16), fom those of Sulzberg Baden in the Bernese Oberland (17) and of Odderade in Schleswig Holstein (18). Without doubting the very cold nature of the brief episode which ended the Eemian interglacial palynological observations indicate a complex process of climatic deterioration that was therefore also relatively long (covering a period of several centuries). None of these observations confirm the abrupt climatic change described by G. Woillard (19) in one of the samples taken at La Grande Pile - where, incidentally, sedimentological studies indicate that the transition periods are under-represented (20).

2.3 Comparison with ocean data

Reconstruction of summer and winter surface water temperatures in the North Atlantic Ocean has been carried out as part of the CLIMAP programme to reconstruct oceanic conditions during the last interglacial period (21). Pujol and Duplessy (22) have put forward a synthesis of available data to reconstruct the conditions existing in this ocean approximately 115 000 years ago - half-way through the transition to the glacial period. At that time, an ice sheet measuring approximately 15 million cubic kilometres had developed in the north of the northern hemisphere, but no direct geological data are available to show the limits of this ice sheet. The North Atlantic Ocean was only chilled along its western fringe ; in the central and eastern part of it, along the European coasts, temperatures were at least as high as during the preceding

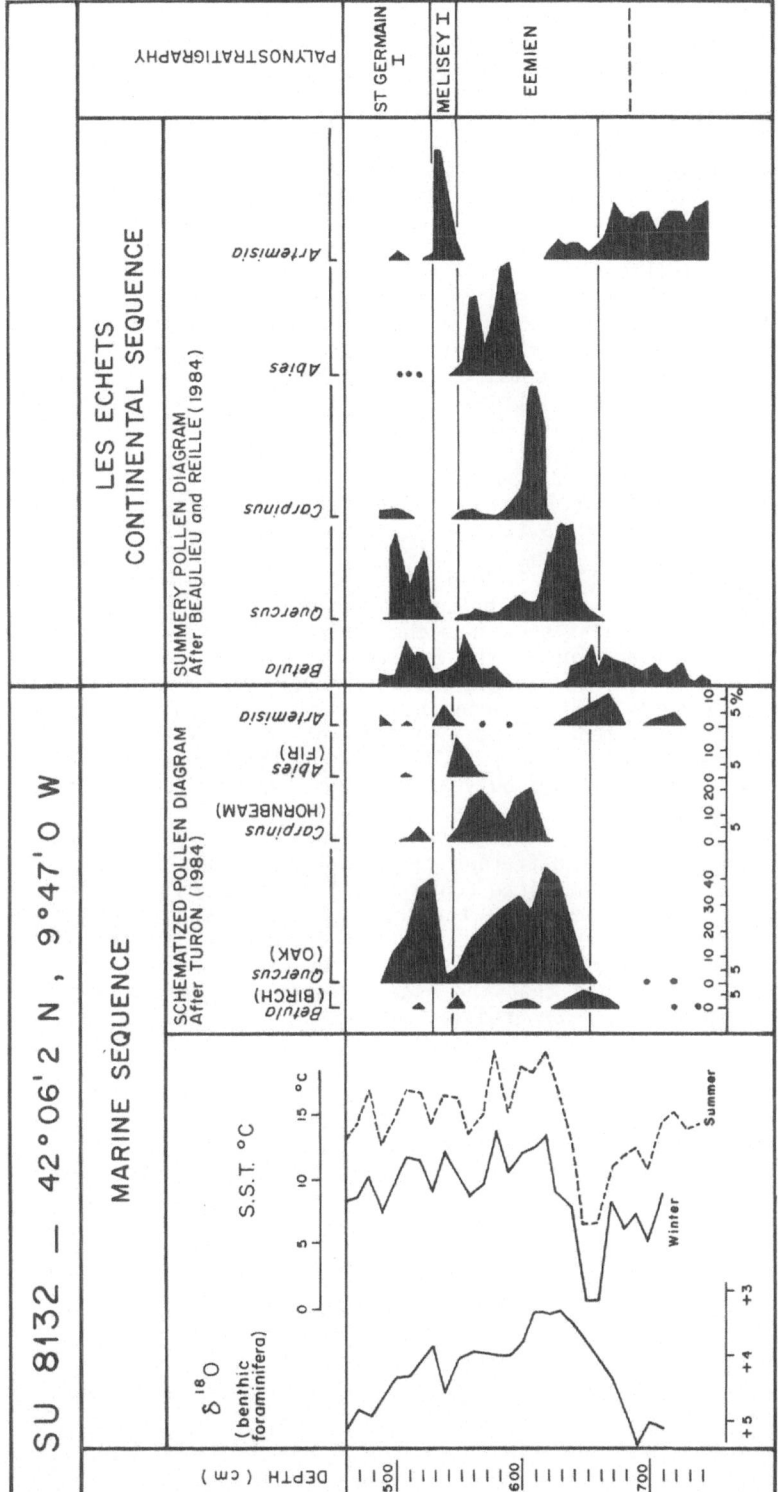

Figure 2 : Comparison of the Les Echets pollen sequence and the oceanic sequence of core SU81/32 during the last interglacial period and the onset of the glacial period.

- 33 -

interglacial period. The beginning of the glacial period was therefore characterized by a considerable thermal disparity between the western and eastern edges of the North Atlantic - considerably more than is the case today (21, 22).

The marine cores which contain sediments laid down during the last interglacial period are generally extracted far out to sea because the accumulation of erosion products along the edge of the continents is largely responsible for burying the 120 000 year old deposits under a sedimentary cover that is too thick to be pierced by the corers currently available. The pollen content of these cores is therefore very low because of the dilution associated with increasing distance from the continent. It is for this reason that we undertook, in 1981, an oceanographic campaign along the coast of the Iberian peninsula in the hope of finding sedimentary accumulations of moderate thickness that would be likely to yield samples rich in pollen and dating from the last interglacial period. It was as a result of this campaign that the cor No SU81-32 (42°6'N, 9°41'W, 2 280 m) made it possible for the first time to provide a direct correlation between marine isotopic stages and continental pollen sequences (23).

Figure 2 shows that pollens within the SU81-32 core underwent the same evolution as that observed on the continent both at Les Echets and at La Grande Pile. These results show that the marine isotopic stage 5e corresponds only to a fraction of the Eemian as observed on the European continent, and that the ice sheets began to cover the surface of northern hemisphere continents in the second half of the Eemian a time at which, in Europe, the development of the hornbeam - oak association was occuring under stable temperatures and hydrological conditions over the continent between 40°N and 50°N. It was only after a time lag of several thousands of years that the ice age reached Europe.

This situation is quite different from that described by Heusser and Shackleton (24) for the American continent. These authors noted in a core taken from the Pacific Ocean off the Oregon coast that the vegetation on the neighbouring continent had varied in accordance with the isotopic curve. When the history of climatic change in North American and Europe is compared, it is clear that the establishment of the ice age in the northern hemisphere was a progressive phenomenon which did not affect all continents at the same time. Our observations, which agree well with the simulation carried out by Royer et al (2), suggest that ice sheet development began in Canada at time in which Europe and the North East Atlantic Ocean - at least as far north as 50°N - continued to experience the typical temperatures of interglacial conditions. It was only in the final phase of the development of the ice age that cold and drought invaded the European continent.

3. THE LAST DEGLACIATION

The ice age, which began 110 000 years ago, lasted for approximately 90 000 years. During this time there was a series of less significant oscillations between glacial advances and periods of moderate increases in temperature. The ice sheets in the northern part of the northern hemisphere reached tneir maximum extent 18 000 years ago, at which time 50 million cubic metres of ice formed ice sheets covering the north of

America and Europe. The next 10 000 years would see these ice sheets disappearing, leaving only a few vestiges such as the Greenland ice cap. How is it possible to explain such a sudden disappearance of these enormous quantities of ice ?

3.1 Material and methods

The last deglaciation is represented by a large number of geological deposits both marine and continental in origin. In contrast to deposits more than 100 000 years old, these sediments can be directly dated using the carbon-14 method provided that they contain sufficient carbon. Although this makes it easier to correlate continental samples found a great distance apart, those marine cores which contain sufficient sediment to show the details of climatic transition are low in fossils because these sediments essentially comprise the accumulation of erosion products from the continent. For this reason, only the CH73-139C core could be dated using the carbon-14 method. For the other cores, we resorted to isotopic stratigraphy, assuming that the major isotopic transitions in the north east Atlantic ocean were of the same age as that measured in the CH73-139C core.

The reconstruction of climatic conditions over Europe and the adjcent Atlantic has therefore been carried out by comparing continental pollen series, the pollen series found in ocean cores and estimates of surface sea temperatures deduced from the analysis of foraminiferal fauna.

3.2 The onset of deglaciation

The isotopic curves (figures 3) show that after the glacial maximum no significant deglaciation occurred until after 15 000 $\pm$ 800 years BP. The ice sheets then melted in two quite distinct phases separated by a period of little overall change which lasted from 13 000 years BP to about 10 000 years BP.

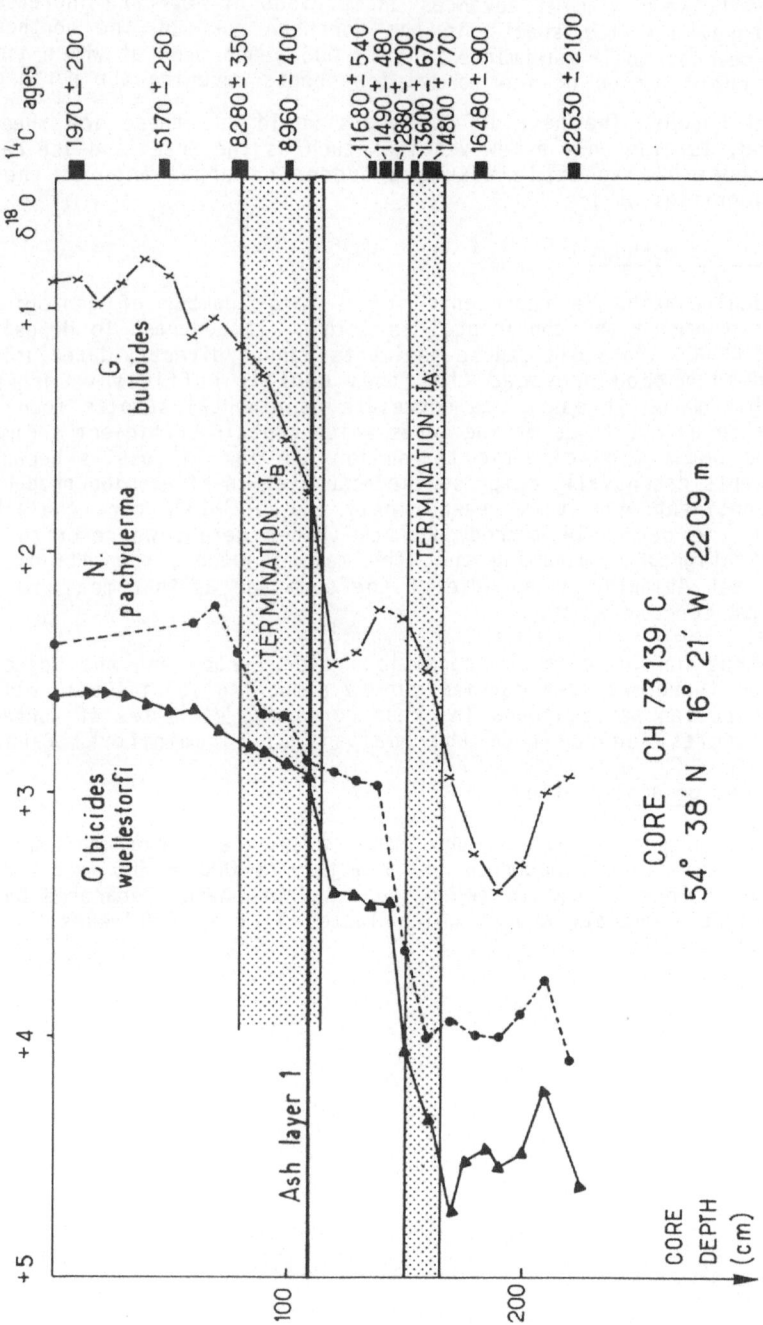

Fig. 3 Variations in the $^{18}O/^{16}O$ ratio in foraminifera during deglaciation

Analysis of foraminiferal fauna in cores extracted between 36°N and 54°N (Figure 4 and reference 4) in the Atlantic Ocean show that when the ice was beginning to melt, surface water temperatures were as low as they had been at full glaciation.

Similar observations can be made on the continents. In the Pyrenees, whether on the northern piedmont (Biscaye deposit, reference 25) or halfway up the mountains (the Freychinède deposit, reference 26), there is a clear increase between 20 000 and 13 000 years BP in sagebrush (Artemisia) - a typical steppe-land grass. In south-east France, between 20 000 and 15 000 years BP , it has not proved possible to attribute any change in pollen diagrams to a significant climatic fluctuation (27). At about 15 000 years BP, all medium-altitude sites show an increase in the proportion of Artemisia, chenopodiaceae and caryophyllaceae (10, 15, 27, 28, 29) which no doubt indicates colonization of previously bare ground by regional steppe formations. This expansion of Artemisia can also be found in Italy (28) and in Greece (29). It can also be observed (5) that absolute pollen frequencies increase notably only after 13 500 years BP.

A principal component analysis and a regression from present day conditions have made it possible to show that in the case of the Tenagi Philippon site, the maximum expansion of Artemisia is associated with temperatures that were 5°C and 10°C lower than present day summer and winter temperatures respectively. At the same time, spring, autumn and winter precipitation was lower by a factor of 2 (29). The fact that increasing quantities of Artemisia coincides with the onset of deglaciation therefore suggests that this was the maximum period of aridity on the European continent, which tallies with the small quantity of maritime moisture reaching the European continent (3).

The temperature changes that took place on the American continent were scarcely different from those observed in Europe. The first signs of climatic improvement date from 14 000 years BP, but it was only 1 500 years later that warmer conditions become general (31): the improvement of conditions is associated with the corresponding decrease in the size of the Laurentide ice sheet.

In their entirety, these results indicate that the decrease in the volume of continental glaciers was not associated with any heat inflow via either the atmosphere or the ocean in the northern part of the northern hemisphere. A fall in precipitation could have contributed to 'drying out' the ice sheets but is not in itself a powerful enough mechanism to cause the rapid, though discontinuous, disappearance clearly demonstrated by the isotopic study of ocean (figures 3, 4 and 6). It is therefore more likely the inherent instability of ice sheets (32) and the summer disappearance of the pack ice that linked them together (33), which were the major mechanisms leading to the destruction of ice in the high northern latitudes.

3.3. Climatic oscillations in Europe during the deglaciation

As early as 1973, Ruddiman and Mc Intyre (34) showed that during the transition from a glacial climate to that of the Holocene, the retreat of the polar front separating cold waters from temperate ones took place at different times in different regions of the Atlantic Ocean. The same authors also discovered the existence of a short phase, at about 10 000 years BP, in which the polar front again advanced southwards reaching a position near the one occupied at full glaciation. Duplessy et al (4) have shown that these changes in the position of the frontal zones were accompanied by changes in summer and winter temperatures of surface water of as much as 14°C off the French coast. In addition, comparison of the fossil fauna (which yields estimates of sea water temperature) and pollens included in the same sediments provided strong support for the hypothesis that the advance of the polar front in the Atlantic coincided with the well known Younger Dryas cold phase in northern Europe.

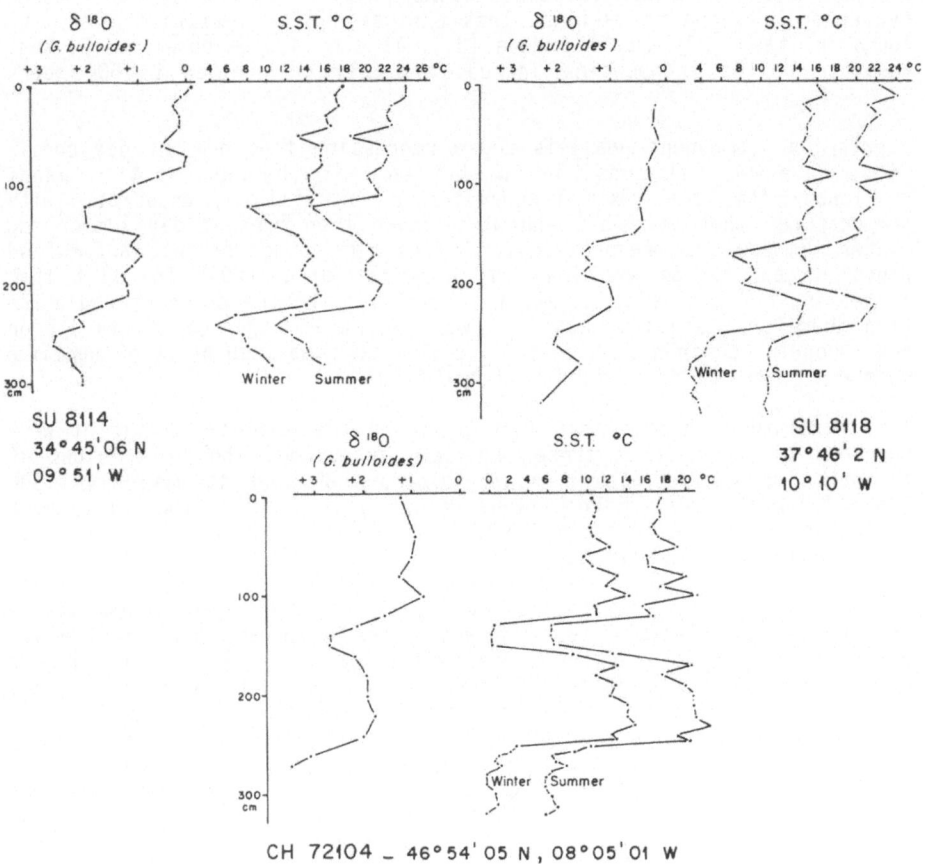

Fig. 4 Details of the last deglaciation ($18^0/16^0$ and estimates of surface sea temperatures) in cores extracted off the European coast.

We have therefore recommended the detailed study of continental and oceanic sequences covering the period 13 500 years BP to 10 000 years BP in order to determine the extent of variations in temperature and their geographical distribution across the north east Atlantic Ocean and Europe during this period of great climatic instability.

As general rulem there is evidence everywhere of a net increase in pollens in the period 13 500 years BP to 13 000 years BP (figure 5). At the same time, the majority of French sites demonstrates a major expansion of juniper (Juniperus) or pine (Pinus) which appears as almost pure stands in the Maritime Alps. Taken together, these results confirm a very significant climatic improvement which gave rise to the well-known Bölling and Alleröd interstadial periods. The colder Younger Dryas period, after 11 000 years BP, is a major event which can be traced everywhere both by changes in the sediments of lake beds, the thinning out of the pine forest and the resurgence of all kinds of steppe vegetation.

The SU81/47 core (Figure 6) provides a pollen diagram showing the same climatic oscillations in southwest France and northern Spain and therefore in good agreement with previous work in this zone (4). The first indications of a warmer climate appear about 14 000 years BP with the more widespread appearance of the birch (Betula), which approximately coincides with the midpoint of the first phase of isotopic deglaciation (termination I_A). The significant appearance of oak pollens, dated in continental sediments at about 13 000 years BP (cf the synthesis found in reference 4), marks the beginning of the Bölling and Alleröd interstadial periods and was synchronous with the presence of warm water along the European coast. The reduction in the oak cover and the virtual disappearance of birch which characterizes the Younger Dryas can be observed in oceanic sediments containing the most cold-tolerant fauna. The second warming and deglaciation phase in the isotopic curve (termination I_B) is, on the continent, accompanied by a clear reduction in herbaceous plants and by expansion of the forest cover.

4. CONCLUSIONS

The temperature of surface water off the European coast determines the character of the continent's vegetation. The vegetation's response to deteriorating climate is very rapid, perhaps taking less than a century, so that continental pollen spectra provide readings that synchronize with increasingly cold oceanic water both when the earth as a whole moved into the ice age (transition between the isotopic stages 5e and 5d) and when there were very frequent climatic oscillations such as those observed during the last deglaciation.

During periods of climatic improvement, when temperature and humidity conditions encourage the development of forest, only a small number of kinds of vegetation are an immediate indication of this improvement - those, such as the birch and the pine, which had survived in 'refugia' within their main range.

At the start of the ice age, the climates of Europe and America evolved differently because there were considerable contrasts between their

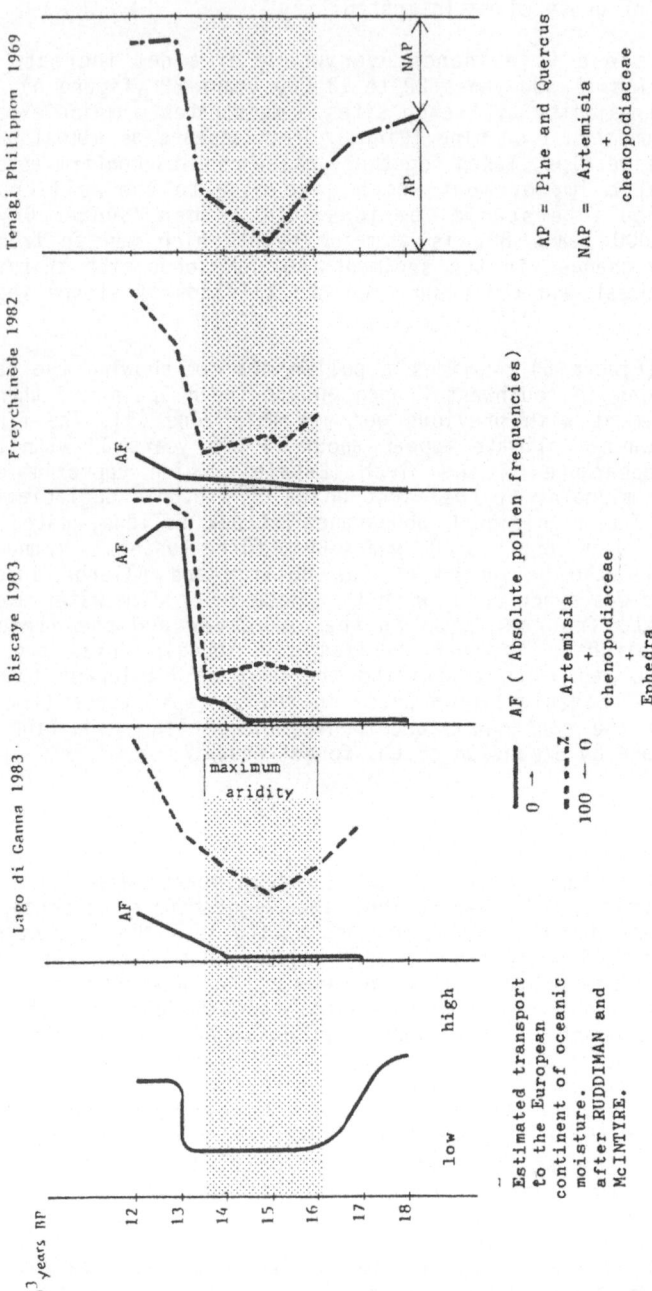

Figure 5 : Comparison of moisture transport onto the European continent (3), absolute pollen frequencies and the percentage of Artemisia + chenopodiaceae + Ephedra compared to those of trees (Pinus and Betula) in continental sequences.

- 40 -

respective fringes of the Atlantic ocean : the western edge of the Atlantic was the first to be chilled whereupon insolation conditions then allowed the establishment of a cold and humid climate over North America and the development of ice sheets. In Europe, by contrast, continued high temperatures in the North East Atlantic contributed to the maintenance of an interglacial climate for several thousand years. Glaciation, and probably the development of the Scandinavian ice sheet, did not appear until the ice age was well established on the American continent and cold water had reached the European coast.

Deglaciation, which began when both the continents and the northern oceans were experiencing low temperatures, was a discontinuous process in which the inherent instability of ice sheets must have played a major role. In the European area, this climatic transition was marked by major fluctuations (over $10^{0}C$) in temperature within just a few centuries. The precise chronology of these climatic fluctuations, which is necessary if models of the dynamic processes involved are to be constructed, still has to be established and will require direct experimental measurement.

ACKNOWLEDGMENTS

This work was supported by the following contracts CLI-004-F (M. Van Campo), CLI-005-F (J. Moyes), CLI-012-F (A. Pons) and CLI-085-F (J.C. Duplessy).

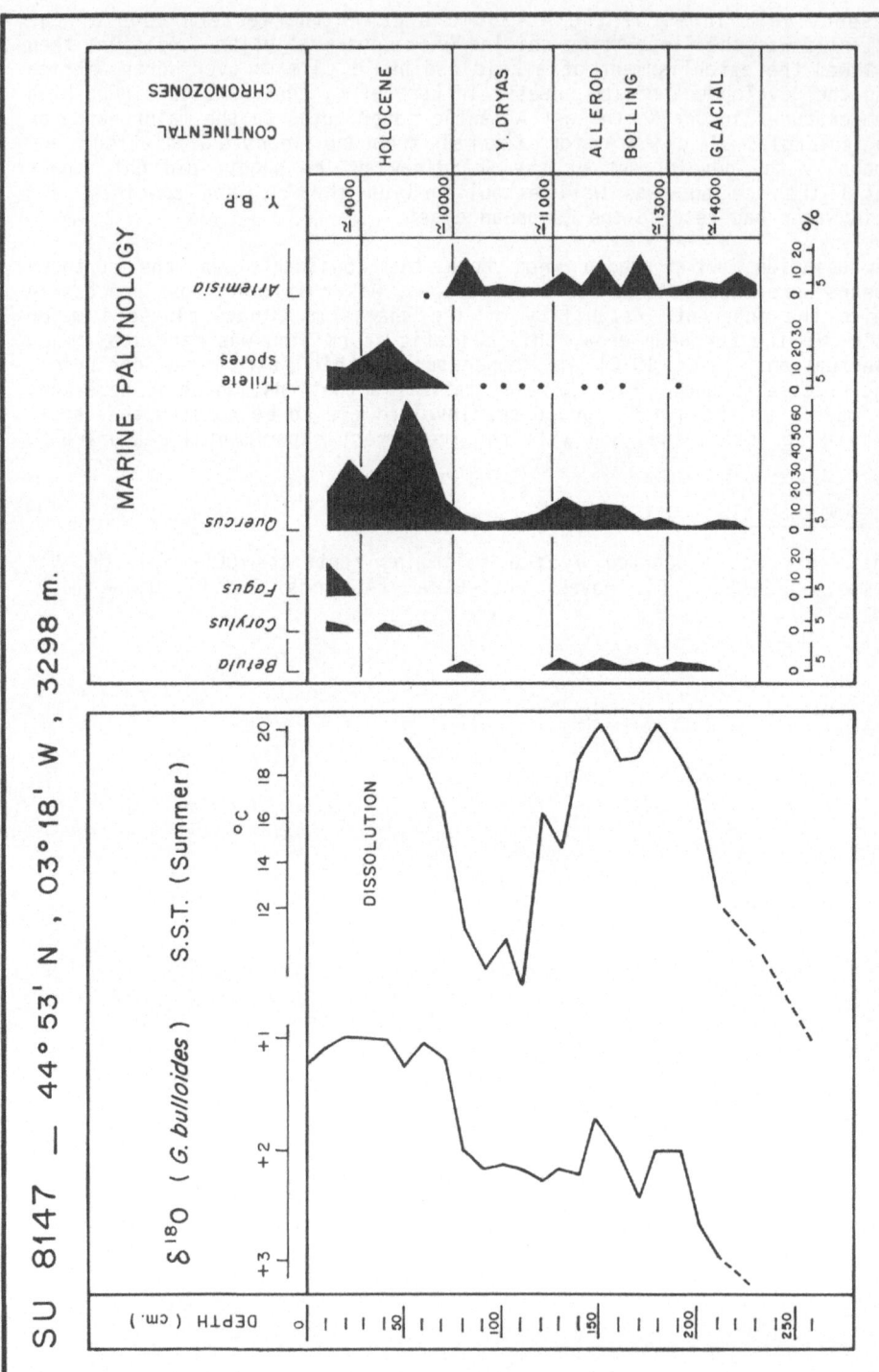

Figure 6 : Variations in the isotopic composition of foraminarera, of surface water temperatures and of pollens in the SU81-47 core extracted in the Bay of Biscay.

- 42 -

REFERENCES

1. BERGER, A., IMBRIE, J., HAYS, J., KUKLA, G , and SALTZMAN, B. (1984).
   Milankovitch and climate. D. Reidel Publishing Company.
2. ROYER, J.F., DEQUE,M. and PESTIEAUX, P. (1983). Nature. Vol. 304, 43–45
3. RUDDIMAN, W.F. and McINTYRE, A. (1981). The North Atlantic ocean during
   the last deglaciation. Palaeogeogr. Palaeoclimat. Palaeoecol. Vol. 35,
   145–214
4. DUPLESSY, J-C., DELIBRIAS, G., TURON, J-L., PUJOL, C. and DUPRAT, J.
   (1981). Deglacial warming of the northeastern Atlantic Ocean : correlation with the paleoclimatic evolution of the European continent.
   Palaeogeogr. Palaeoclimat. Palaeoecol. Vol. 35, 121–144
5. FRENZEL, B. (1973). Climatic fluctuations of the Ice Age. The Press of
   Case Western Reserve University
6. VAN DER HAMMEN, T., MAARLEVELD, G.C., VOGEL, J.C. and ZAGWIJN, W.H.
   (1967). Stratigraphy, climatic succession and radiocarbon dating of the
   last glacial in The Netherlands. Geol. Mijnb. Vol. 46, 79–95
7. FRENZEL, B. (1980). Klima der Letzten Eiszeit und der Nacheiszeit in
   Europa. Veröff. Joachim Jungius-Ges. Wiss. Hamburg. Vol. 44, 9–46
8. LALOU, C. (1982). Sediments and sedimentation. In: Uranium series dis-
   equilibrium. M. Ivanovich and R.S. Harmon Eds. Clarendon Press Oxford,
   431–458
9. WOILLARD, G. (1978). Grande Pile peat bog: a continuous pollen record
   for the last 140 000 years. Quaternary Research. Vol.9, 1–21
10. BEAULIEU,J-L. de and REILLE, M. (1984). The pollen sequence of Les Echets
    (France): a new element for the chronology of the Upper Pleistocene.
    Géographie Physique et Quaternaire. Vol. 38, 3–9
11. IMBRIE, J. and KIPP, N.G. (1971). A new micropaleontological method for
    quantitative paleoclimatology: application to a late Pleistocene Carib-
    bean core. In: The Late Cenozoic Glacial Ages. K.K. Turekian, Ed. Yale
    University Press. 71–181
12. PUJOL, C. (1980). Les foraminifères planctoniques de l'Atlantique Nord
    au Quaternaire. Ecologie - Stratigraphie - Environnement. Thesis,
    University of Bordeaux (France)
13. SHACKLETON, N.J. and OPDYKE, N.D. (1973). Oxygen isotope and paleoma-
    gnetic stratigraphy of equatorial Pacific core V28-238: Oxygen isotope
    temperatures and ice volumes of a 10^5 and 10^6 year scale. Quaternary
    Research. Vol. 3, 39–55
14. DUPLESSY, J-C. (1978). Isotopes studies. In: Climatic Change. Gribbin,
    J. Ed. Cambridge University Press, 46–67
15. BEAULIEU, J-L. de. and REILLE, M. (1984). A long upper Pleistocene pol-
    len record from Les Echets, near Lyon, France. Boreas. Vol.13, 111–132
16. WOILLARD, G. (1975). Recherches palynologiques sur le Pleistocène dans
    l'est de la Belgique et dans les Vosges lorraines. Acta Geographica
    Lovaniensia. Vol.14, pp.118
17. WELTEN, M. (1981). Verdrängung und Vernichtung der anspruchsvollen Ge-
    hölze am Beginn der letzten Eiszeit und die Korrelation der Frühwürm-
    Interstadiale in Mittel- und Nordeuropa. Eiszeitalter u. Gegenwart.
    Vol.31, 187–202
18 AVERDIECK, F.R. (1967). Die Vegetationsentwicklung des Eem-Intergla-
    zials und der Frühwürm-Interstadiale von Odderade/ Schleswig-Hostein.
    Fundamenta B. Vol.2, 101–125
19. WOILLARD, G. (1979). Abrupt end of the last interglacial s.s. in north-
    east France. Nature. Vol. 281, 558–562

20. SERET, G. (1982). Rather long duration of the transient climatic events in the "Grande Pile" (Vosges-France). In: Palaeoclimatic Research and Models. A. Ghazi, Ed., D. Reidel Publishing Company 139-144
21. CLIMAP Project Members (1984). The last Interglacial Ocean. Quaternary Research. Vol. 21, 123-224
22. PUJOL, C. and DUPLESSY, J-C. (1982). The ocean surface during the last interglacial to glacial transition : a review of the available data. In: Palaeoclimatic Research and Models. A.Ghazi, Ed., D. Reidel Publishing Company 145-151
23. TURON, J-L. (1984). Direct land/sea correlations in the last interglacial complex. Nature. Vol. 309, 673-676
24. HEUSER, L.E. and SHACKLETON, N.J. (1979). Direct marine-continental correlation : 150,000-year oxygen isotope-pollen record from the North Pacific. Science. Vol. 204, 837-839
25. MARDONES, M. and JALUT, G. (1983). La tourbière de Biscaye (alt.409m), Hautes-Pyrénées) : approche paléoécologique des 45.000 dernières années. Pollen et Spores. Vol. 25, 163-212
26. JALUT, G., DELIBRIAS, G., DAGNAC, J., MARDONES, M. and BOUHOURS, M. (1982). A palaeoecological approach to the last 21.000 years in the Pyrénées : the peat bog of Freychinède (alt.1350m, Ariège, South France) Palaeogeogr. Palaeoclimat., Palaeoecol. Vol. 40, 321-359
27. BEAULIEU J-L. de, PONS, A., REILLE, M. (1984). Recherches pollenanalytiques sur l'histoire de la végétation des monts du Velay, Massif Central, France. Diss. Bot. Vol. 72 (Festschrift Welten), 45-70
28. SCHNEIDER, R. and TOBOLSKI, K (1983). Palynologische und stratigraphische Untersuchungen im Lago di Ganna (Varese, Italien). Bot. Helv. Vol. 93, 115-122
29. WIJMSTRA, T.A. (1969). Palynology of the first 30 metres of the 120m deep section in northern Greece. Acta Bot. Neerl. Vol. 18, 511-527
30. SABATIER, R. and VAN CAMPO, M. (1984). L'analyse en composantes principales de variables instrumentales appliquée à l'estimation des paléoclimats de la Grèce, il y a 18.000 ans. Bull. Soc. Bot. Fr. Vol. 131, Actual. Bot. (1), in press
31. DUPLESSY, J-C. and RUDDIMAN, W.F. (1984). La fonte des calottes glaciaires. La Recherche. Vol.15, 806-816
32. DENTON, G.H. and HUGHES, T.J. (1981). The last great ice sheets. John Wiley & Sons
33. GROUSSET, F. and DUPLESSY, J-C. (1983). Early deglaciation of the Greenland sea during the last glacial to interglacial transition. Marine Geology. Vol.52, M11-M17
34. RUDDIMAN, W.F. and McINTYRE, A. (1973). Time-transgressive deglacial retreat of Polar waters from the North Atlantic. Quaternary Research. Vol.3, 117-130.

CLIMATIC HISTORY FROM ICE CORE STUDIES IN GREENLAND

DATA CORRECTION PROCEDURES

W. DANSGAARD, H.B. CLAUSEN, D. DAHL-JENSEN, N. GUNDESTRUP and C.U. HAMMER
Geophysical Institute, Department of Glaciology
University of Copenhagen
Haraldsgade 6, DK-2200 Copenhagen, Denmark

Summary

Greenland ice cores contain a broad spectrum of informations on past environmental conditions in the North Atlantic region. However, the ice
flow pattern, which is particularly complicated in South Greenland, dic-
tates correction of ice core data for thinning of annual layers since
the time of formation, and/or for deviating conditions upstream from the
drill site. Measurements on the changing geometry of the deep drill hole
at Dye 3; on the surface and bedrock topographies; and on the surface
velocities upstream, have been used as inputs to an ice flow model that
has led to a deeper understanding of the ice dynamics that is needed for
a proper correction proceedure. The corrected data series show that the
annual precipitation in South Greenland has varied considerably less
through the last several millenia, than suggested by the raw data. The
interpretation of stable isotope data series is discussed, and examples
of Pleistocene environmental data series are presented.

1. Introduction

The American-Danish-Swiss joint effort, Greenland Ice Sheet Program
(GISP, 1971-81; 1), emerged from the first successful ice core drilling to
bedrock at Camp Century, NW Greenland (Fig. 1). Numerous studies on this
and other ice cores, recovered later on, have demonstrated that the great
ice sheets are rich sources of information on past environmental conditions.

The main objective of GISP was to extract this information by drilling
and analyzing new ice cores, including a new one to bedrock at a more favorable location than Camp Century. The ice sheet was surveyed by airborne
radio-echo sounding equipment, which showed that most of the bedrock in
central and North Greenland is favorably flat, whereas the ice in South
and East Greenland rests on a mountainous bedrock (2).

Three ice cores were drilled to 400 m depth by a thermal drill (3),
and a number of cores up to 110 m length were drilled by other techniques
(4,5) at the locations shown in Fig. 1, for the purpose of picking out
the areas that meet a series of requirements, which together define a
favorable location for deep ice core drilling (see note 18 in 6).

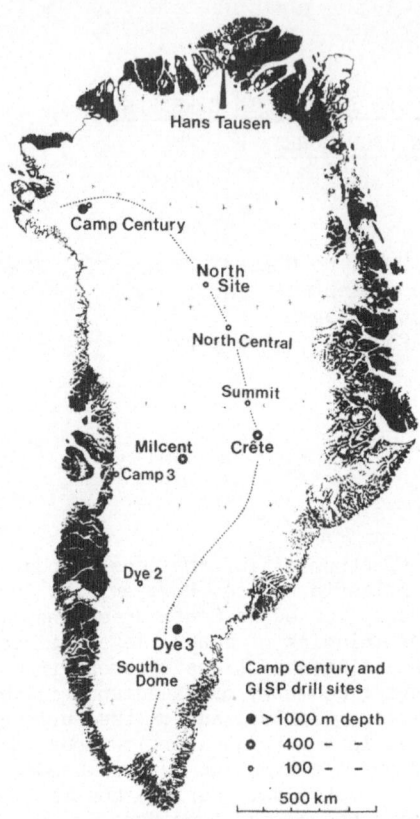

Fig. 1. *Camp Century and GISP drill sites on the Greenland ice sheet. The dotted curve shows the main ice divide that encloses the West Greenland discharge area. The deep drill site at Dye 3 in South Greenland is located some 45 km from the ice divide, but the deepest part of the ice core was deposited much farther to the South, probably close to the South Dome.*

Within the map:

Hans Tausen

Camp Century

North Site

North Central

Summit

Milcent · Crête

Camp 3

Dye 2

Dye 3
South Dome

Camp Century and GISP drill sites

● >1000 m depth
◉ 400 - -
○ 100 - -

500 km

As a result of these efforts, Central Greenland appeared to be most suitable. However, the Dye 3 station in South Greenland was chosen for logistic reasons, and because the deep drilling had to be performed by a new and untested technique (7). The GISP field activities culminated in the summer of 1981, when the new drill penetrated the ice sheet and reached bedrock 2037 m below the surface.

In parallel with the drilling, extensive stratigraphic, chemical and physical studies were performed on the ice cores, and 67,000 samples were cut for successive stable isotope analysis. The initial results were presented at a GISP symposium in 1982 (8), and quite a few reports have been published since then (cf. e.g. 9,10 and 11).

However, the Dye 3 site had some disadvantages from a scientific point of view, in so far as there is always some surface melting in the summer time. The wet snow collects disproportional amounts of CO_2 from the atmosphere, and the percolating melt water refreezes in the deeper and colder snow thereby disturbing the stratigraphy.

Furthermore, Dye 3 is far from the ice divide, which means that the deep part of the ice core was originally deposited far upstream, as explained in the text to Fig. 2. Hence, the present thicknesses of annual layers in the ice core can only be interpreted in terms of past accumulation rates at

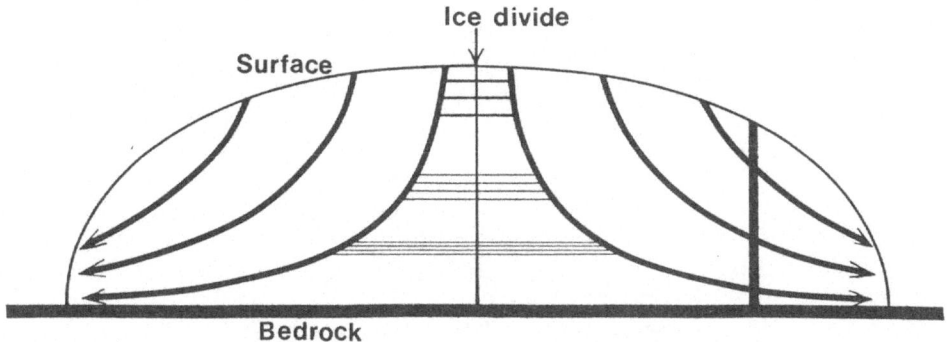

Fig. 2.   *Cross section perpendicular to the ice divide of an idealized ice*
          sheet resting upon a plane horizontal bedrock. The thickness of the
          ice is exaggerated by two orders of magnitude relative to the
          horizontal dimension. The arrows show internal ice flow lines. The
          thin, horizontal lines symbolize annual layers that are stretched
          and   therefore get thinner, as they sink into the ice sheet. The
          annual layers in an ice core (heavy vertical line), drilled far
          from the ice divide, were deposited the farther upstream from the
          drill site, the deeper their present position in the ice core.

the drill site, if they are corrected for the total thinning of the layers
since the time of deposition, and for the geographical variation of the
accumulation rate upstream from the drill site. Data on the composition of
the annual layers in the ice core (isotopic; impurity concentrations) only
need correction for possible upstream effects.

   Finally, the bedrock topography upstream from the Dye 3 drill site is
hilly, and the ice flow pattern therefore complicated, which makes it difficult to apply a proper correction proceedure to ice core data.

   The first step in this proceedure is to design a model that describes
the ice flow in the area upstream from the drill site. This implies extensive field work for the purpose of collecting data on the variations of the
ice thickness upstream; the mechanical properties of the ice at various
depths; and, for verification of the flow model, data on the surface velocities upstream.

   The next step is to collect data on the upstream accumulation rates,
and concentrations of isotopes and impurities.

   Finally, the ice flow model is used to compare ice core data with the
corresponding surface data at the upstream locations, where the individual
ice core layers were deposited in the past.

   This is one of the main objectives of the EEC climatological program
in Greenland 1983-85. The present paper is a midway summary of results
hitherto obtained. They are supplementary to other results obtained prior
to 1983, and they will be presented together in order to give an overall
picture of the situation. The cores and snow samples collected during the
1984 field season have not been taken into account.

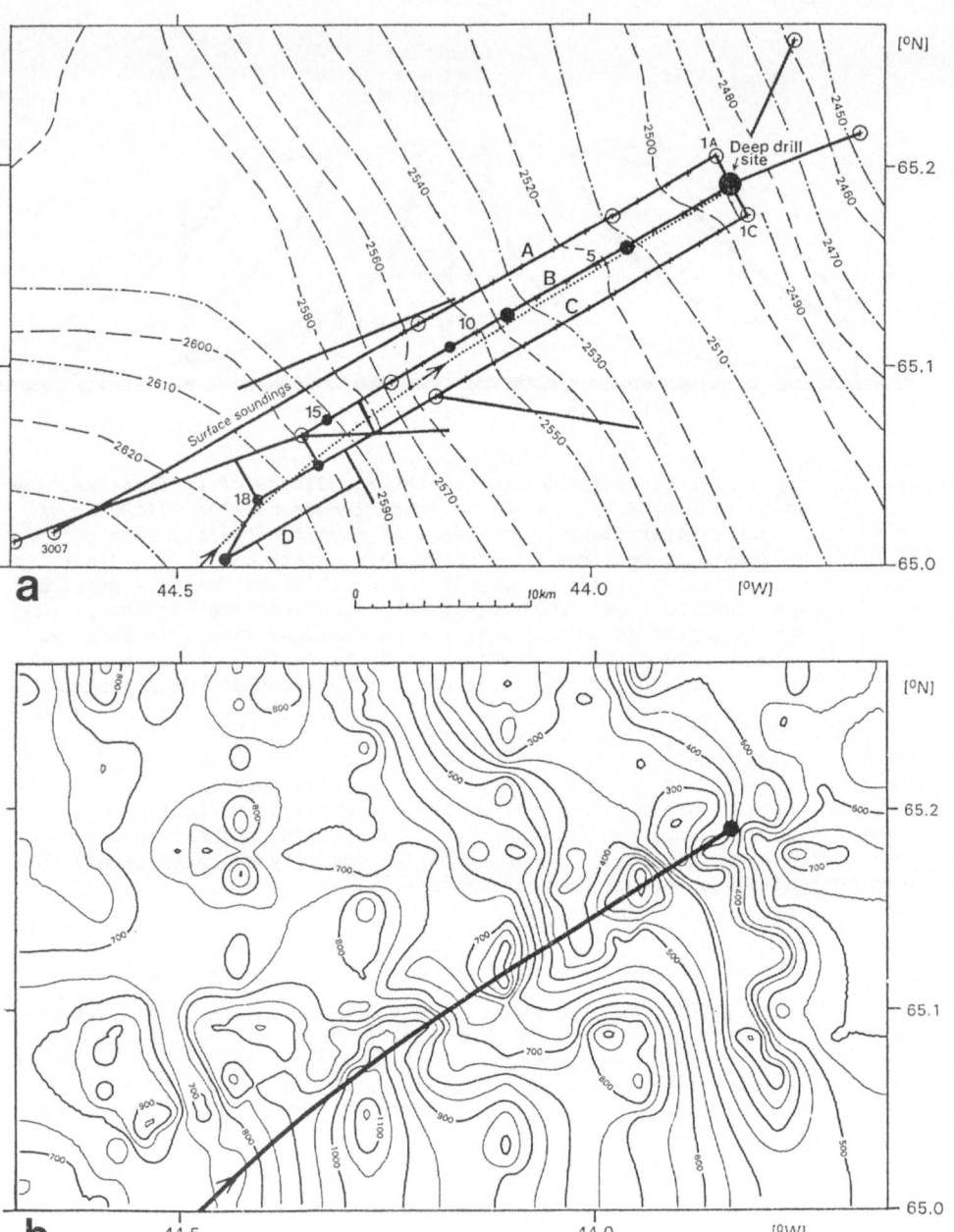

Fig. 3a: *Contours of surface elevations in m, and surface tracks (full lines) upstream from the deep drill site at Dye 3, South Greenland. The ice is estimated to flow in northeasterly direction along the dotted curve. The filled circles are drill sites, where shallow cores were recovered to depth from 14 to 174 meters.*

Fig. 3b: *Contours of bedrock elevations in m. The heavy curve is the projection of the flow line through the drill site.*

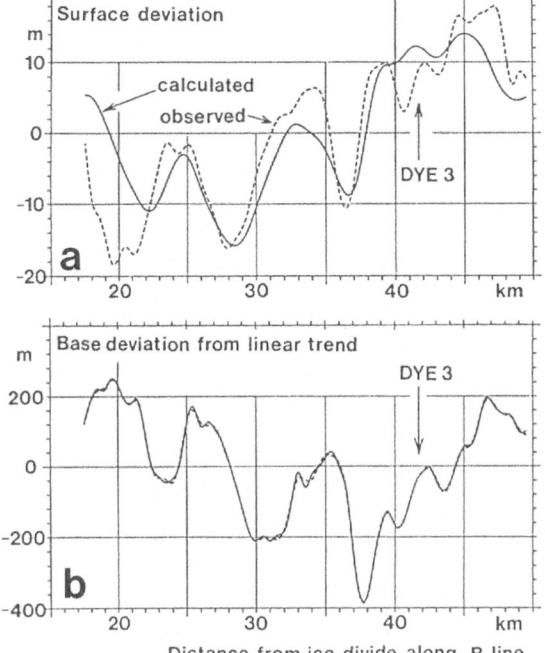

Fig. 4a:
Observed and calculated surface elevation along the
B-line, given as deviations
from a minor linear trend.
The undulations are 20
times smaller than the
hills on the bedrock, which
cause them (from 13).

Fig. 4b:
The rough bedrock topography
along the B-line, shown as
deviations from a minor
linear trend (from 13).

2. Measurements upstream from Dye 3

2.1. Surface and bedrock topographies.

The ice thickness is an important parameter in ice flow modelling. It appears as the difference between the surface and bedrock elevations.

Airborne radio-echo soundings in the 1970'es (2) mapped the ice thickness in South Greenland along flight lines in a rather coarse network (8 km spacing), and these measurements, supplemented by radio-echo soundings along the A, B and C lines in Fig. 3a (12), has been compiled into the bedrock topography map shown in Fig. 3b.

The heavy curve in Fig. 3b is the projection of the flow line through the drill site (filled circle), as estimated from the surface elevation contours shown in Fig. 3a. These contours are based partly on the airborne soundings, and partly on barometric elevation measurements along the heavy lines.

Fig. 4 shows the surface and bedrock topography perturbations from minor linear trends along the B-line. The bedrock is hilly. Notice that 5 km from Dye 3 the bedrock elevation changes by more than 400 m over a distance of 2 km, corresponding to a 20 % change of the ice thickness.

The bedrock hills cause 20 times smaller undulations of the surface. They appear with a phase shift depending on the wavelength (13).

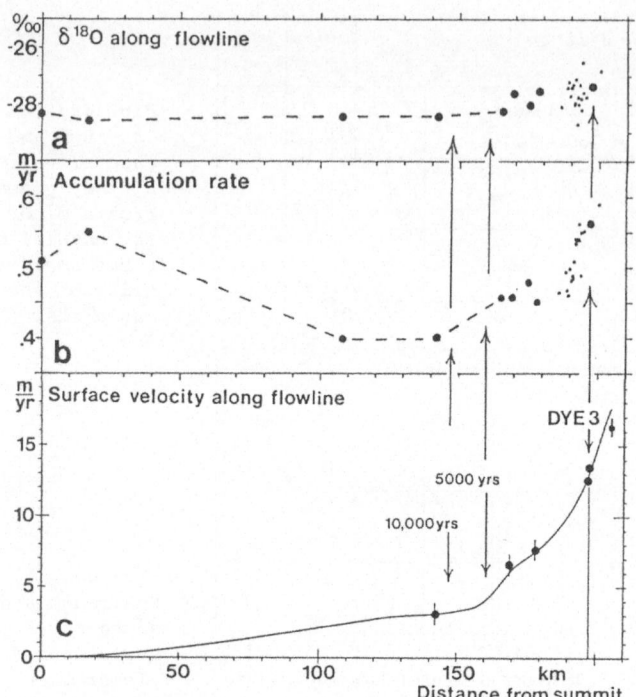

Fig. 5a: $\delta^{18}O$ of the snow pack plotted as a function of the distance from the summit of South Dome along the flowline through Dye 3.

Fig. 5b: Mean accumulation rates in meters of ice equivalent per year through the 9 year period 1966-74.

Fig. 5c: Surface velocities measured by repeated satellite positioning (dots), and calculated by ice flow modelling (curve). The arrows indicate the location of Dye 3 and the calculated sites of deposition of 5000 and 10,000 yr old ice in the deep core.

2.2. $\delta^{18}O$ of the snowpack.

Detailed $\delta^{18}O$ profiles were measured along several firn cores (from 5 to 174 m length) drilled or augered close to the estimated flowline. In all cases the annual layers were identified by the seasonal $\delta^{18}O$ oscillations, in some cases they were verified by analyses of other seasonally varying parameters.

The mean $\delta^{18}O$ value of the total snow accumulation in the 9 year period 1966-1974 is plotted in Fig. 5a as a function of the distance along the estimated flowline from the summit of South Dome (see Fig. 1), which is assumed to be the site of deposition of the oldest ice in the Dye 3 deep ice core.

The arrows indicate the Dye 3 drill site, and the sites of deposition of 5000 and 10,000 yr old ice in the deep core, according to flow model calculations (see section 4).

The $\delta^{18}O$ difference between Dye 3 and South Dome is 1.0 ‰, as expected from the difference between the mean annual temperature of the snow-pack, 1.7°C, and the isotopic temperature effect in Greenland, 0.62 ‰ per °C (14).

The scatter close to Dye 3 seems to be connected to the considerable surface undulations in this area, perhaps because of a varying summer to winter snow accumulation ratio from place to place. Anyhow, the scatter indicates that the $\delta^{18}O$ profile along the younger part of the deep core needs considerable correction.

2.3. Accumulation rates.

The mean accumulation rate in meters of ice equivalent per year through the same period is plotted in Fig. 5b.

The scatter of the data close to Dye 3 does not reflect high uncertainty, but rather real short-distance variability due to the undulations in the surface topography and the consequent relatively high accumulation rates in the valleys.

2.4. Surface velocities.

The surface velocity was measured by repeated satellite positioning at one location downstream; two locations close to the drill site; and three upstream locations close to the flowline. These measurements are listed in Table 1, and plotted in Fig. 5c.

Table 1.

Station	Initial Position	km upstream	Period	Velocity m/yr	Direction degr. E of N	Ref.
Helen	65.25918°N 43.74307°E	-9	1980-83	16.27±0.6	65.4±4	
1	65.18781°N 43.82862°E	0	1972-83	12.46±0.11	61.2±0.5	(15,16)
1A	65.20153°N 43.84021°E	0	1980-81	13.44±0.7	63.8±4	(17)
11.5A	65.11869°N 44.19756°E	19	1980-81	7.57±0.7	52.5±4	(17)
16B	65.06366°N 44.34052°E	28	1981-84	6.60±0.5	43.2±4	
3008	64.84977°N 44.65324°E	55	1980-81	3.06±0.7	-12±20	(17)

3. Deep hole measurements

The deep hole at Dye 3 was drilled by GISP in 1979-81 (6). It is 2037.6 m long, and 130 mm in diameter. The hole is filled with a fluid to counteract the outside pressure.

In 1983, three independent borehole surveys were made using different tools to measure the temperature profile along the hole (Fig. 6); the inclination, azimuth and diameter of the hole; and the pressure in the fluid (18).

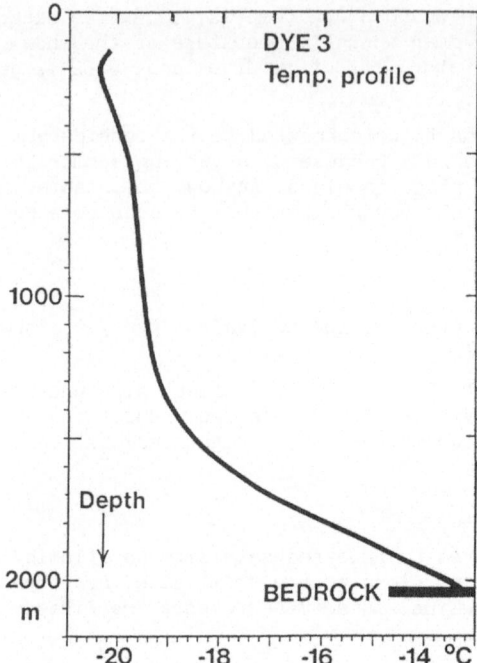

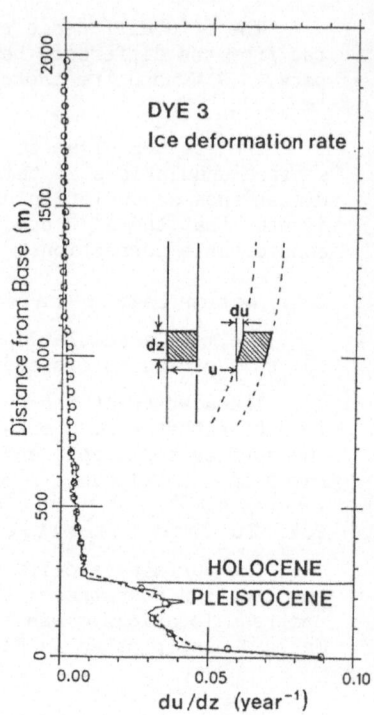

Fig. 6. Temperature profile measured
along the deep hole in 1983. The minimum
at 200 m depth is a remain of the little
ice age. The increase at great depths is
caused by the geothermal heat flux (18).

Fig. 7. Deformation rate du/dz
along the deep core. Pleistocene ice is obviously much softer than Holocene ice (13).

Comparison with similar measurements made in 1981 leads to the full
curve in Fig. 7 that shows a profile of the ice deformation rate, du/dz,
through the entire ice sheet, u and z being the horizontal velocity and
the distance from the bedrock, respectively (13).

The small change in du/dz from surface and down to z = 242 m is to be
ascribed to the generally increasing temperature downwards, according to
the Arrhenius formula.

The remarkable shift in du/dz at z = 250.8 m coincides with drastic
changes in essentially all other ice core parameters, e.g. $\delta^{18}O$ (6), trace
element (19) and dust concentrations (20), and mechanical properties
measured directly on the ice core (9).

Since the age of the ice at z = 250.8 m is some 10,400 years (6), the
shift mentioned above is believed to reflect the drastic change of the
environment at the Pleistocene to Holocene transition.

It is concluded that the ice deposited in Greenland during the last
glaciation is much softer than ice deposited in post-glacial time, probably
due to much higher concentrations of particularly continental dust (18).
The same conclusion has been drawn regarding the ice cap on the Canadian
Devon Island (21).

80 % of the surface velocity at Dye 3 is due to shear deformation in
the Pleistocene ice. This is extremely important for ice flow modelling.

Throughout the Holocene the slow warming of the ice sheet, and the thinning of the layer of soft Pleistocene ice have contributed to changing ice flow parameters. The surface and deep hole observations show that the central part of the ice sheet in South Greenland is very close to being in balance at present (16).

4. Ice flow modelling

As mentioned in the introduction, an ice flow model is needed for correction of deep ice core data on annual layer thicknesses and $\delta^{18}O$ to make them representative of annual accumulation rates and isotopic composition at the site and time of deposition.

The flow models designed for this purpose (22,13) consider, as a first approximation, a "basic" ice flow between smoothed surface and sub -surface boundaries of the ice sheet. The second approximation model includes the mechanical ice flow properties measured in the deep hole, and the observed details of the bedrock topography.

The validity of the first approximation model appears from Fig. 5c, cp. the fit between measured and calculated surface velocities along the flow line from South Dome to Dye 3 (N. Reeh, in preparation).

The validity of the second approximation model appears from Fig. 4. The dashed curve in Fig. 4b is a 12 component Fourier representation of the full curve showing the measured deviations of the bedrock elevations from a minor linear trend.

This representation is used as an input, and the calculated surface undulations are shown by the full curve in Fig. 4a. They compare well with the observed ones, but for the ends of the interval considered, where a good fit cannot be expected.

Combined, two models of this kind are used later for correction purposes, see sections 5.3. and 5.4.

5. Oxygen-18 profile along the deep ice core

According to the simple Rayleigh condensation model, the isotopic composition (oxygen-18 or deuterium concentration in the δ scale) of a given polar snow fall depends on several parameters, of which the most important one is cooling of the precipitating air mass, since the last substantial uptake of water vapor from the ocean.

As the seasonal or longer term climatic conditions in the source area of the vapor are considerably more stable than those at high latitudes, the isotopic composition of the polar snow is strongly influenced by the temperature at the site and time of snow deposition (23,14). With some reservations (24,25) $\delta^{18}O$ profiles along polar ice cores therefore exhibit seasonal oscillations (26,27) and changes due to general climatic temperature variations (14,28).

However, the δ profiles may also be influenced by other parameters, such as
1. the summer to winter precipitation ratio, which is, for example, different on the West and East coasts of Greenland, and which may furthermore change at a given location, if the climatic conditions change;
2. ice sheet surface elevation changes, which cause surface temperature variations that are not necessarily connected to general climatic change.

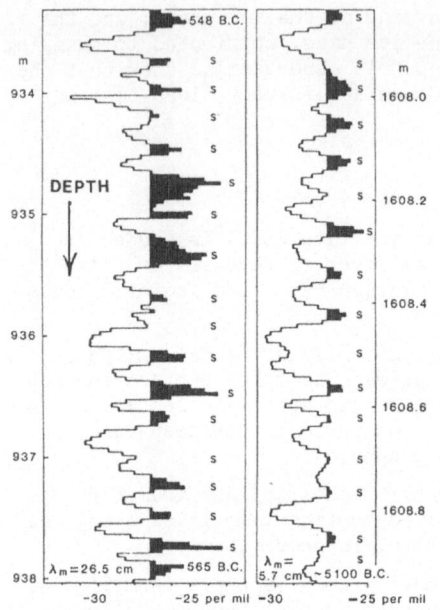

Fig. 8. Seasonal $\delta^{18}O$ variations in
two core increments from about 936
and 1608 m depth. The annual layer
thickness λ_m are 26.5 and 5.7 cm,
respectively. Summer layers are
marked by s. The curve to the left
spans the period 548-565 B.C. The
deepest increment was deposited
approximately 5100 B.C., but it
is not yet absolute dated (6).

3. changing sea-ice cover, i.e. changing distance to the open ocean, which
   may affect the degree of cooling of precipitating air masses, but not
   necessarily the mean air temperature on the ice sheet, at least not to
   the same degree.

In general, it should be kept in mind that the isotopic composition
of a given ice core increment may give a hint as to the climatic conditions
that prevailed during the relatively short periods of snowfall when the
material was deposited, but it tells nothing about the much longer periods
of dry weather conditions inbetween.

5.1. Seasonal $\delta^{18}O$ variations.

The isotopic composition of the snow, as well as the impurity concentrations, vary in an annual cycle (6,29,30,31). Therefore, detailed and
continuous profiles of these parameters may be used for identification of
the individual annual layers in an ice core and, hence, for absolute dating
by counting the layers downwards from surface.

In the Dye 3 core, refreezing of percolating meltwater cause occasional
disturbances in the stratigraphy. Nevertheless, the annual layers are identifiable in an $\delta^{18}O$ profile many millenia back in time, as shown in Fig. 8.

5.2. Absolute dating.

Up till now, the absolute $\delta^{18}O$ dating of the Dye 3 ice core has been
extended back to more than 6000 B.P.

The accuracy of this dating is a few years per thousand, to judge from
the fact that strongly acid ice was found in the layers dated at 1816 A.D.,
1783 A.D. and 1107 A.D., in agreement with the expected fall-out of strong
acids from the well-known great volcanic eruptions of Tambora (1815), Laki
(1783) and Hekla (1104-1105) (32).

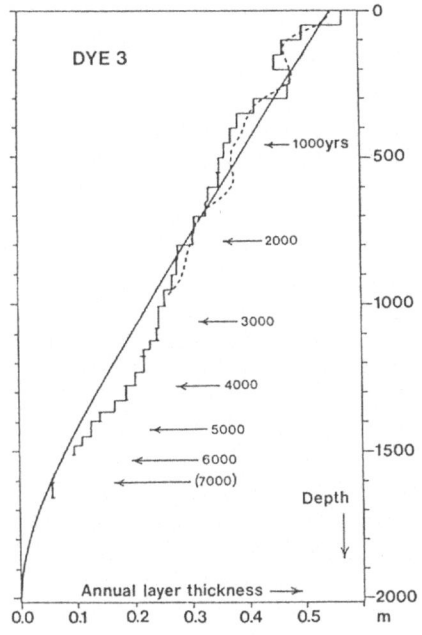

DYE 3

←1000yrs   ⌐500

←2000   ⌐1000

←3000

←4000

←5000   ⌐1500

←6000

←(7000)

Depth

Annual layer thickness →   ⌐2000

0.0   0.1   0.2   0.3   0.4   0.5   m

Fig. 9. Full curve: Annual layer thickness profile predicted by a first approximation ice flow model. Dashed curve: Prediction by a second approximation model. Both models assume no temporal change of the upstream accumulation pattern. Step curve: Measured mean annual layer thickness profile along the deep core. Figures: Absolute ages.

5.3. Annual precipitation record.

Once the individual layers are identified, their thicknesses may be used as a basis for establishing a record on annual precipitation in the past.

The step curve in Fig. 9 shows measured mean annual layer thickness in time intervals ranging from 81 to 278 years, plotted against the depth of their present position in the Dye 3 ice core.

The age of the layers is added on the right hand side of the curve. The general trend reflects the thinning of the layers as they approach the bedrock.

However, the thickness of the layers needs correction for vertical strain since the time of deposition, and for deviating accumulation rate at their sites of deposition, as described in section 1.

The full smooth curve in Fig. 9 shows the result of ice flow calculations with the first approximation model described in section 4. The curve predicts the annual layer thickness in the core with the assumption that the upstream accumulation rates and the ice sheet thickness have not changed in time.

Comparison with the measured step curve suggests that a few hundred years ago (little ice age), and a thousand years ago, the accumulation rate was 10 % lower than now, and that in the period 6000 to 2600 yrs ago the accumulation rate was more than 10 % higher than now (40 % higher 4000 yrs ago).

However, a second approximation calculation (dashed curve in Fig. 9), which takes the observed perturbations in upstream accumulation rate and ice sheet thickness into account, shows that most of the former variations are due to upstream flow and accumulation effects (*22*).

Whether or not the relatively high thickness of 6000 to 2700 yr old annual layers can also be ascribed to upstream effects, is being investigated.

At present, it seems that the annual precipitation in the Dye 3 area has remained surprisingly constant at least through the last 1400 yrs, at it has previously been shown to be the case in Central Greenland (33).

The annual accumulation record based on seasonal $\delta^{18}O$ cycles can hardly be extended far beyond 8000 yrs B.P., because the $\delta^{18}O$ cycles are expected to be obliterated by diffusion in the ice (25).

Furthermore, the correction proceedure gets the more uncertain, the farther from the drill site the ice was deposited (although the new knowledge about the ice deformation rate (Fig. 7) is a substantial step forward). This will be a problem no matter which seasonally varying ice core parameter, one uses for annual layer identification, and beyond 10,000 yrs B.P. it will be further complicated by our lack of knowledge about the flow conditions in the Pleistocene ice sheet.

5.4. Isotope-temperature record

With the reservations mentioned in the beginning of section 5, a smoothed $\delta^{18}O$ record reveals past climatic changes in terms of mean annual temperatures.

5.4.1. The last millenium

A smoothed $\delta^{18}O$ profile along the upper 650 m of the Dye 3 deep core is shown in Fig. 10 on an absolute time scale.

However, according to Fig. 5a this record needs considerable correction for the irregular $\delta^{18}O$ pattern in the snow at least 10 km upstream, i.e. the area where the youngest 1000 annual layers in the core were deposited.

The dashed curve in Fig.10 is a preliminary result of a correction based on combined flow models of the types described in section 4. However, more surface data are needed for making a reliable correction of the isotope record prior to 1400 A.D.

5.4.2. The Pleistocene record

The deepest 250 m of the Dye 3 dep core spans most of the last glaciation, and the bottom 25 m of silty ice undoubtly originates from a period warmer than the present interglacial, probably the Eemian, cp. Fig.11.

The time scale to the left is essentially derived from comparison with dated deep sea core records, but it has only been extended to 90 ka B.P., i.e. 90,000 yrs before present, because the core is hardly continuous beyond this date (6).

The Pleistocene record needs no correction, because the $\delta^{18}O$ variability is much higher than possible upstream effects.

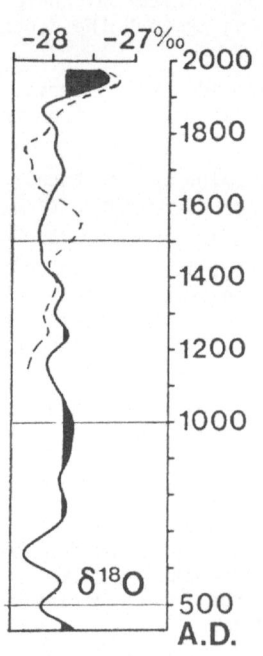

Fig. 10. Solid curve: $\delta^{18}O$ profile along the upper 650 m of the Dye 3 deep ice core, spanning the last 1500 yrs. Dashed curve: *Same after correction for upstream effects back to 1150 A.D.*

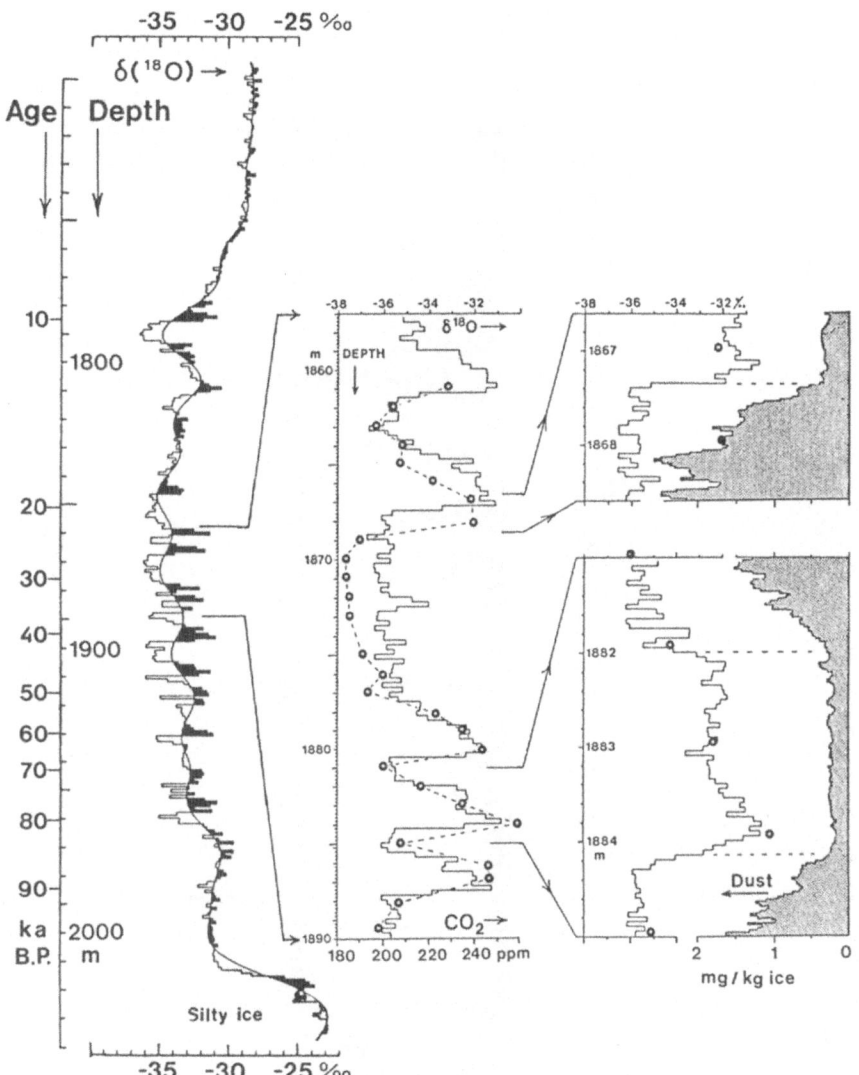

Fig. 11. Left section: *Continuous δ¹⁸O profile along the deepest 340 m of the Dye 3 ice core, provided with a depth scale and an approximate time scale.* Mid section: *Details of the 1857 to 1890 m increment. CO_2 concentration analyses of air inclusions (from 11).* Right section: *Further details of some of the drastic δ¹⁸O shifts, compared with continuous dust concentration profiles.*

The abrupt shifts in δ, some of which are shown in detail in the mid and right sections of Fig. 11, may be caused by sudden retreats, and less sudden advances of the ice cover in the North Atlantic Ocean, apparently with a periodicity of 2550 years (*10*).

The concentrations of CO_2 (*11*) and dust oscillate essentially in phase and antiphase with δ, respectively, but it is still doubtful, if the CO_2 analyses are interpretable in terms of atmospheric concentrations.

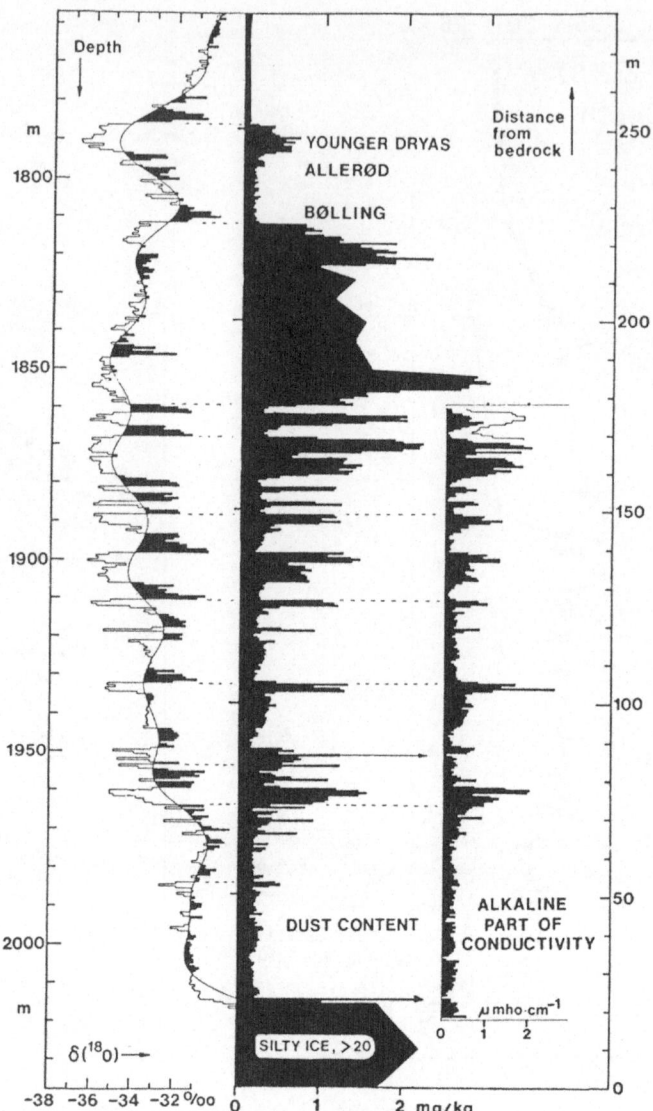

Fig. 12. Left section: *The δ profile along the deepest 280 m of the Dye 3 ice core.* Mid section: *Insoluble particle concentration.* Right section: *The alkaline part of the conductivity, which is mainly due to $Ca(HCO_3)_2$ (from 20).*

At least, it cannot be excluded that the CO_2 analyses on Pleistocene ice in Greenland may be affected by the alkalinity of the ice, which varies in antiphase with δ, cp. Fig. 12 (from 20).

Fig. 12 also shows that the antiphase variation of δ and insoluble particle (dust) concentration is a general feature in ice deposited during the glaciation. A possible explanation is that, in periods of colder climate (lower δ values), there was less precipitation, and higher flux of dust into the Arctic due to higher storminess in areas covered by loess.

ACKNOWLEDGEMENTS

This work was done under contract No. CLI-067-DK(G) with the European Economic Communities. Further contributions were provided by the University of Copenhagen, the Danish Commission for Scientific Research in Greenland, the Danish Natural Science Research Council, the Royal Danish Air Force, and the U.S. National Science Foundation, Division of Polar Programs.

REFERENCES

1. GISP, Greenland Ice Sheet Program, Science Plan, 2'nd Ed. (1976). Univ. Nebraska, Lincoln, Nebraska, U.S.A.
2. GUDMANDSEN, P. (1976). Studies of ice by means of radio echo sounding. Report 162, Electromagnetic Institute, Techn. Univ. of Denmark, Lyngby.
3. UEDA, H.T. and GARFIELD, D.E. (1969). The USA CRREL drill for thermal coring in ice. J. of Glaciol., 8, 311-314.
4. RUFLI, H., STAUFFER, B. and OESCHGER, H. (1976). Lightweight 50 m core drill for firn and ice. Proc. Symp. Ice-core drilling (J.F. Splettstoesser ed.), Univ. of Nebraska Press, Lincoln, Nebraska, 139-153.
5. JOHNSEN, S.J., DANSGAARD, W., GUNDESTRUP, N., HANSEN, S.B., NIELSEN, J.O. and REEH, N. (1980). A fast lightweight core drill, J. Glaciol., 25, 169-174.
6. DANSGAARD, W., CLAUSEN, H.B., GUNDESTRUP, N., HAMMER, C.U., JOHNSEN, S.J., KRISTINSDOTTIR, P.M. and REEH, N. (1982). A new Greenland deep ice core, Science, 218, 1273-1277.
7. GUNDESTRUP, N., JOHNSEN, S.J. and REEH, N. (1983). ISTUK, a new deep core drill system, Proc. 2'nd Internat. Symp. Drilling Technol., published by USA CRREL, Hanover, New Hampshire.
8. LANGWAY, C.C., DANSGAARD, W. and OESCHGER, H. (eds.) (1984). Proc. Symp. GISP 1971-1981, Am. Geophys. Union Meeting, Philadelphia May-June 1982, to be published, A.G.U. Special vol., in press.
9. SHOJI, H. and LANGWAY, C.C. (1984). Flow behavior of basal ice as related to modeling considerations, Ann. of Glaciol., 5, 141-148.
10. DANSGAARD, W., JOHNSEN, S.J., CLAUSEN, H.B., DAHL-JENSEN, D., GUNDESTRUP, N., HAMMER, C.U. and OESCHGER, H. (1984). North Atlantic Climatic Oscillations revealed by deep Greenland ice cores, in Climate Processes and Climate Sensitivity (J.E. Hansen and T. Takahashi eds.), Am. Geophys. Union, Geophys. Monogr. 29, Maurice Ewing Volume 5, 288-298.
11. STAUFFER, B., HOFER, H., OESCHGER, H., SCHWANDER, J. and SIEGENTHALER, U. (1984). Atmospheric CO_2 concentration during the last glaciation, Ann. of Glaciol., 5, 160-164.
12. OVERGAARD, S. and GUNDESTRUP, N. (1984). Bedrock topography of the Greenland Ice Sheet in the Dye 3 area. Proc. GISP Symp., Am. Geophys. Union Meeting, Philadelphia, May-June 1982, AGU Spec. Vol. in press.
13. DAHL-JENSEN, D. (1984). The flow properties at Dye 3, South Greenland derived from borehole tilting measurements and perturbation modelling, J. of Glaciol., in press.
14. DANSGAARD, W., JOHNSEN, S.J., CLAUSEN, H.B. and GUNDESTRUP, N. (1973). Stable isotope glaciology, Medd. om Grønland, 197, nr. 2, 53 pp.
15. MOCK, S.J. (1976). Geodetic positions of borehole sites of the Greenland Ice Sheet Program, USA CRREL report 1976-41, Hanover, New Hampshire.

16. REEH, N. and GUNDESTRUP, N. (1984). Mass balance of the Greenland Ice Sheet at Dye 3. Submitted to J. Glaciol.
17. DREW, A.R. and WHILLANS, I.M. (1984). Measurement of surface deformation of the Greenland ice sheet by satellite tracking (1984), Ann. of Glaciol., 5, 51-56.
18. GUNDESTRUP, N. and HANSEN, B.L. (1984). Borehole survey at Dye 3, South Greenland, J. of Glaciol., in press.
19. SHOJI, H., and LANGWAY, C.C. (1983). Volume relaxation of air inclusions in a fresh ice core, J. Phys. Chemistry, 87, 4111-4114.
20. HAMMER, C.U., CLAUSEN, H.B., DANSGAARD, W., NEFTEL, A., KRISTINSDOTTIR, P. and JOHNSON, E. (1984). Continuous impurity analyses along the Dye 3 deep core, Proc. GISP Symp., Am. Geophys. Union Meeting, Philadelphia May-June, 1982, A.G.U. Special volume, in press.
21. KOERNER, R.M. and FISHER, D.A. (1979). Discontinuous flow, ice texture and dirt content in the basal layers of the Devon Island ice cap, J. of Glaciol., 23, 209-221.
22. REEH, N., JOHNSEN, S.J. and DAHL-JENSEN, D. (1984). Dating the Dye 3 deep ice core by flow model calculations, Proc. GISP Symp., Am. Geophys. Union Meeting, May-June, 1982, A.G.U. Special volume, in press.
23. PICCIOTTO, E., DE MAIRE, X. and FRIEDMAN, I. (1960). Isotopic composition and temperature of formation of Antarctic snows, Nature, 187, 857-858.
24. JOHNSEN, S.J., DANSGAARD, W., CLAUSEN, H.B. and LANGWAY, C.C. (1972). Oxygen isotope profiles through the Antarctic and Greenland ice sheets, Nature, 235, 429-434.
25. JOHNSEN, S.J. (1977). Stable isotope homogenization of polar firn and ice. Internat. Union of Geodesy and Geophysics, IASH, Commission on Snow and Ice, Grenoble 1975, IAHS-AISH publication No. 118, 210-219.
26. EPSTEIN, S. and BENSON, C. (1959). Oxygen isotope studies, Trans. Am. Geophys. Union, 40, 81-84.
27. HAMMER, C.U., CLAUSEN, H.B., DANSGAARD, W., GUNDESTRUP, N., JOHNSEN, S.J. and REEH, N. (1978). Dating of Greenland ice cores by flow models, isotopes, volcanic debris, and continental dust, J. of Glaciol., 20, 3-26.
28. DANSGAARD, W., JOHNSEN, S.J., REEH, N., GUNDESTRUP, N., CLAUSEN, H.B. and HAMMER, C.U. (1975). Climatic changes, norsemen and modern man, Nature, 255, 24-28.
29. HAMMER, C.U. (1980). Acidity of polar ice cores in relation to absolute dating, past volcanism, and radio-echoes, J. of Glaciol., 25, 359-372.
30. HERRON, M.M. (1982). Impurity sources of F^-, Cl^-, NO_3^- and SO_4^{2-} in Greenland and Antarctic precipitation, J. Geophys. Res., 87, 3052-3060.
31. RISBO, T., CLAUSEN, H.B. and RASMUSSEN, K.L. (1981). Supernovae and nitrate in the Greenland ice sheet, Nature, 294, 637-639.
32. HAMMER, C.U., CLAUSEN, H.B. amd DANSGAARD, W. (1980). Greenland ice sheet evidence of post-glacial volcanism and its climatic impact, Nature, 288, 230-235.
33. REEH, N., CLAUSEN, H.B., DANSGAARD, W., GUNDESTRUP, N., HAMMER, C.U. and JOHNSEN, S.J. (1978). Secular trends of accumulation rates at three Greenland stations, J. of Glaciol., 20, 27-30.

AIR-SEA INTERACTION DURING THE LAST 100 YEARS AND TIME SCALES OF CLIMATIC FLUCTUATIONS

H. FLOHN and K.-H. WEBER

Meteorologisches Institut, Universität Bonn (FRG)

Abstract

In the first chapter some contributions to maritime climatology and air-sea interaction are given. They include time-series of fluxes of sensible heat and evaporation, comparisons of maritime wind observations and geostrophic winds, large-scale changes of Atlantic circulation patterns and the role of equatorial upwelling/downwelling for the exchange of water vapour and carbon dioxide.

These examples have been selected according to their role for changes of the CO_2 content under natural conditions and for a rational interpretation of the enigma of "abrupt" climatic changes. At the interannual scale much smaller counterparts ("jumps") do occur.

Introduction

Oceans cover 71 percent of the earth and control climate due to their storage capacity of heat and water. Physical understanding of recent climatic fluctuations depends, to a large extent, on our knowledge of air-sea interaction and its evolution since the beginning of instrumental observations. PALTRIDGE and WOODRUFF (1981) suggested that the global temperature trend of the sea surface temperatures (SST) is delayed with respect to air temperature (AT) indicated by northern continental stations. FOLLAND et al. (1984) analysed a global set of SST and maritime AT (MAT) data, applying some corrections. A comparison between MAT and AT (his Fig. 2d) at the northern hemisphere seems to indicate similarly a delay, at least between 1900 and 1940; no critical discussion can be given now. These marine observations now become increasingly important, because of the need for long homogeneous records at a global scale and for heat flux studies; the latter should be a key in a geophysical interpretation of climatic fluctuations.

Air-sea interaction plays also a key role in another newly arisen problem. Increasing evidence shows climatic fluctuations of the geological past did not happen at a 10^4-10^5 year scale, as expected from the earth's orbital fluctuations, but at a much shorter scale in the order of a century (FLOHN 1979, 1984). In European physical laboratories - at Bern (OESCHGER et al. 1983a, 1983b) and Grenoble (LORIUS and RAYNAUD 1983) - a surprising discovery was made: such "abrupt" changes are accompanied by simultaneous variations of the CO_2 content of the atmosphere. A possible working hypothesis (FLOHN 1983) interprets this coincidence as caused by the air-sea exchange of CO_2 and H_2O; at any rate, it emphasizes the role of the greenhouse effect for climatic change, be it natural or man-made.

A) Air-Sea Interaction and Atmospheric Circulation

1) Sea surface temperature and air temperature

        Maritime observations include wind, state of the sea, SST and less
frequently humidity of the air. Techniques of measuring SST and MAT varied
with time, and new techniques - measuring SST in buckets or at the intake
of cooling water in the engine room, MAT in screens or with sling psychro-
meter - led to systematic errors which are difficult to quantify and to
correct. While intake measurements of SST tend to be 0.4-0.6°C higher than
bucket measurements - see, among many others, BARNETT (1984), RAMAGE (1984)
-, recent evaluations seem to indicate that the correction is smaller, e.g.
(FOLLAND et al. 1984) +0.3°C for data before 1940. German observations
(WEBER 1984 and earlier results) show however, that the difference SST-MAT
which is so important for the flux of sensible heat (and to some degree also
of latent heat), varies also systematically before 1920. At a representative
field in the tropics (5-10°N, 70-80°E) this difference drops from 0.25°
(1902-13) to 0.16° (1922-37), but rises again to 0.55° (1958-67), using all
available observations. While the diurnal temperature variations of SST
remain constant, that of MAT during the middle period is definitely higher
(Fig. 1) than during the other periods. This reflects a change in measuring
MAT: thermometers in screens gave higher daytime values due to the solar
heating of the ship and bad ventilation. Together with the uncertainty of
wind speed estimates this yields unrealistic long series of the flux of
sensible heat H (RAMAGE 1984); no absolute calibration is possible. The
same is true for evaporation LE. However, it is possible to evaluate spatial
and time variability of these parameters from series after 1948, especially
in well-documented areas along the main shipping routes.

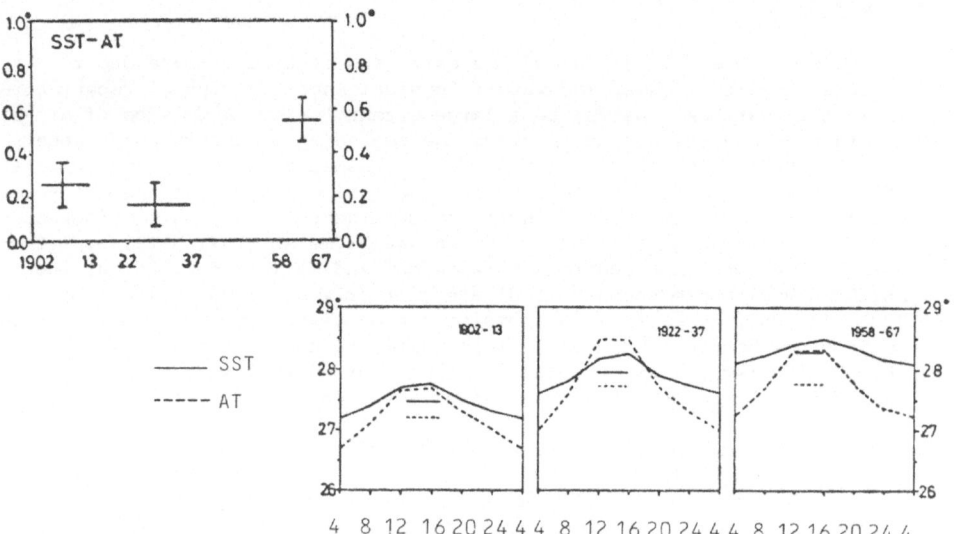

Fig. 1.   SST and MAT in the field 5-10°N, 70-80°E (Wof Sri Lanka). Above :
          average difference SST-MAT; below ; diurnal variation of SST and
          MAT.

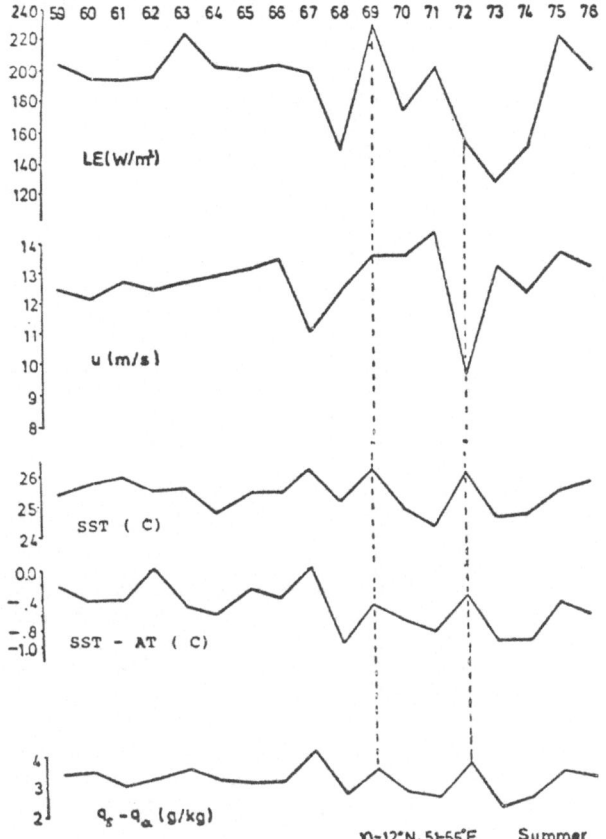

Fig. 2 Time series of LE, wind speed u, SST, SST-MAT and $q_s - q_a$ off the Somali coast ($10-12^{\circ}N$, $51-55^{\circ}E$), summer 1959-1976 (WEBER). Note the anomalies of the Niño year 1972.

2) Time series of sensible heat flux and evaporation

In most tropical and subtropical latitudes H is positive and relatively small, compared to evaporation; the Bowen ratio is usually of the order 0.05-0.10, in temperate latitudes of the order 0.2-0.3. In wintertime along east coasts of continents, where very cold arctic air is advected over warm water, very large fluxes of sensible heat occur. Negative values can occur during spring and summer along coasts with prevailing advection of warm continental air above cool water: this suppresses convection. The same is true for areas with upwelling cool water, including the equatorial Pacific and Atlantic. The effective difference is that between oceanic skin temperature and true air temperature just above the interface; SST and MAT give only a rough estimate which nevertheless proves to be quite sensitive. As an example the annual variation of air-sea exchange parameters are given for an area off the southern coast of the Arabian peninsula (Fig. 2), along the shipping route to the Persian Gulf, with great seasonal variations. In this area the reversal of SST-MAT occurs between May and September, with negative values of H; SST drops nearly $6^{\circ}C$ between May and August. The time variability of seasonal values of H and LE at the ocean surface is surprisingly high (Table 1).

Table 1: Interannual Variability (IAV) of evaporation (LE) and sensible heat flux (H), Watt/m^2

		Mean	Max.	Min.	IAV	
Weather Ship C	LE	151.0	192.3	118.1	23.1	
(53°N,35°W, 22 years)	H	22.3	38.8	5.1	9.1	
October–March	LE+H	173.3	213.1	134.2	23.0	
Field 0.4°S,28–34°W	LE	93.4	114.0	81.0	8.8	
13 years,	H	3.1	6.0	1.0	3.1	
June–October	LE+H	96.5	117.0	82.0	9.8	
Light Ship Elbe 1	LE	75.7	92.7	65.1	7.7	
(54°N,8°E, 25 years)	H	19.0	26.0	11.0	3.6	
September–December	LE+H	94.6	117.5	76.5	11.4	

   While at low latitude oceans only the sign of H is important, LE often provides the largest term in the surface energy budget. However, reliable ship measurements of relative humidity or dew-point are rather rare. In the southern North Sea, both E and H depend on season and wind direction; a long marine fetch is only given with wind directions between about 300 and 360°. Here E becomes negative between March and June, but limited to the landborne wind direction between 030° and 270° (Fig. 3); the field of H is similar, but the area with negative (downward) flux of H is larger, slightly displaced towards a later month. In this investigation (KATZSCHNER 1978) the bulk aerodynamic formula is used with a constant drag coefficient of 1.3 x 10^{-3}. Unfortunately the correlation between the thermal stability SST-MAT, which is available in the Historical Sea Surface Temperature Data since 1861, and the saturation deficit q_s-q_a (q = specific humidity) is positive but not very high: at the equatorial Atlantic it reaches, under purely oceanic conditions, 0.62 resp. 0.72, in the eastern Pacific west of the Galapagos 0.49 and in the coastal upwelling region off western Africa only 0.29. Any attempt to estimate q_s-q_a (and then LE) from the available data cannot lead to really reliable values. This is the more true since the correlation between wind speed and SST-MAT is small, partly even negative.

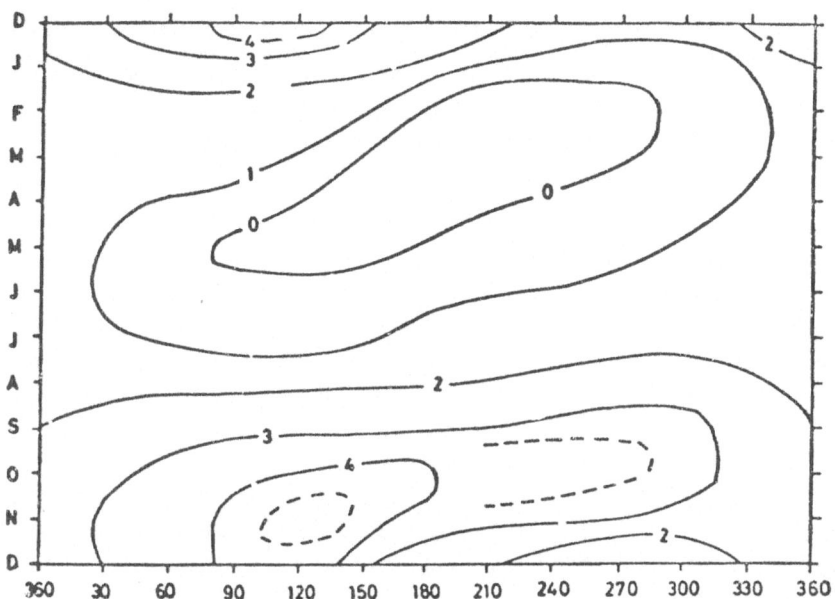

Fig. 3 Evaporation in mm/d at Lightship Elbe 1 in the German Bight
        (54°N, 8°E) as a function of season and wind direction (270° = W);
        1 mm/d equivalent to LE = 28.5 W/m².

        In equatorial upwelling regions LE is drastically reduced. At tropi-
cal oceans the daily amount of E varies around 4-5 mm/d; in the equatorial
areas of the eastern Pacific and Atlantic it can drop, in years with undisturbed upwelling and SST ∿23°, to values near or below 1 mm/d (HENNING-
FLOHN 1980); see also Chapter A 5.

        Since upwelling is coupled with thermal stability (SST<MAT), the rela-
tive humidity at the boundary layer increases to values around 90 percent,
to be compared with average values around 78 percent. This fact should not
be interpreted as indicating higher evaporation: indeed E is low, but the
water vapour is trapped in the boundary layer, while tropospheric subsidence
leads to a minimum of precipitable water in the atmospheric column above
(GRODY et al. 1980).

3) Ocean surface winds and geostrophic winds

        Few studies are available comparing, above the oceans, observed surface
winds and geostrophic winds derived from the sea-level pressure field. This
gap can be filled at least over the North Atlantic, where fairly reliable
daily pressure maps are available since 1881. Three comparable fields have
been selected:

        HSSTD field 34: 46-52°N, 23-30°W
          "   "   44: 40-46°N, 20-30°W
          "   "   62: 27-32°N, 17-25°W

        Here the boundaries of these fields are fairly similar to the 5°x10°
grid of the pressure fields. From these examples, some tentative results
shall be outlined. While field 34 is situated right at the North Atlantic

Table 2: Comparative annual results of maritime surface winds (m) and geostrophic winds (g) at the northern Atlantic Ocean. Average positions given in degrees latitude and longitude; periods a-d see
text, q = constancy (v_{scal}/v_{result}) in percent, dd = direction of
resultant winds (v), velocity in m/s.

Period and central position		Marine observations			geostrophic data	
		v_{scal}	q	dd_m	dd_g	v_{result}
Field 34	a	9.48	34	241.6°	255.2	5.51
m: 48.3°N,26.1°W	b	9.04	43	257.8	256.7	6.09
g: 47.5°N,25.0°W	c	9.18	40	259.2	256.8	6.16
	d	9.57	41	255.7	255.9	6.22
Field 44	a	8.69	26	257.5	261.7	3.68
m: 42.8°N,24.7°W	b	8.17	28	268.2	261.9	5.05
g: 42.5°N,25.0°W	c	7.81	31	265.6	262.6	5.32
	d	8.01	33	263.6	260.7	4.71
Field 62	a	6.67	48	31.5	46.3	3.59
m: 29.5°N,19.5°W	b	5.74	55	37.1	40.0	4.67
g: 30.0°N,20.0°W	c	5.85	58	37.2	44.1	4.30
	d	6.26	58	31.3	42.4	3.92

main shipping route (British Channel to American east coast), the position
of field 44 is just south of it. Field 62, along the shipping route from
Europe to South America, lies off the coast of Morocco, i.e. in the NE-
trade region.

Only a few (preliminary) results shall be given here (Table 2), limited to the annual values. Since during both world wars observations are
lacking, we distinguish four different periods: 1885-1899, 1900-14 (or 13),
1920-38 and 1946-60, denoted as a, b, c and d; the number of pairs varies
between 14 and 19. The resultant (marine) surface winds show a weak tendency
to rotate clockwise from period a to period b and c; an even weaker tendency of an anticlockwise rotation can be observed towards period d. However,
the standard deviation of the u- and v-components indicate that this rotation remains below the significant level. The resultant geostrophic wind
fails to coincide with these changes; here the permanent smoothing of the
pressure field during the analysis renders it less sensitive against minor
changes. The scalar speed of the (marine) wind tends to show highest values
during period a, and decreases during periods b and c; during period d a
weak increase is observed. But these variations could also be an artifact,
caused by the different calibrations of the Beaufort scale (RAMAGE 1984).

Correlating the u- and v-component of both data sets, the correlation coefficient tends to increase towards period d. This could be interpreted as a gradual improvement of the reliability of the pressure analyses; the greatest deviations of individual seasons and years occur before 1900. 53 percent of these ccf's are 0.80 or higher; only in the field 62 (trades) statistically insignificant ccf's occur (35 percent). The surface resultant wind is distinctly weaker than the geostrophic resultant, in field 62 82 percent of the latter. The constancy (steadiness) of the winds increases distinctly around 1900; at field 62 it is fairly constant and high during summer, but quite variable during the remaining seasons (Fig. 4).

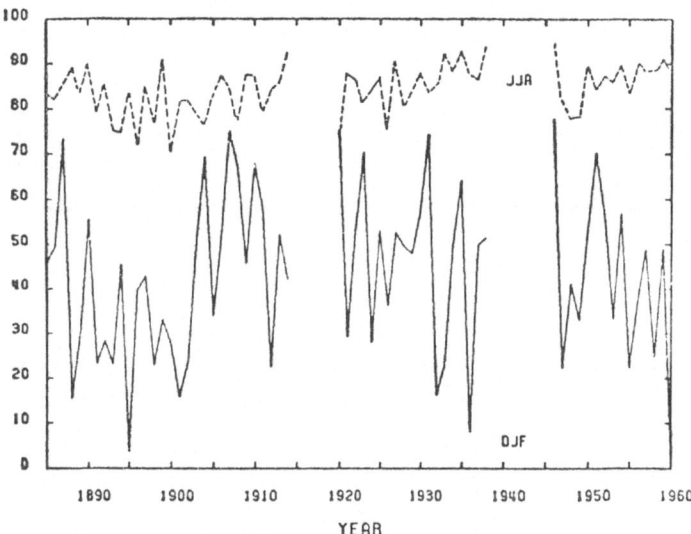

Fig. 4 Steadiness of winds v_{RES}/v_{scal}) in percent at extreme seasons (summer = JJA, winter = DJF) off the NW coast of Africa (Lat. 27-32°N, Long. 17-25°W). Data source: Historical Sea Surface Temperature Data (HSSTD) (K.-H. WEBER 1984).

4) Variations of the position of the North Atlantic "centers of action"

Sea-level pressure data at 5° Lat x 10° Long. grid points, available at the North Atlantic between Lat. 20°N and 70°N since 1881, serve as base for an investigation of the variability of the latitude of the two main centers of action, the Icelandic Low (IL) and the Azores High (AH). Zonal averages are used for the section Long. 10°W-70°W; position and central pressure of both centers have been derived for each individual month using polynomials (for details see R. GLOWIENKA 1984) as well as the meridional pressure gradient. The ccf's between pressure and latitude of IL (AH) are negative (positive) and statistically significant for most months. Those between the position of both centers are positive (between 0.46 and 0.70), between the pressure negative, as found also by WALLACE and GUTZLER (1981).

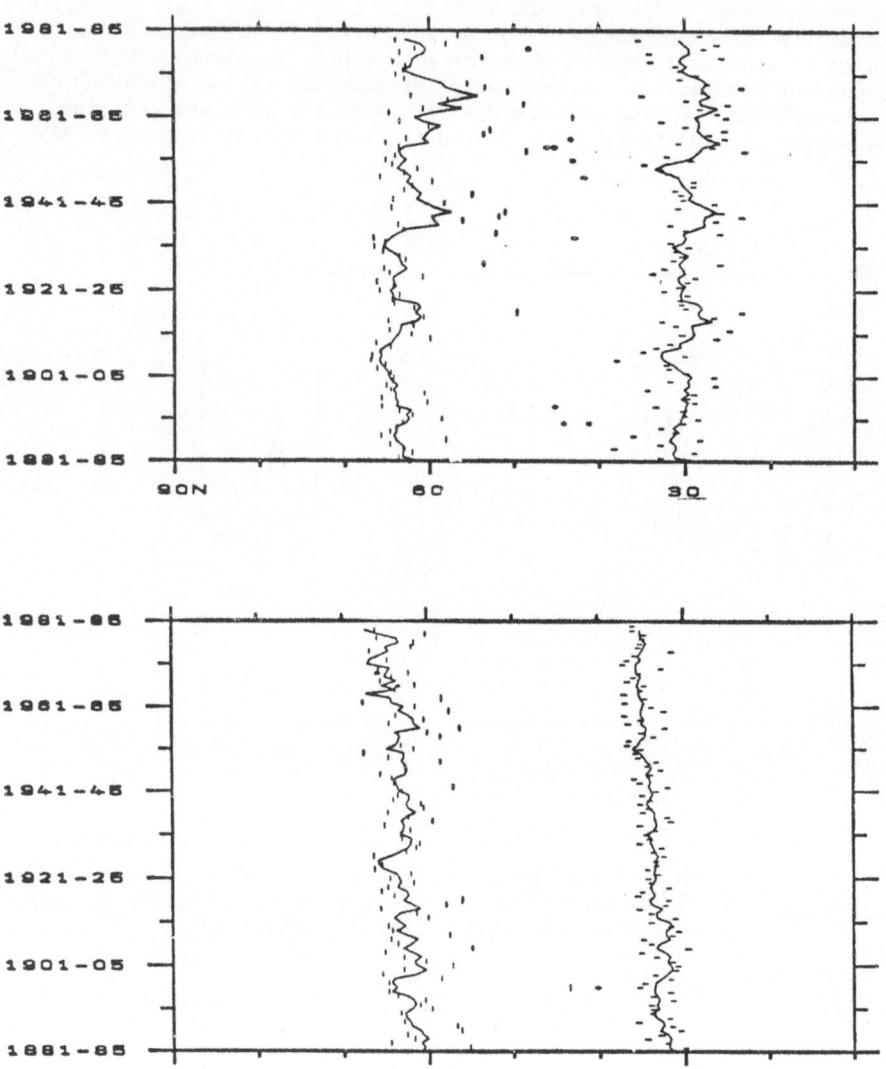

Fig. 5 Position of Icelandic Low (left) and Azores High (right), indivi-
       dual years (dashes) and 5-year averages (solid lines) for the period
       1881-1983, averaged over Long. 10-70°W, January (above) and July
       (below). Note. the opposite trend (R. GLOWIENKA 1984).

Of high interest are the (linear) trends of the positions (Fig. 5).
During summer both centers are shifted northward; the displacement rates
for IL (AH) July are 0.046° (0.034°) Lat. per year. In contrast to this
both centers are shifted southward during winter; the January displacement
rates are resp. -0.044° (-0.022°) Lat. per year. A total displacement around
400 kms during 103 years is certainly not negligible. While the shifts
are significant at the 99 percent level, the associated pressure changes
are insignificant, except the IL pressure during December, with a trend
of +0.053 mb/year.

The meridional pressure gradient varies strongly, especially during
winter: in January the extreme 5-year-averages are 0.35 and 0.90 mb/°Lat.

5) Parallel variations of H_2O and CO_2 ?

Recent studies in upwelling areas (HENNING-FLOHN 1980, WEBER-FLOHN
1984) have shown that saturation deficit and evaporation drop sharply with
decreasing SST. Surprisingly enough, this has a global effect on tropospheric temperatures: PAN and OORT (1983) found a significant ccf up to
+0.65 between SST at the equatorial Pacific and the temperature of the
entire mass of the Northern Hemisphere atmosphere 3-7 months later. At
least during northern winter, the H_2O content of most of the global
atmosphere, especially in the tropics, was significantly higher in cases
with "warm" SST at the key region than in the "cold" cases.

Regarding CO_2, several investigators (NEWELL et al. 1978, ANGELL 1981)
found significant positive correlations between the growth rate of CO_2
after removal of the seasonal fluctuations, and the SST anomalies in the
equatorial Pacific. BACASTOW and KEELING (1981) demonstrated a negative
correlation between the CO_2 growth rate and the Southern Oscillation Index,
which itself is highly negatively correlated with SST in the eastern
equatorial Pacific. In five El Niño years (with warm water at the key
region: 1958, 1963, 1965, 1969, 1972) the annual increase of CO_2 was 1.04
ppm, while in five "cold" years (1960, 1964, 1967, 1971, 1974) it dropped
to 0.57 ppm. Selecting periods of 6 months length with " warm" or "cold"
SST at Christmas Island (1.5°N, 157°W), WEBER and FLOHN (1984) found in
12 "warm" periods a CO_2 increase of 1.06 ppm/6 months, in 10 "cold" periods
an increase of only 0.38 ppm; the difference is significant at the 95 percent level. Similar differences had been found when selecting "dry" cases
(with suppressed rainfall during upwelling) and "wet" cases (with high
rainfall during El Niño); the correlation between SST anomaly and rainfall
index is 0.70. These interannual differences of CO_2 exchange through the
air-sea interface reach ∿1.4 ppm or ∿3 Gt (= 10^{12} kg) Carbon per year, i.e.
they are of the same order as the annual storage of fossil CO_2 in the air
(2.6 Gt).

While the H_2O exchange through the air-sea interface occurs with a
phase change, following physical laws, the CO_2 exchange is also a biological
process (NEWELL et al. 1978, BAES 1982): in the upwelling nutrient-rich
water primitive algae consume CO_2 from the air and from the CO_2-super-
saturated water, while the nearly sterile warm water of the tropical mixing
layer is deficient of nutrients and thus releases CO_2 into the atmosphere.
Recent investigations of the bioproductivity of the upwelling oceans indicate an annual CO_2 production by photosynthesis in the right order of magnitude (WEBER-FLOHN 1984). With a global model of the oceanic primary
productivity VIECELLI (1984) estimates the increase of atmospheric CO_2
content with a 1 percent productivity loss to 0.5-2.5 percent (depen-

ding on turbulent mixing), which is also of the right order of magnitude (Gt/year). CHAVEZ et al. (1984) estimate the primary productivity under cool conditions to 2 Gt/a, under El Niño conditions to about one third.

These investigations indicate a parallel variation of the water vapour and carbon dioxide contents of the atmosphere. Both gases contribute to the greenhouse effect of the atmosphere - indeed H_2O can be more efficient than CO_2 (RAMANATHAN 1981), especially in the tropics. This parallel course enhances the CO_2 effect; however, it will be difficult to find convincing evidence for a gradual increase of the atmospheric H_2O content too. But this is likely if the global SST increase of about $0.6^{\circ}C$ since the beginning of this century (FOLLAND et al. 1984) is correct. It would also be diffi- cult to find convincing evidence for an increase of global precipitation. The problem of maintenance of the global water budget belongs to the most difficult, but also the most urgent questions: it is a real challenge for future climate research.

B) Abrupt Climatic Changes (scale 1-100a) Now and in the Past

1) Occurrence of climatic "jumps" in recent data series

Detailed studies of long climatic series have indicated that climatic changes - statistically described as a more or less linear trend - in some cases occur stepwise, with a short and rather rapid transition between two different modes, each of them including relatively minor fluctuations. Two quite dramatic examples have been observed in the runoff series of the Nile (1899) and of Lake Victoria (1962). The mean annual Nile runoff at Aswan fluctuated, during the period 1870-1898, around an average of 109.8 km^3/a (standard deviation 13.5 km^3/a), while during 1899-1928 the equivalent figures were 83.7 and 13.9 km^3/a respectively. This means a decrease of about 25 percent, twice the standard deviation; KRAUS (1955) has demonstra- ted some parallel trends, while RIEHL et al. (1979) and HASSAN (1981) dis- cussed longer time-series. The Aswan record has been found to be homoge- neous; thus this drop reflects a drastic change in the summer rainfall regime of Ethiopia, which contributes about 80 percent of the annual discharge. In recent decades the Nile runoff has been increasingly manipu- lated for irrigation purposes.

Even more striking is the abrupt change of the level of Lake Victoria, caused by the rainy season 1961/2 which brought 200-600 percent of the nor- mal value in large areas of eastern Africa. Lake Victoria rose about 2.5 m and remained high: the runoff of the White Nile rose from 20.8 km^3/a for the period 1900-61 to 41.2 km^3/a for 1962-78, with standard deviations of 4.6 and 4.7 km^3/a respectively (FLOHN and BURKHARDT 1984). Possible physi- cal interpretations have been discussed elsewhere (FLOHN 1983a).

Similar infrequent cases exist. We may mention the rainfall over the Sudan-Sahel belt of western Africa, where drought conditions last nearly uninterrupted since 1968. Using normalized anomalies from 20 stations (Lat. $11-18^{\circ}N$, west of $9^{\circ}S$, ca. $2 \times 10^6 km^2$), LAMB (1984) indicates, during the wet period 1950-67, only one year with weak negative anomalies, during the dry period 1968-83 only one year with weak positive anomalies. Multi-year droughts are well-known in this area, but the present duration (including 1984) seems to be exceptionally long.

Another example has been given by an evaluation of the temperature of the 500/1000 mb layer above the central Arctic, averaged over the whole area north of Lat. 75°N (about 8.8 x 10^6 km^2). These data (SCHAIRDEL-FLOHN 1983) indicate a rather abrupt cooling between 1961 and 1963, mainly during the summer months. The average deviation of the temperature of the lower troposphere is +0.48°C (1949-61) and -0.42°C (1962-76); the standard deviations of the two periods are 0.43°C and 0.35°C. Only two years of each period indicate a weak opposite deviation. Surface temperatures show a similar drop, while the sea-ice area apparently has increased by about 0.5x10^6 km^2; its correlation with tropospheric temperature is -0.54, significant at the 99 percent level.

These "jumps" have in common that the difference of the averages of consecutive periods (in the order of 15-30 years) reaches or exceeds twice the standard deviation of the individual periods. They occur at a regional (not local) scale and represent areas of several million km^2 (Lake Victoria only about 10^5 km^2). At present, speculations about their origin are not meaningful: spatial coherence and cohesiveness in time are hitherto unknown.

2) Occurrence of abrupt changes in the geological past

Recent investigations at continental sites have revealed that in a number of cases during the Pleistocene a change of annual or summer temperatures of 3 K or more evolved in a time-span of about 100 years. These abrupt changes are much shorter than expected as a consequence of orbital variations (10^4-10^5 years); they comprehend half or more of the difference glacial/interglacial. Table 3 gives a list of such events; for comments and references see FLOHN 1984).

Even more challenging cases have been found by physical investigations of a recent ice core (Dye 3 in southern Greenland, Lat. 65°N). DANSGAARD et al. (1982) have evaluated the $\partial^{18}O$ record which represents very nearly surface temperature, while OESCHGER et al. (1983a) and OESCHGER and STAUFFER (1983b) have measured the CO_2 content entrapped in air bubbles with an improved technique. Here we refer to OESCHGER-STAUFFER's investigations comparing both records at a depth of 1860-1900 m, equivalent to an age of 27-40 ka (10^3 years) ago. In a 30 m section (equivalent to about 10 ka, Fig. 6) they found non-periodic simultaneous fluctuations of both $\partial^{18}O$ and of CO_2 (between about 260 ppm and 190 ppm). The accuracy of the measurements has been discussed by BARNOLA et al. (1983), the age difference between the bubble air and ice by SCHWANDER and STAUFFER (1984). Most convincing is the high-resolution investigation of a 3 m section (OESCHGER-STAUFFER Fig. 8) equivalent to about 1 ka: here the transition from a "cold" level with an average CO_2 content of 190 ppm to a "warm" level (256 ppm) took not more than 100 years (32 cm). Fig. 7 gives a time-independent diagram of CO_2 data versus simultaneous $\partial^{18}O$ values (ccf = 0.95). A linear regression equation (WEBER-FLOHN 1984) reads

$$CO_2 \text{ (ppm)} = 617 + 11.7 \, \partial^{18}O \text{ (per mil)}$$

The same procedure for the much longer time series of Fig. 6 gives

$$CO_2 \text{ (ppm)} = 537 + 9.47 \, \partial^{18}O \text{ (per mil)}$$

with a ccf of 0.82 (due to the lower resolution and the less accurate fixation of simultaneous points). The $\partial^{18}O$ difference of the two different climatic periods in Fig. 7 (5.2 per mil) is equivalent - using DANSGAARD's (1964) data - to a local temperature difference of 8 K. With other techniques similar results have been reached for the well-documented cases a), b) and d) of Table 3.

Since at present the CO_2 fluctuations affected mostly the northern hemisphere, spreading towards the South Pole station in 1-2 years, such data stored nearly 2 kms deep in the Greenland ice are indeed indicative for the global climate. The most intriguing facts are the occurrence of "abrupt" climatic changes in the "human" 100^a-scale and the simultaneous participation of CO_2 fluctuations: both facts now force hesitating climatologists to revise some of their fundamental textbook ideas.

3) The "greenhouse effect" as a key for an abrupt climatic change?

What processes could have been responsible, under purely natural conditions, for CO_2 fluctuations of a comparable rate (60-80 ppm per 100 years) as those now produced by man? Based mainly on our present understanding of seasonal variations of the atmospheric/oceanic circulation, a working hypothesis has been proposed (FLOHN 1981, 1982, 1983c; WEBER-FLOHN 1984). It starts from the observed fact that the variable inter-annual changes of the atmospheric CO_2 content are correlated with the sudden reversals of oceanic upwelling/downwelling (El Niño); see Chapter A 5. The coincidence between glacials and global dryness and between interglacials and global wetness (SARNTHEIN 1978) suggests that the (under present conditions) observed parallel course of CO_2 and H_2O may have been a quite general feature of climatic history at scales between 1 and 10^4 years. The interglacial/glacial difference between global annual precipitation/evaporation has been (conservatively) estimated to 20-25 percent of the present value (FLOHN 1983c).

These fluctuations during the last ice age, in earlier interglacials or glacial/interglacial transitions (Table 3) indicate that in rare cases (ca. 0.1-0.5 per ka) abrupt climate fluctuations as described above occur under purely natural conditions.

In Chapter a it has been shown that the sudden transition between up-welling and downwelling - occurring in a few weeks or months - are responsible for large-scale changes in the H_2O and CO_2 fluxes between sea and air. Both gases are mainly responsible for the greenhouse effect: this effect contributes also to the difference between glacial and interglacial (OESCHGER et al. 1983a) and to the stadial-interstadial fluctuations with a (non-periodic!) time-scale in the order of 2-3 ka. Fluctuations of this scale are also observed during the Holocene as glacier variations in the Alps (GAMPER and SUTER 1982), Scandinavia (KARLEN 1976, 1979) and other high mountains. They are also confirmed at deep-ocean cores; SCHNITKER (1982) has proposed a climatic feedback mechanism of this time-scale based on the slow deep-sea currents transporting bottom water from its source in the subarctic Atlantic to the other oceans.

Our proposed feedback (FLOHN 1983c) - see also the more general view of SIEGENTHALER and WENK (1984) - is based on the observed seasonal and/or interannual fluctuations of the El Nino-type: these experiences are only expanded in time. We start from a hypothetical cooling (or warming) event caused e.g. by a cluster or a lull of heavy volcanic eruptions; their thermal effect shall be concentrated - as a consequence of the geometry of solar rays at high latitudes and of a longer residence time of aerosol

in the polar stratosphere - in polar regions. In equatorial regions
temperature changes are distinctly smaller and (due to the extension of
oceans) slower. This differential cooling (or heating) corresponds with an
expansion (or contraction) of the circumpolar vortex (Ferrel circulation),
in principle similar to the seasonal variations of each year. This is
also equivalent to the equatorward (poleward) shift of the subtropical
anticyclonic belt with a strengthening (weakening) of the tropical easterlies. This causes increasing upwelling, correlated (under natural conditions)
with suppressed evaporation, precipitation and CO_2 release into the atmosphere and further radiative cooling, or in the case of warming equatorial
downwelling (Nino) with increased evaporation and CO_2 release, increased
precipitation and tropospheric warming due to radiative processes as well
as release of latent heat (PAN and OORT 1983).

Conclusions

The ideas expressed in Chapter B on the occurrence of abrupt climatic
changes and their correlation with the air-sea exchange processes of greenhouse gases are based on a transfer of recent experiences at the interannual time-scale to a larger scale of 10^2 (and finally 10^3-10^4) years.
Similarly the observed interannual climatic "jumps", at a regional scale,
seem to be an analogue to these "abrupt" changes which most probably are
of a global scale. Thus we may emphasize two points: Understanding of
climatic changes at the time-scale near 100 years is not possible without
due account of the key role of the oceans and of the time variability of
the composition of the earth's atmosphere. And: the CO_2 climate problem
is not only a recent environmental issue: it is of a quite general nature
within the climatic evolution of our planet. This perception greatly under-
lines the need for further research. This is especially necessary at the
field of paleoclimatology: only here is a possibility to verify model
results which otherwise remain in a vacuum. Perhaps more essential are the
fundamental questions, which nature puts at the disposal of far-sighted
model-designers: we need their answers.

During the last years, several authors (e.g. BUDYKO 1982) suggested
a key role of CO_2 variations for the evolution of the global climate
since primordial times. This follows early ideas pointed out by Sv.
ARRHENIUS (1896) and Th. CHAMBERLIN (1897). Indeed the "acryogenic" climate
of the Mesozoic and Early Cenozoic - both poles ice-free - can probably
not be understood without the assumption of a higher atmospheric CO_2
level. It may be possible that the gradual transition to a unipolar
glaciated earth during the Late Cenozoic(existing between 14 and about 3
million years ago) and finally into the formation of a more or less permanent Arctic sea-ice cover, are both correlated with changes in the
atmosphere's composition. But here it is impossible to enter into the
geophysical and geochemical background of climatic evolution. Climatologists are now beginning to widen their time horizon from 300 years of
instrumental observations into the geological past. In this view the
anthropogenic growth of the CO_2 level is not new - we have to look
seriously into the very long history of the evolution of our climate and
to learn its lesson.

REFERENCES

J.K. ANGELL (1981): Monthly Weather Review 109, 230-243

Sv. ARRHENIUS (1896): Philos. Mag. (5th Ser.), 41, 236-275

R.B. BACASTOW, C.D. KEELING (1981): WMO/UNEP/ICSU Conference on Analysis and Interpretation of CO_2 data. Bern, 109-112

C.F. BAES jr. (1982): in W.C. Clark (Ed.) Carbon Dioxide Review 1982, Oxford Univ. Press, p. 187-204

T.B. BARNETT (1984): Monthly Weather Review 112, 303-312

J.M. BARNOLA et al. (1983): Nature 302, 410-413

M.J. BUDYKO (1982): The Earth's Climate, Past and Future. International Geophysical Series 29, 307 pp.

Th. CHAMBERLIN (1897): Journal of Geology 5, 653-683

F.P. CHAVEZ et al. (1984): Trop. Ocean/Atmos. Newsletter 28, 1-2

W. DANSGAARD (1964): Tellus 16, 436-468

W. DANSGAARD et al. (1982): Science 218, 1273-1277

H. FLOHN (1979): Quaternary Research 12, 135-149

H. FLOHN (1981): Phys. Blätter 37, 184-190

H. FLOHN (1982): Journ. Meteor. Soc. Japan 60, 268-273

H. FLOHN (1983a): Wiss. Ber. Meteor. Inst. Univ. Karlsruhe 4, 17-34

H. FLOHN (1983b): in A. Ghazi (Ed.) Palaeoclimatic Research and Models. D. Reidel, Dordrecht p. 17-33

H. FLOHN (1983c): in A. Street-Perrott et al. (Eds.) Variations in the Global Water Budget. D. Reidel, Dordrecht, p. 403-418

H. FLOHN (1984): in N.A. Mörner, W. Karlén (Eds.): Climatic Changes on a Yearly to Millennial Basis. D. Reidel, Dordrecht, p.521-531

H. FLOHN, Th. BURKHARDT (1984): in press

C.K. FOLLAND et al. (1984): Nature 310, 670-673

M. GAMPER, J. SUTER (1982): Geographica Helvetica 37, 105-114

R. GLOWIENKA (1984): in press (Contrib. Atmos. Phys.)

N.C. GRODY et al. (1980): Journ. Appl. Meteor. 19, 986-996

F.A. HASSAN (1981): Science 212, 1142-1145

D. HENNING, H. FLOHN (1980): Contrib. Atmos. Phys. 53, 430-441

W. KARLEN (1976): Geografiska Annaler 58A, 1-34; 61A(1979), 11-28

L. KATZSCHNER (1978) Dipl. Thesis Univ. Bonn (unpublished)

E.B. KRAUS (1955): Quart. Journ. Roy. Meteor. Soc. 81, 198-210

P. LAMB (1984): in press (Z. f. Gletscherkunde und Glazialgeologie)

C. LORIUS, D. RAYNAUD (1983): in W. Bach et al. (Eds.): Carbon Dioxide. D. Reidel, Dordrecht, p. 145-177

R.E. NEWELL et al. (1978): Pure and Applied Geophysics 116, 351-371

H. OESCHGER et al. (1983a) in: A. Ghazi (Ed.) Palaeoclimatic Research and Models. D. Reidel, Dordrecht, p. 95-107

H. OESCHGER, B. STAUFFER (1983b): Paper at Global Carbon Cycle Symposium, Knoxville, Tenn., Nov. 1983

Y.H. PAN, A.H. OORT (1983): Monthly Weather Review 111, 1244-1253

G. PALTRIDGE, S. WOODRUFF (1981): Monthly Weather Review 109, 2427-2434; cf. also 112 (1984), 1093-1095

C.S. RAMAGE (1984): Journ. Climate Appl. Meteor. 23, 187-193

V. RAMANATHAN (1981): Journ. Atmos. Sci. 38, 918-930

H. RIEHL et al.(1979): Monthly Weather Review 107, 1546-1553

M. SARNTHEIN (1978): Nature 271, 43-46

B. SCHAIRDEL, H. FLOHN (1983): Paper at IAMAP/UGGI Symposium Hamburg, August 1983

D. SCHNITKER (1982): Palaeogeography, Palaeoclimatology, Palaeoecology 40, 213-234

J. SCHWANDER, B. STAUFFER (1984): Nature 311, 45-47

U. SIEGENTHALER, Th. WENK (1984): Nature 308, 624-626

J.A.VIECELLI (1984): Climatic Change 6, 153-166

J.M. WALLACE, D.S. GUTZLER (1981): Monthly Weather Review 109, 784-812

K.-H. WEBER (1984): Final Report EC-Contract (unpublished)

K.-H. WEBER, H. FLOHN: Bonner Meteorologische Abhandlungen (in press)

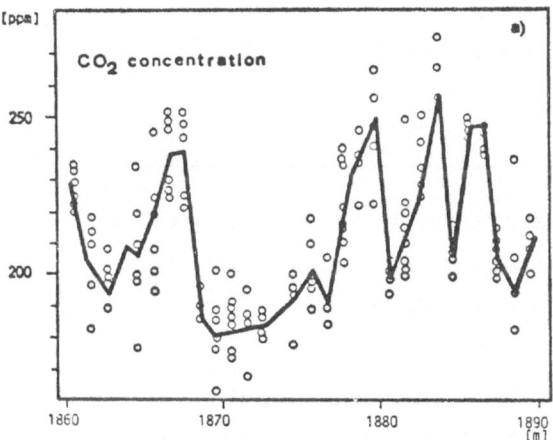

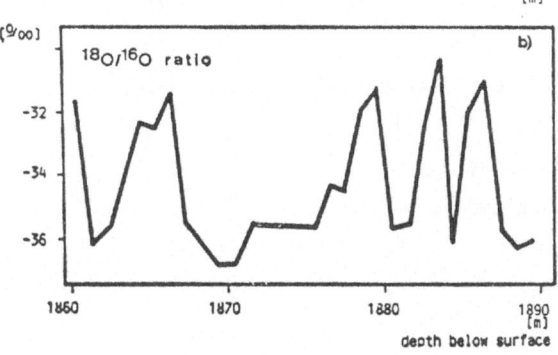

Fig. 6: CO_2 concentration and $\partial^{18}O$ ratio at 1860-1890 m depth (approximately 29-38 ka BP), Dye 3, Lat. 65°N, Greenland (OESCHGER-STAUFER 1983). Circles = individual CO_2 measurements, distance between averaged values ~ 300a

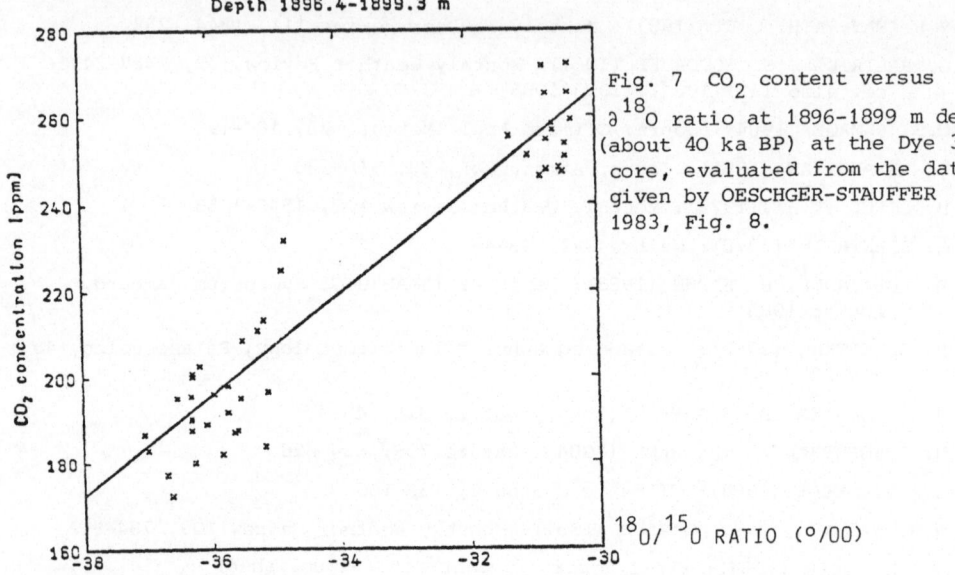

Fig. 7 CO_2 content versus $\partial^{18}O$ ratio at 1896-1899 m depth (about 40 ka BP) at the Dye 3 core, evaluated from the data given by OESCHGER-STAUFFER 1983, Fig. 8.

Table 3 **Abrupt Paleoclimatic Events**
(Time Scale ~100a)

ΔT		Time BP
W	Transition Y. Dryas - Preboreal	~10.2 ka
C	* Alleröd - Y. Dryas	10.8
?	Onset Moist Period (Lat. 15 - 35 °N)	12
W	Transition O. Dryas - Bølling	12.8
C	* Stages 5a / 4	73
C	* * 5c / 5b	95
C	* * 5e / 5d	115
C	Cold Episode Holstein Interglacial	190 ?
	(Stage 7, 9 ?)	
C	* * Stage 19	700
	(Matuyama - Brunhes)	

MODELLING THE ASTRONOMICAL THEORY OF PALEOCLIMATES IN THE TIME AND
FREQUENCY DOMAIN

An example of the relationship between long term and
short term climate changes

A. BERGER, P. PESTIAUX
Université Catholique de Louvain
Institut d'Astronomie et de Géophysique G. Lemaître

SUMMARY

A simple multiple regressive spectral model based upon monthly insolations shows that the next cool stage is expected to occur 5 kyr AP and
that a natural decrease of the atmospheric CO_2 would follow the present
cooling trend.
    Spectral analyses and a paleoclimatic data bank have allowed to per-
ceive the overall physical meaning of the isotopic spectra collected from
deep-sea cores. The response of the climatic system to the astronomical
forcing can not be isolated from the physical processes (bioturbation,
variable sedimentation rates, dissolution and ^{18}O fractionation) which
transform the original climate signal into proxy data like the $\delta^{18}O$ recorded in foraminifera burried in deep-sea cores. They all must be taken
into account before any astronomical calibration of the original geological data based upon too restrictive hypotheses is used.

1. Introduction

    Among the longest astrophysical and astronomical cycles which might
influence climate (and even among all forcing mechanisms external to the
climatic system itself) only those involving variations in the elements of
the Earth's orbit have been found significantly related to the long-term
series available in the geological record.
    Over the last 700 kyr record, $^{18}O/^{16}O$ isotope ratios and other clima-
tic proxy data revealed clear periodicities of 100 kyr, 40 kyr, 23 and 19
kyr. According to the now-respected Milankovitch theory (1,2) ice ages are
driven at these periodicities by changes, respectively, in the eccentricity of the Earth's orbit, the tilt and the precession of the Earth's rotation axis.
    In lights of all the results obtained so far, the relevance of this
astronomical theory can be assessed for modern climate research including
the study of climate variability and prediction at the human time scale.
The repercussion of this research comes directly from its results (e.g.,
validation of climate models responding to an exact astronomical forcing)
but also from the transfer of techniques elaborated for its own purpose
and largely applicable to other climate studies (e.g., spectral analysis,
insolation models ...).

2. The long-term and short term variations of insolation

As the climate system is thermally driven by solar insolation, there is a real interest to check whether or not there are some relationships between insolation parameters and the climate at the global scale. The difficulty which then first arises is to determine which are the most realistic latitudes and during which periods of the year. The adjustment may be tested in the time and/or the frequency domains. This is why the different kinds of insolation which are supposed to be used for modelling the climate or for simulating the climatic variations, have been carefully analysed as far as their computation, accuracy and spectrum are concerned (3,4).

These insolations depend upon (i) 4 astronomical parameters which all affect the total energy received by the planet : the astrophysical solar constant, the length of the tropical year and of the day, the secularly variable mean distance from the Earth to the Sun (defined by the eccentricity, e), (ii) and 2 other ones which re-distribute differently the energy among the latitudes and months : the climatic precession, $e \sin \tilde{\omega}$, and the obliquity, ε. This dependance is summarized in table I.

Table I. Insolations as a function of astronomical parameters (++ means stronger dependancy).

	ε	$e \sin \tilde{\omega}$
Mid-month insolation at equinoxe		+
at solstice	+	++
Half-year astronomical seasons		
- total insolation	+	
- length		+
- mean in polar latitudes	++	+
in equatorial latitudes		+
Caloric seasons polar latitudes	+	
equatorial latitudes		+
Meteorological seasons (astronomical definition)		
- total insolation	+	
- length		+
Meteorological seasons (monthly mean)	+	++

The accuracy of the long-term variations of the astronomical elements and of the insolation values and the stability of their spectrum have also been analysed by comparing 7 different astronomical solutions and 4 different time spans (0-0.8 Myr BP, 0.8-1.6 Myr BP, 1.6-2.4 Myr BP and 2.4-3.2 Myr BP). The general conclusions are, for the accuracy in time, that improvements are necessary for periods further back than 1.5 Myr BP. About the stability of the frequencies, the fundamental periods (around 40, 23 and 19 kyr) do not detoriate with time over the last 5 Myr but their relative importance for each insolation and even astronomical parameter is a function of the period considered.

Superimposed on these long-term insolation variations, there is a "secular" (decadal and century time scale) trend which amplitude varies

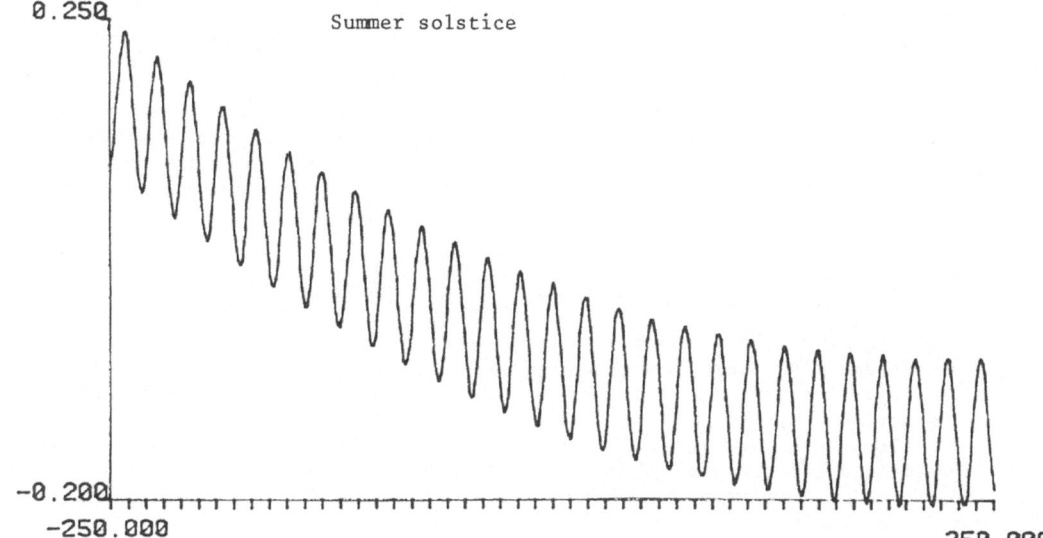

0.250

Summer solstice

-0.200

-250.000   250.000

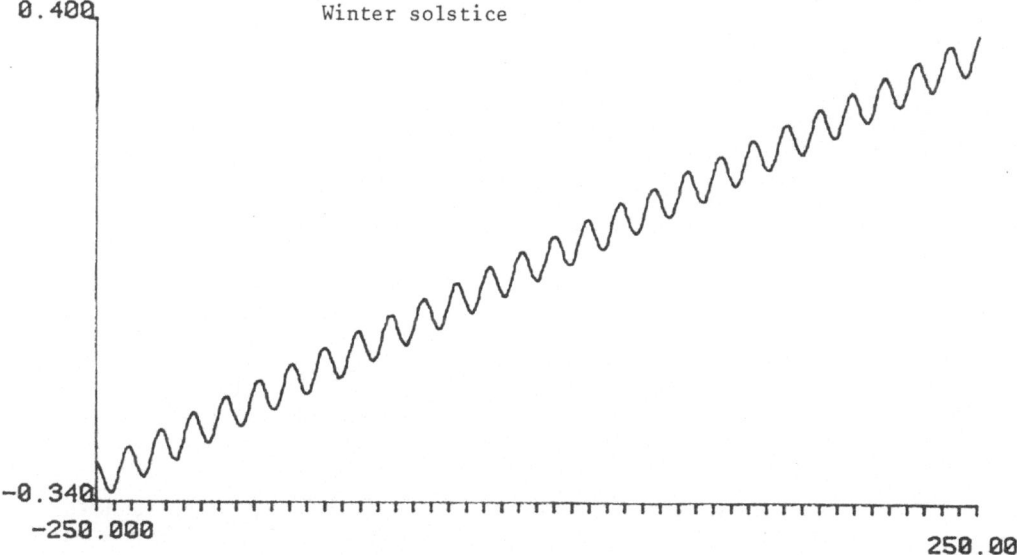

0.400

Winter solstice

-0.340

-250.000   250.000

Figure 1   Short term variations of the deviations (in cal cm^{-2} day^{-1}) of
60°N insolation from their 1950 A.D. values for solstices (1950 A.D.
values for summer solstice : 984 cal cm^{-2} day^{-1}; and for winter
solstice : 50 cal cm^{-2} day^{-1}).

between hundredths to tenths of one per cent. These short term fluctuations are more or less similar to those associated to solar activity as measured by satellites, the total radiative output of the Sun varying at the 0.1–0.3% level (5).

This is why a computation similar to the one done for the whole Quaternary (6), was completed at much shorter frequencies over a period of 500 years centered 1950 (Fig. 1). The quasi- periodicities are typically those related to the seasonal cycle (1, 0.5 yr) and to the motion of the moon (18.6, 9.3 yr), with no Jupiter component (11.9 yr) as found in Borisenkov et al. (7).

3. Prediction of natural climate and CO_2 for the next 60 kyr

Not only these fundamental frequencies are alike but the climatic series are also phase locked and strongly coherent with orbital variations (8). Moreover, most recent models outputs compare favorably with data of the past 400 kyr (9, 10).

To objectively support the correlations found between the orbital parameters and the past climate, spectral multivariate analyses have been performed (1) between $\delta^{18}O$ deep-sea cores data (11) and the zonal monthly mean insolations (12).

The close agreement between the simulated and the geo-ecological paleoclimates (87% of the climatic variation is explained) authorizes the prediction of the future natural climate. This, however, will only materialize if man's impact on land and the atmosphere has not yet modified the mechanism of climate change and does not do so in the future. The first cold peak will arrive 4 kyr AP, and the models foresee an improvement peaking at about 15 kyr AP, followed by a cold interval centered around 23 kyr AP. Major glaciation, comparable to the stage four of the last glacial cycle, is indicated at 60 kyr AP (Fig. 2).

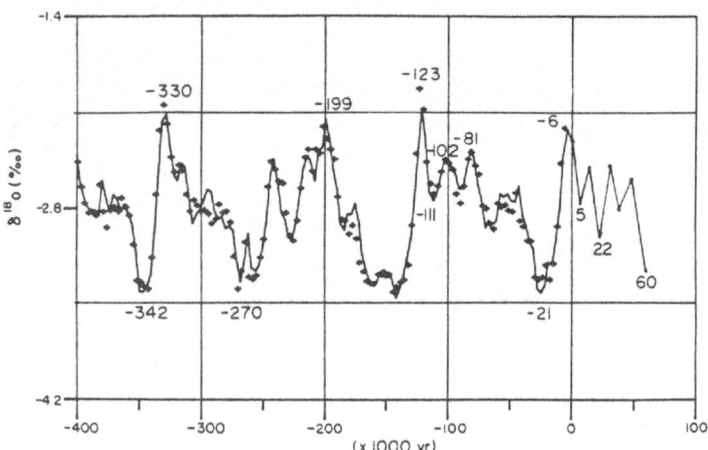

Figure 2   Long-term climatic variations over the past 400 kyr, and prediction for the next 60 kyr. Crosses represent the $\delta^{18}O$ deep-sea cores data from Hays et al. (11). Full line is the climate simulated by the auto-regressive spectral multivariate insolation model.

These predictions are far from being only of academic interest. Indeed, the effect on the global mean temperature of the 20th century CO_2

increase in the atmosphere due to man's activities must be viewed in the frame of the glacial-interglacial cycle of the past 150 kyr (13). Consumption of the bulk of the world's known fossil fuel reserves would plunge our planet into a super- interglacial. This man-made perturbation of our present climate would be totally different if exerted in a different climatic background. As we are still in an interglacial phase, the CO_2-induced global mean temperatures expected in the next century(ies) could reach levels several degrees higher than those experienced at any time in the last million years. Moreover, the slow long-term cooling towards the next "ice age" would have to wait until this warming had run its course, more than a thousand years from now, if ever, depending whether the climate system is transitive or not.

On the other hand, there seems to be a direct link between the long-term variations of the atmospheric CO_2 (14) and the astronomical paleoclimates. Following the results of Shackleton and Pisias (15) or the hypothesis by Broecker and Peng (16), a natural decrease in the atmospheric CO_2 concentration during the next 5 kyr must be expected, but it will most probably not exceed 50 ppmv. The rate at which this natural lowering of the CO_2 content in the atmosphere will occur is nevertheless very important and requires further investigation. This is why a 2-D seasonal model of the ocean-atmosphere-cryosphere-biosphere system is presently built to take into account both the astronomical and the CO_2 effects. (17).

4. Frequency analysis of paleoclimatic data

There are numerous justifications to a spectral approach for paleoclimate modeling (18). Its purpose is here to use a combination of different spectral analysis techniques (19) in order to extract an optimal information about the continuum and the peaks characterizing the paleoclimatic variability at periodicities ranging from 10^3 to 10^5 years.

Fourteen of the most accurate existing oxygen isotope records have been selected from the paleoclimatic data bank (PACDATA : (20)). They have been classified in three groups covering respectively more than 700 kyr for L3, 400 kyr (the last 10 oxygen isotope stages) for L2 and less than 125 kyr (the last Interglacial-Glacial cycle) for L1 (18). The chronologies of these records have been built either by absolute dating, or/and by isotopic stratigraphy without making any a-priori assumption of a relationship between a master curve and the analysed isotopic record.

The Blackman-Tukey Spectral Analysis has been first applied for its confidence in the amplitude estimation and its good statistical properties. Then, the Maximum Entropy Spectral Analysis by Least Squares has been used in order to increase the frequency resolution of the statistically significant spectral peaks. An experimental bandwidth has also been added to the statistical one, in order to take into account the uncertainty in the chronology. These upper and lower limits associated to each selected spectral peak have been used to define different frequency bands of increased paleoclimatic variability.

For the 6 cores of L3, three such distinct frequency bands are apparent over the last 800 kyr : B6(200-80 kyr), B5(70-30 kyr), and B4(30-18 kyr), which mean and standard deviation are respectively 103-24, 42-8 and 23-4 kyr, all the spectra (Fig. 3) presenting peaks that can thus be associated to the 100, 41 and 23 kyr astronomical quasiperiodicities. The Evolutive Maximum Entropy Spectral Analysis applied to the whole records (data window of 400 kyr; Fig. 4) exhibits a satisfying stationarity up to 600 kyr; further back in time, the location of the peaks and their ampli-

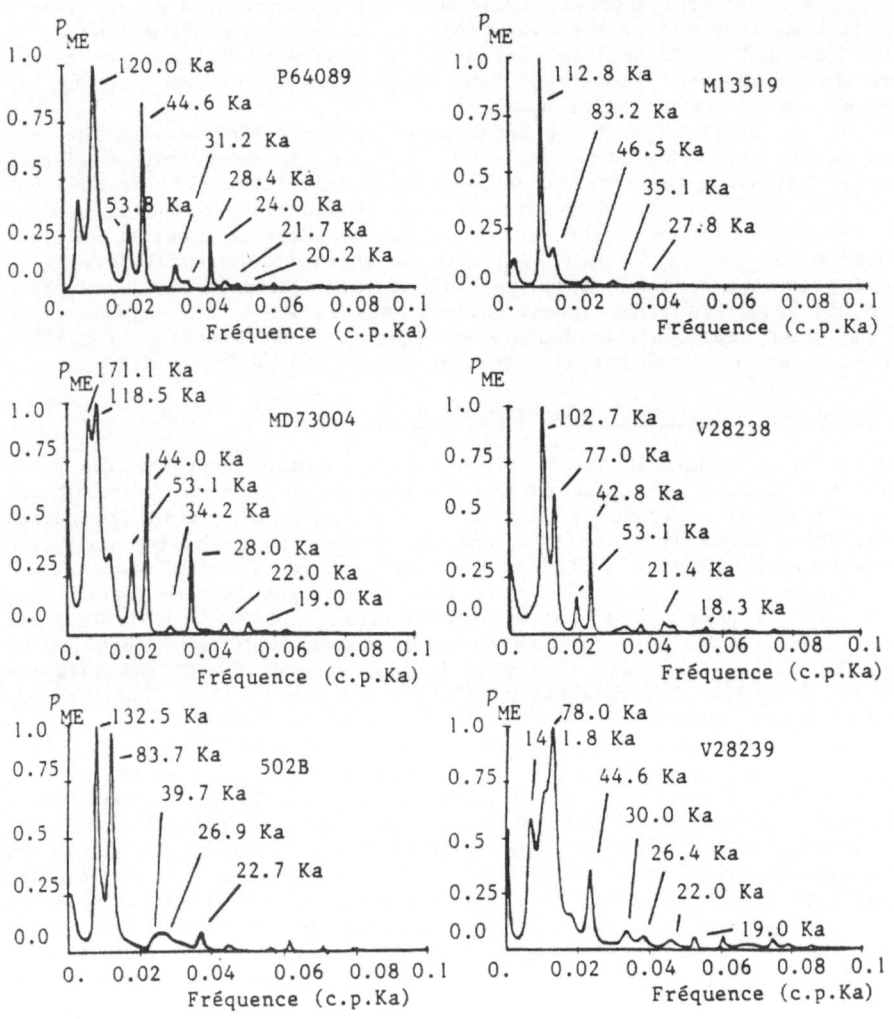

Figure 3   Power spectrum density for 6 deep-sea cores (L3 category) using th Maximum Entropy model over the last 800 kyr.

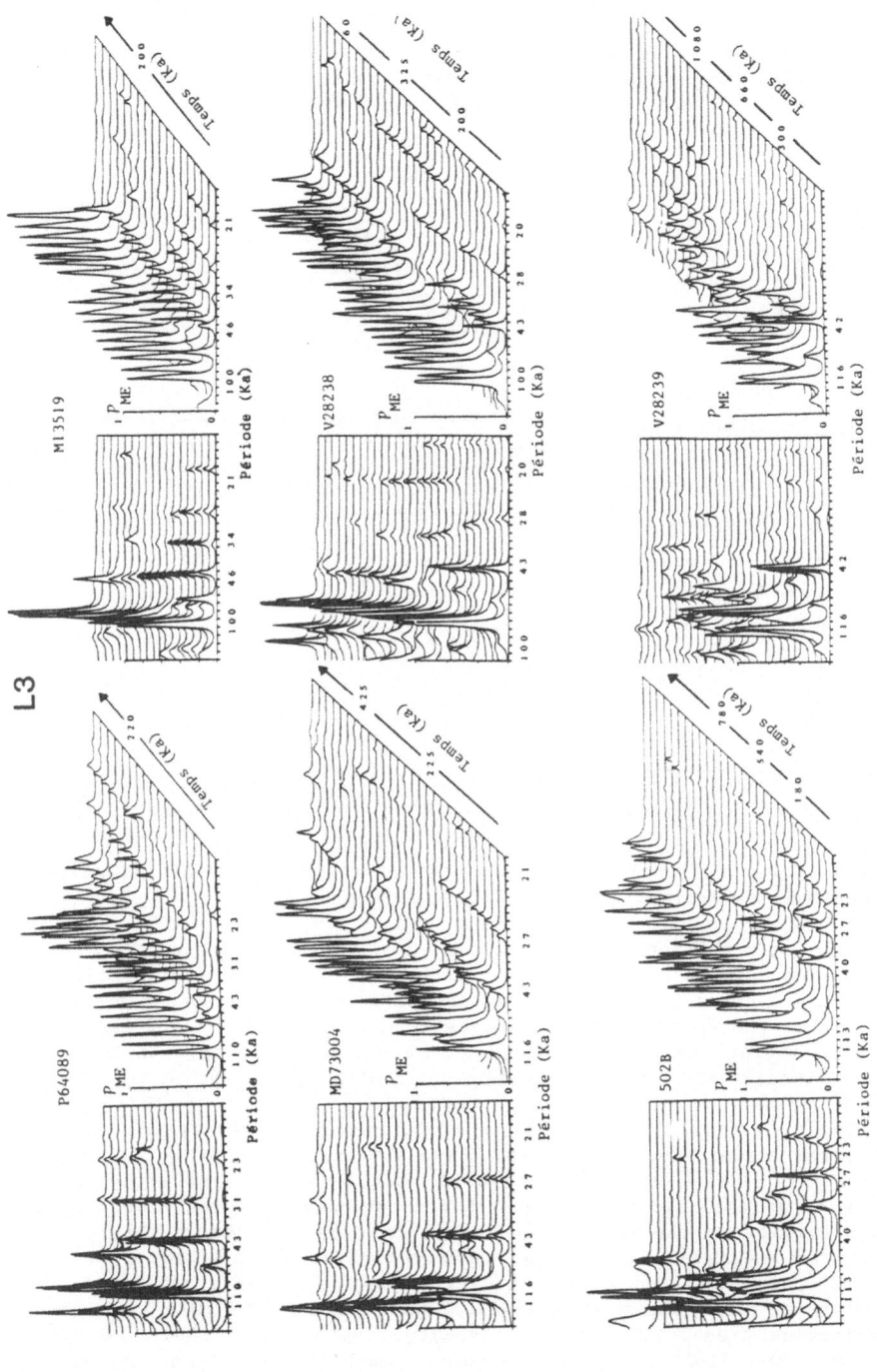

Figure 4 The Evolutive Maximum Entropy Spectral Analysis applied to 6 deep-sea cores covering the last 2 million years.

tude change strongly, probably in relation to climatic non-stationarities affecting also the sedimentation rate and therefore the chronology.

The Maximum Entropy Power Spectra of the 4 L2 deep-sea cores (Fig. 5) only present a poor frequency resolution characterized by a few number of peaks in the range of the astronomical frequencies. All the peaks cannot be distinguished from the mean periodicities which are calculated from the 4 cores : 120 ± 18 kyr, 41 ± 14 kyr and 23 ± 8 kyr.

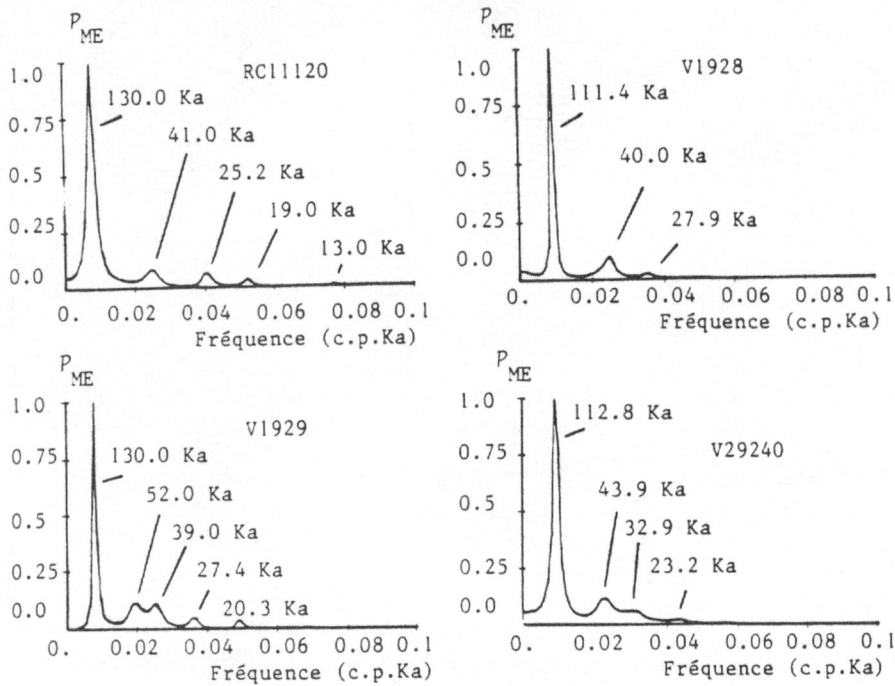

Figure 5 Power spectrum density for 4 deep-sea cores (L2 category) using the Maximum Entropy model.

High sedimentation rate and refined sampling are necessary to identify high frequency paleoclimatic variability. Following these two criteria, four deep sea cores (L1 group) have been selected in the Indian Ocean. For each of these cores two foraminifera have been analysed, one of them being a surface living species. All these cores are strongly influenced by the monsoon circulation and therefore by the hydrological cycle. In all the records, a high frequency variability is superimposed to the glacial-interglacial oscillation implying a strong non stationarity in the mean. If spectral analysis was misused on these original non stationary records, the shape of the power spectrum would have been reddened artificially, leading to an amplification of the low frequencies and therefore masking any variability in the high frequencies. In order to remove this effect, it has been necessary to detrend each record by fitting an appropriate piecewise polynomial and substracting it from the original record.

Starting with the most reliable dated record (MD77191), the quasi-periodicities around 2.3 kyr and 1.9 kyr, together with their associated upper and lower limits, suggest the existence of a first frequency band of

paleoclimatic variability. This (B1) will include the quasiperiodicities
ranging from 1.3 kyr to 3.2 kyr. The spectral peaks of the other records
which have a large part of their extreme possible range in this first frequency band, are considered as belonging to it (Fig. 6).

Two other frequency bands have been defined in a similar way, one
including quasiperiodicities ranging from 3.5 kyr to 6.5 kyr (B2) and the
other those from 7 kyr to 14 kyr (B3). All these quasiperiodicities
refined in frequency resolution and belonging to the same frequency band
have been averaged to give mean quasi periodicities of 10.2 kyr, 4.6 kyr
and 2.3 kyr with corresponding standard deviations of 1.2 kyr, 0.3 kyr and
0.2 kyr.

5. Reconstruction of a paleoclimatic spectrum from proxy oxygen isotope data

The oxygen isotopic ratio recorded in the foraminiferal shells taken
from the deep sea sediments is the result of successive transformations of
the original climatic signal.

First, it is well accepted that the oxygen isotopic ratio of the
shell is related both to the oxygen isotopic ratio and to the temperature
of the medium in which this shell lived (21). If, for example, the
Epstein formula is used to convert a benthic and a planctonic isotopic
record into temperature, the resulting power spectrum will thus be a com-
binaison of the two initial spectra.

Furthermore, the shells can be submitted to the dissolution resulting
from the corrosivity of the water. This dissolution can be local in time
or cyclic, with opposite effects during glacials and interglacials. From a
simulation using the core V28238, a significant modification of the power
spectrum was demonstrated to occur only for a "Pacific" type of dissolution.

The uncontrolled variations in the sedimentation rate must also be
taken into account. Using the same deep-sea core stretched and compressed
in the limits of accuracy of the dating levels, it was shown that the
spectral peaks shift, their width is increased and their amplitude
reduced.

Finally, the sediment is mixed up to a variable depth by the bottom
ocean infauna and this bioturbation might thus have a drastic low pass
filter effect on all the isotopic records taken from cores which have a
sedimentation rate lower than 5 to 10 cm/kyr.

6. Modelling the shape of the paleoclimatic spectra

The only known deterministic inputs to the climate system acting at
the scale of the Quaternary paleoclimatic variations are the insolation
parameters with quasiperiodicities of 41 kyr, 23 kyr, 19 kyr and a very
weak 100 kyr. To which extent can the typical power spectrum of paleoclimatic variables (Fig. 7) be explained by such a deterministic input
and/or stochastic one corresponding to a white noise ?

The shape of the continuum and its logarithmic slope can provide indications about the formulation of the basic equations governing the
underlying dynamical climatic system. For example, the Power Spectral
Density of the output of some simple linear differential systems have a
logarithmic slope in -2. Hasselman (22) has studied the integration of the
meteorological fluctuations by such a linear system having two different
characteristic times. This model was also used by Kominz in 1979 (23) to
discuss the spectral slope of paleoclimatic data. In this respect, it is

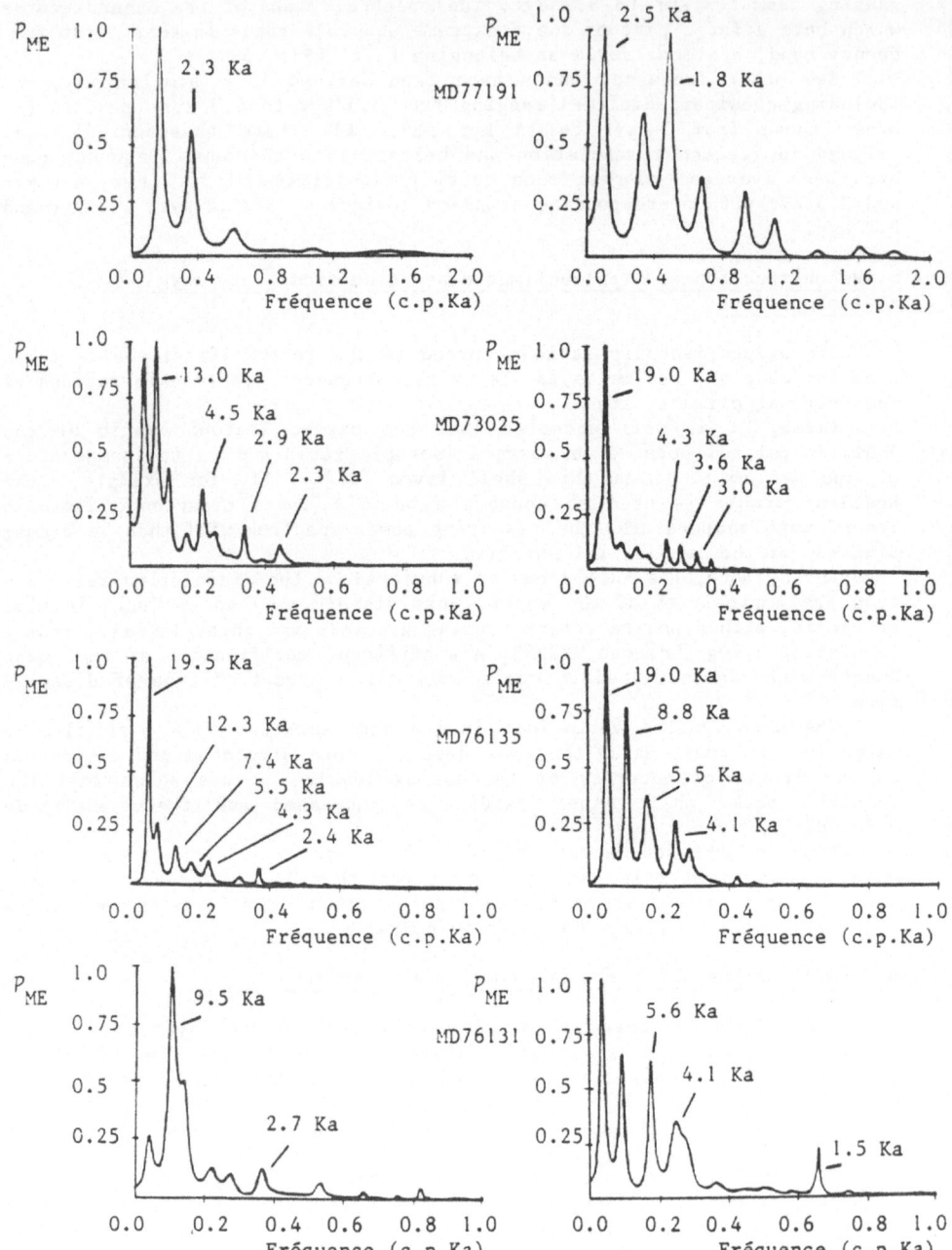

Figure 6   Power spectrum density for 4 deep-sea cores (L1 category;
   left: surface planctonic species, right : planctonic intermediate
   layers or benthic) using the Maximum Entropy model.

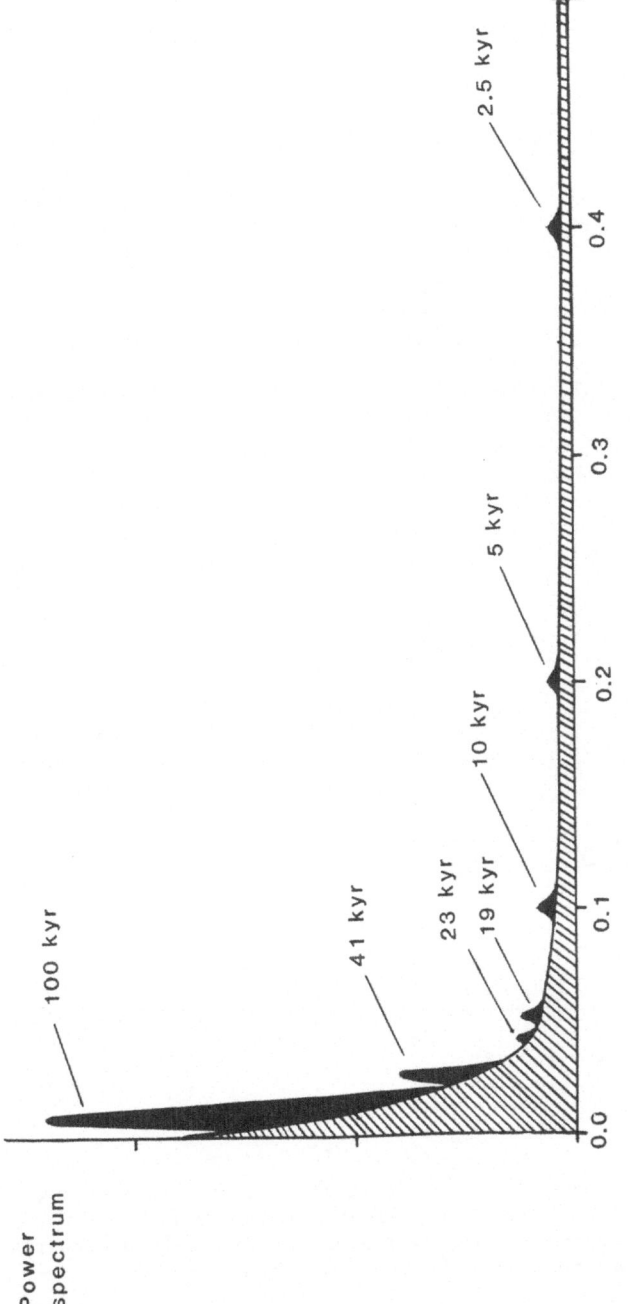

Figure 7 Typical shape of the Power Spectrum of Paleoclimatic Variations, with the continuous background and the superimposed peaks of "astronomical" origin.

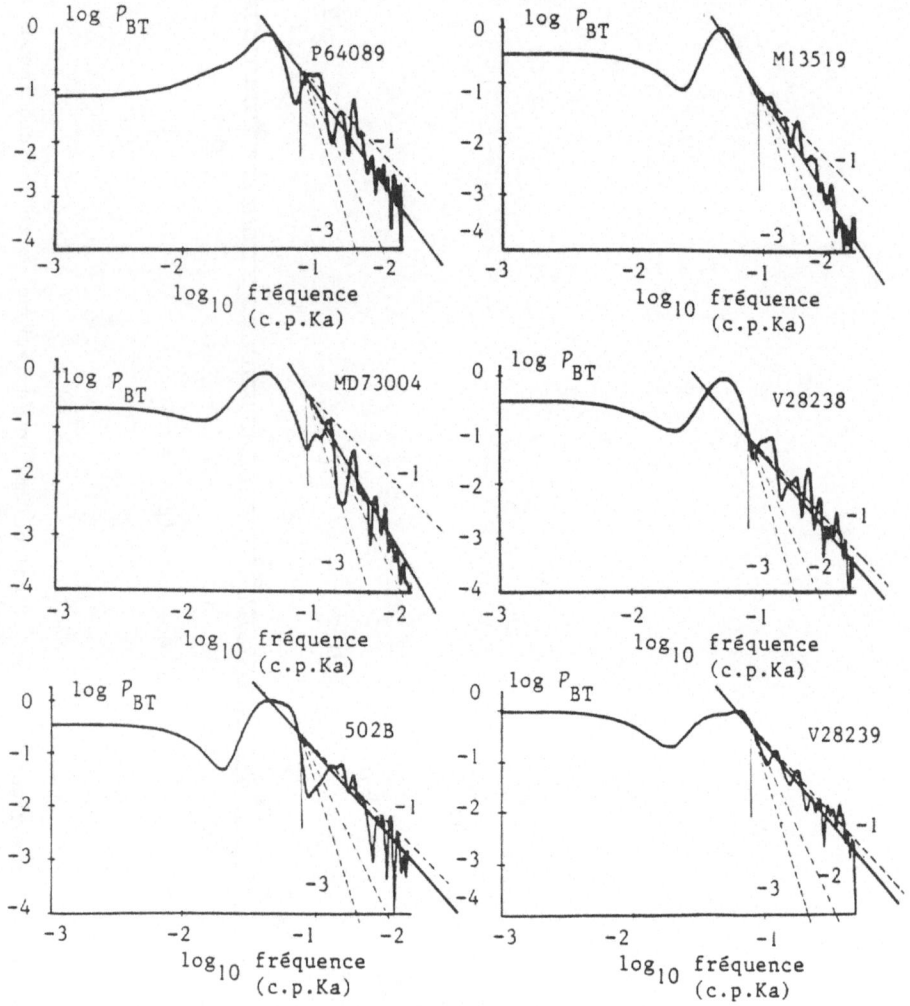

Figure 8   Blackman-Tukey Power spectrum density in a logarithmic scale
for the L3 deep-sea cores. The slopes of the dashed lines are respectively -1, -2 and -3. The estimated slopes of the decreasing part
of the spectra are respectively -1.3, -1.4, -1.6, -1.1, -1.1 and
-1.1.

worthwhile to point out that the logarithmic spectra of the oxygen isotope data may differ from core to core. Moreover, the long and medium records have slopes of the order of −1 (Figs 8 and 9), whereas the Blackman-Tukey spectra transformed in a logarithmic scale have a −1.8 slope for the planctonic species while it is even steeper for the deeper living species (Fig. 10).

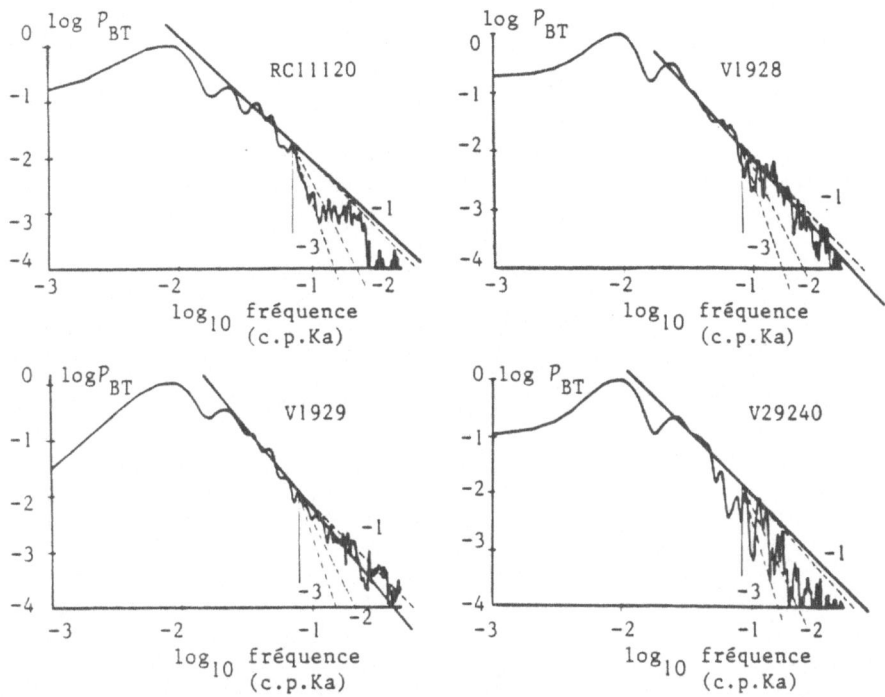

Figure 9 Blackman-Tukey Power spectrum density in a logarithmic scale for the L2 deep-sea cores. The slopes of the full lines are respectively −1.1 (or −2.7), −1.1, −1.2, −0.9 (or −2.0).

In order to illustrate how the observed spectral shape might be explained by a purely stochastic autoregressive model (a particular case of the linear models just mentionned) or an autoregressive model forced by both insolation and white noise, a set of simulations have been performed. When only the white noise is used as input, the continuum is not neces-sarily constant, nor uniformly decreasing ; spectral peaks can appear in the astronomical frequency bands, but none of them is statistically signi-ficant and they appear at different frequencies from one realization to another. If one adds to the white noise forcing the forcing corresponding to the summer caloric insolation at 55°N, only the astronomical frequen-cies are statistically significant, although small peaks can appear at lower frequencies. This facts together with the presence of clearly iden-tified spectral peaks at 100 kyr, 10 kyr, 5 kyr and 2.5 kyr (Table II) in-dicate that the climate system is highly non linear and that the paleo-climatic variability can hardly be explained by linear models, stochastic or not.

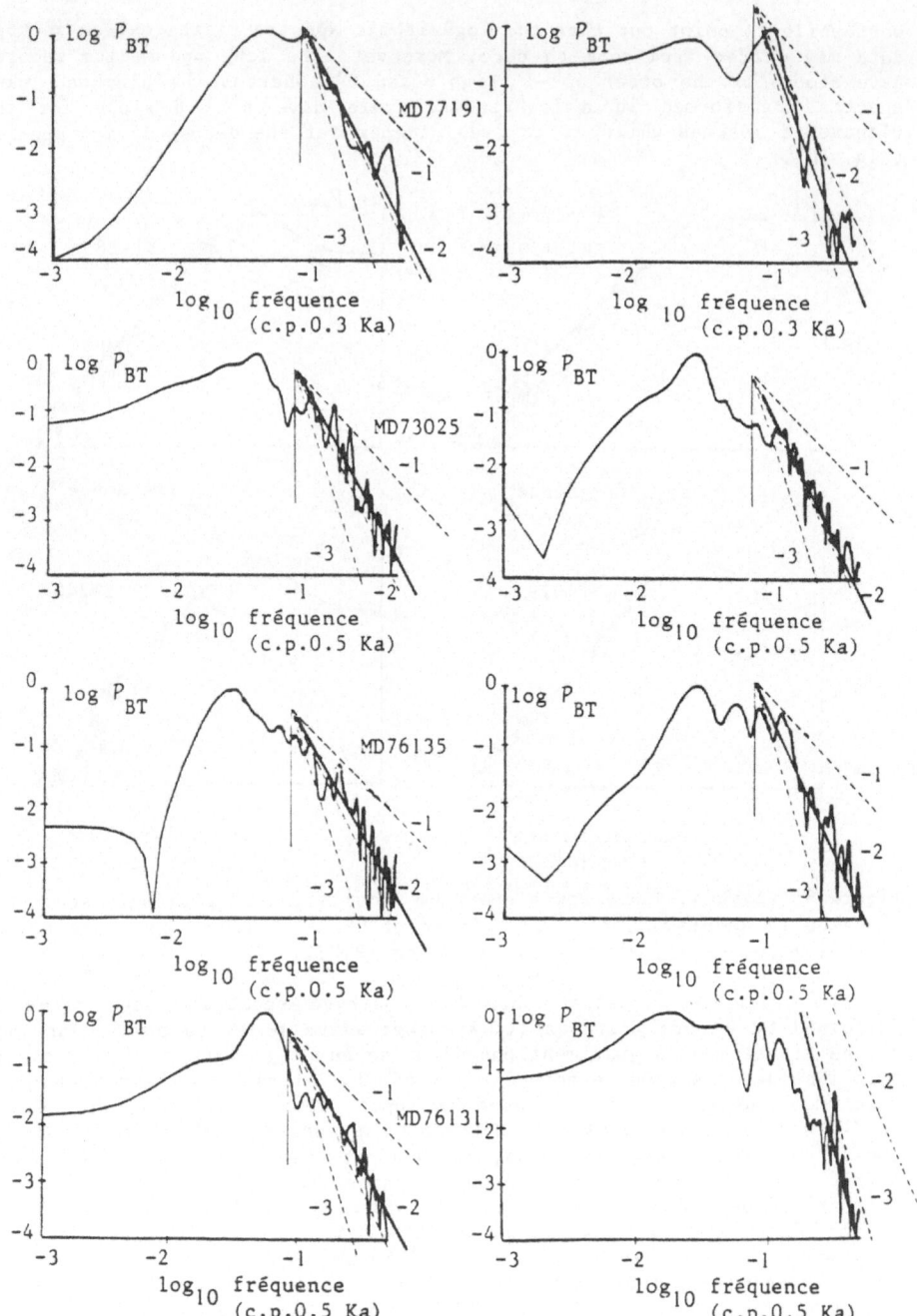

Figure 10   Blackman–Tukey Power spectrum density in a logarithmic scale
   for the L1 deep-sea cores. The slopes of the full lines are respectively −1.9 and −2.5, −1.8 and −1.9, −1.8 and −2.0, −1.8 and −3.0.

Table II. Combination of tones related to the 3 insolation frequencies $\omega_{f1}=1/19$ kyr^{-1}, $\omega_{f2}=1/23$ kyr^{-1}, $\omega_{f3}=1/41$ kyr^{-1}. d_i correspond to the periods detected in the deep-sea cores; n_i, p_i and q_i to the integers used to linearly combining them and O_i the order of that combination. Periodicities found in the results of the non-linear climatic oscillator (24) are also listed.

d_i	n_i	p_i	q_i	O_i	$n_i\omega_{f1}+p_i\omega_{f2}+q_i\omega_{f3}$	Le Treut and Ghil (1983)
13.0	1	0	1	2	12.98	X
12.3	2	0	-1	3	12.37	
10.2	1	1	0	2	10.41	X
9.5	2	0	0	2	9.51	X
8.8	0	2	1	3	8.99	
7.4	1	2	0	3	7.17	X
5.6	0	4	0	4	5.76	
5.5	1	3	0	4	5.47	
4.6	0	5	0	5	4.61	
4.3	4	0	1	5	4.26	
4.1	3	2	0	5	4.02	
3.0	2	4	2	8	3.05	
2.9	2	5	1	8	2.88	
2.8	3	4	1	8	2.81	
2.7	2	5	2	9	2.69	
2.5	5	2	2	9	2.51	
2.3	5	4	0	9	2.29	

An alternative might thus be presented for the explanation of both the detected spectral peaks and the observed continuum. A non-linear climatic oscillator (24) has been selected which uses the energy balance equation for the global ocean-atmosphere system, and an ice sheet evolving under the effect of precipitation and of a viscoelastic adjustment of the earth's crust. An integration was carried out over 2 millions years by forcing it with the mid-month July insolation at 65°N. The latitudinal extention of the ice cap and its isostatic sinkink have been transformed into an ice volume where combination of tones of the forcing frequencies (1/19 kyr^{-1}), (1/23 kyr)$^{-1}$ and (1/41 kyr^{-1}) might be expected (Table II).

Although the model temperature presents only peaks at the forcing frequencies and has a flat log-log power spectrum (Fig. 11), the global ice volume (Fig. 12) has no spectral peak at 41 kyr while an aperiodic continuous background and the 105 kyr and 10.5 kyr periodicities appear clearly. Even more significant is the fact that the high frequencies – corresponding to periods of 13, 10.5 and 7.2 kyr (the maximum resolution of the model being of the order of 5 kyr) – are also detected in L1 geological records.

If the already mentionned transformations leading to the observed oxygen isotope record are now taken into account, the transformation of the global ice volume and the surface temperature into a planctonic oxygen isotope index reintroduces a 41 kyr quasiperiodicity (Fig. 13a,b). If we compress and stretch this record within the classical chronology uncertainties, bimodal peaks are introduced together with harmonics and the logarithmic slope is strongly increased (Fig. 13c,d). Finally, the bioturbation simulated by mixing this record introduces a strong amplification of the low frequency peaks and a further increase of the logarithmic slope (Fig. 14). This show definitively the overall importance of the non-linear phenomena, which transform the original climatic signal, on the spectral shape of paleoclimatic data.

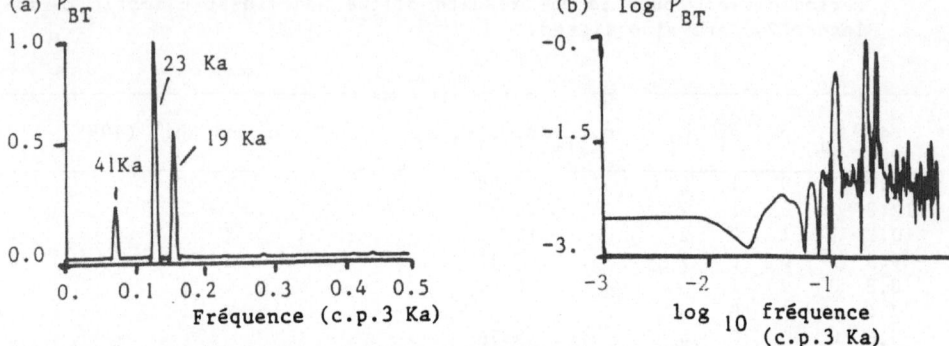

Figure 11 Blackman-Tukey power spectral density (a) of the Le Treut-Ghil (24) model temperature and its log-log transformation (b).

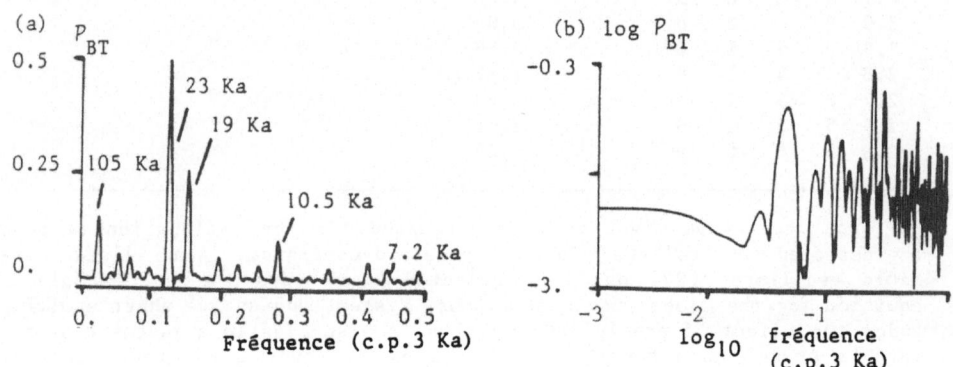

Figure 12 Same as Fig. 11 for the ice volume curve as calculated by Pestiaux (18).

7. Conclusions

The results of the models used, both in time and frequency domains, show the relevance of the astronomical theory to climate studies also at the human time scale.

A carefull analysis of different spectral techniques has demonstrated all their advantages and weaknesses, the Blackman-Tuckey spectrum being used for its statistical properties, the maximum or minimum-Cross Entropy analysis being recommended to refine the spectral peaks.

Spectral analysis of a core with a high sedimentation rate core has shown quasi-periods ranging from 18 to 1 kyr, particularly those related to the astronomical combination tones - 10.2, 4.1 and 2.6 kyr -, and an approximate -2 logarithmic slope as predicted by a linear model stochastically forced. This increase from a -1 to a -2 slope is supposed to be related to a variable sedimentation rate and to the time scale inaccuracies.

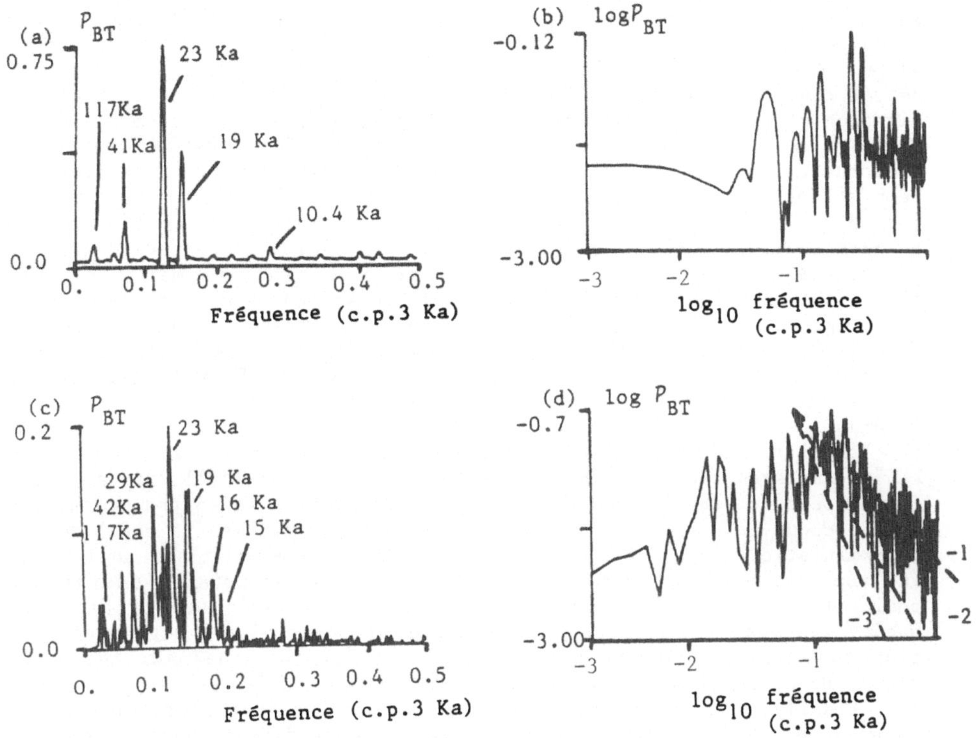

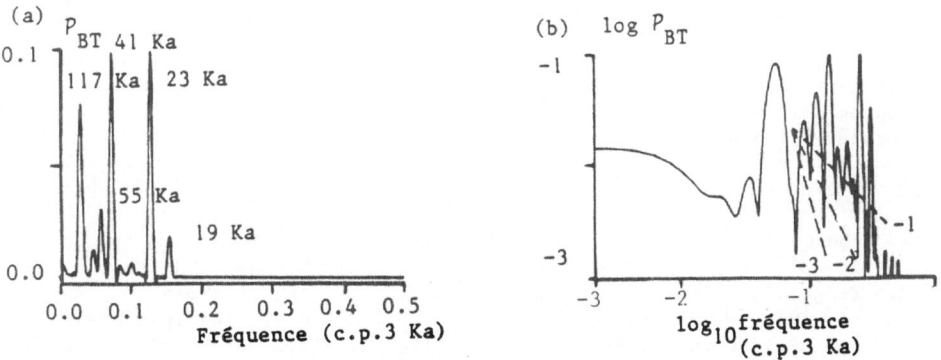

Figure 13 Same as Figs 11 and 12 for the theoretical $\delta^{18}O$ which would have characterized planctonic foraminifera following Epstein formula (a,b). (c,d) gives the same information if the time series is now compressed and stretched within the classical chronology uncertainties.

Figure 14 Same as Fig. 13 if a mixing is now allowed over a depth corresponding to a smoothing period of 10 kyr.

Spectral analysis of deep-sea cores covering the last 2 million years and even only the last 10 oxygen isotope stages show significant spectral peaks associated to the astronomical frequencies, with a -1 logarithmic slope. The Evolutive Maximum Entropy power spectrum analysis allows to detect non stationarities before 600 kyr, most probably linked to the modification of the internal characteristics of the climatic system.

To model these spectra, one must first simulate the effects on the spectrum of the physical transformations that the climatic variables have been subject to, before giving their shape to the isotopic variations recorded in deep-sea cores. It was possible to show that 1) bioturbation reduces the amplitude of the peaks (high frequencies are smoothed out); 2) variable sedimentation rate is by far the most important mechanism which alters the spectrum by introducing double peaks and harmonics and reducing the amplitude of the peaks and of the high frequency background; 3) "Pacific" dissolution shifts the spectral peaks and change their relative importance; 4) surface temperature reconstructed from planktonic and benthic foraminifera has a spectrum which may differ from both original records.

It is highly significant for the astronomical theory that even after having suffered such transformations, isotope records still display significant peaks at the astronomical frequencies. Taking these transformations into account, a non-linear climate oscillator has allowed to conclude to a deterministic origin of the isotopic spectra. Moreover, the analysis of the combinations of tones has clearly shown the importance of the precesional components in a monsoonal climate. Systematic use of high frequency resolution spectral techniques and of cores with a high sedimentation rate would allow to extend these results, obtained for the Indian ocean, to other climatic regions.

Finally, it is thus expected that the paleoclimatic data bank could be enlarged to other reliable cores (land, ice and deep-sea) and that all the techniques applied here will be used to confirm the results found in this study. In order to help resolving the non-stationarity of paleoclimatic data and the transitivity of the climatic system, Walsh analysis and a boolean approach are also underway. A more complete presentation of the results can be found in Pestiaux (18) and in Berger et al. (17).

8. Acknowledgment

This work was supported in part of the Climate Programme of the Commission of the European Communities under Contract CLI-026-B. The authors would like to express their appreciation to J. Backman, J.Cl. Duplessy, C. Emiliani, W. Prell, M. Sarnthein, N. Shackleton and H. Zimmerman for providing the deep-sea cores data. Thanks are due to Mrs N. Materne and F. Mercier for their technical assistance.

9. References

1. Berger, A., 1980. The Milankovitch astronomical theory of paleoclimates a review. Vistas in Astronomy, 24, pp. 103-122.
2. Berger, A., Imbrie, J., Hays, J., Kukla, G. and Saltzman, B. (Eds), 1984. Milankovitch and Climate, Understanding the response to astronomical forcing. D. Reidel Publ. Co., Dordrecht, Holland.
3. Berger, A., 1984. Accuracy and frequency stability of the Earth's orbital elements during the Quaternary. In : "Milankovitch and Climate", op.c., pp. 3-39.

4. Berger, A. and Pestiaux, P., 1984. Accuracy and stability of the Quaternary terrestrial insolation. In : "Milankovitch and Climate", op.c., pp. 83-112.
5. Eddy, J.A., Gilliland, R.L. and Hoyt, D.V., 1982. Changes in the solar constant and climatic effects. Nature 300, pp. 689-693.
6. Berger, A., 1978a. Long term variations of daily insolation and Quaternary climatic changes. J. Atmospheric Sciences 35(12), pp. 2362-2367.
7. Borisenkov, Y.P., Tsetkov, A.V. and Agaponov, S.V., 1983. On some characteristic of insolation changes in the past and future. Climatic Change 5(3), pp. 237-244.
8. Imbrie, J. et al., 1984. The orbital theory of Pleistocene climate support from a revised chronology of the marine ^{18}O record. In: "Milankovitch and Climate", op.c., pp. 269-306.
9. Kukla, G., Berger, A., Lotti, R. and Brown, J., 1981. Orbital signature of interglacials. Nature 290, pp. 295-300.
10. Imbrie, J. and Imbrie, J.L., 1980. Modeling the climatic response to orbital variations. Science 207, pp. 943-953.
11. Hays, J., Imbrie, J. and Shackleton, N.J., 1976. Variations in the Earth's orbit. Pacemaker of the Ice Ages. Science 194, pp. 1121-1132.
12. Berger, A., 1978. Numerical values of mid-month insolation from 1,000,000 YBP to 100,000 YAP (astronomical solution of Berger, 1978). Contribution nr 36, Institut d'Astronomie et de Géophysique G. Lemaître, Université Catholique de Louvain-la-Neuve.
13. Mitchell, 1977. Carbone Dioxide and Future Climate. EDS, Environmental Data Service (NOAA), March 1977.
14. Lorius, C. and Raynaud, D., 1983. Record of past atmospheric CO_2 from tree-ring and ice core studies in Carbon Dioxide, W. Bäch, A.J. Crane, A.L. Berger and A. Longhetto (Eds), D. Reidel Publ. Company, Dordrecht (Holland), pp. 145-178.
15. Shackleton, N.J. and Pisias, N.G., 1984. Modelling the global climate response to orbital forcing and atmospheric carbon dioxide changes. Nature 310, pp. 757-759.
16. Broecker, W.S. and Peng, T.H., 1984. The climate - chemistry connection. In the proceedings of the Ewing symposium. Lamont-Doherty Geological Observatory.
17. Berger, A., Pestiaux, P., Gallée, H., Gaspar, Ph., Tricot, Chr. and Van der Mersch, I., 1984. Modelling the astronomical theory of paleoclimates in the time and frequency domains, an example of the relationship between long-term and short-term climate changes. Scientific Report 1984/6, Institut d'Astronomie et de Géophysique G. Lemaître, Université Catholique de Louvain-la-Neuve.
18. Pestiaux, P., 1984. Approche Spectrale en Modélisation Climatique. Dissertation Doctorale, Institut d'Astronomie et de Géophysique G. Lemaître, Université Catholique de Louvain-la-Neuve.
19. Pestiaux, P. and Berger, A., 1984. An optimal approach to the spectral characteristics of deep-sea climatic records. In : "Milankovitch and Climate", A. Berger et al. (Eds), Reidel Publ. Company, Dordrecht (Holland), pp. 417-446.
20. Berger, A. and Pestiaux, P., 1982. An introduction to the IAG PACDATA Bank. Contribution nr 28, Institut d'Astronomie et de Géophysique G. Lemaître, Université Catholique de Louvain-la-Neuve.
21. Duplessy, J.Cl., 1978. Isotope studies. In : "Climatic Change", J. Gribbin (Ed.), Cambridge University Press, pp. 46-68.

22. Hasselman, K., 1976. Stochastic Climate Models. I. Tellus, 28(6), pp. 473-484.
23. Kominz, M.A. and Pisias, N.G., 1979. Pleistocene climate : deterministic or stochastic ? Science 204, pp. 171-173.
24. Le Treut, H. and Ghil, M., 1983. Orbital forcing, climatic interactions and glaciation cycles. J. Geophys. Res. 88, pp. 5167-5190.

CLIMATE RECONSTRUCTION USING HISTORICAL SOURCES

T.M.L. WIGLEY, G. FARMER and A.E.J. OGILVIE
Climatic Research Unit, University of East Anglia, Norwich, U.K.

Summary

Winter severity indices for Europe are compared and evaluated. New
values are presented for the period 1220-1420 A.D. Overall, the data
show long time scale cooling from c. 1200 to c. 1340, warming to c. 1510
and then cooling into the main Little Ice Age cold period of the 17th
century.

1. Introduction

Historical climatology is concerned with the use of documentary
sources to reconstruct past climate and to study the impact of climatic
fluctuations on society. This paper deals with the use of descriptive
information to reconstruct past temperature fluctuations in Europe since
about 1200 A.D. A more extensive review of historical climatology is given
in the book Climate and History (1). We begin by discussing some of the
problems in using documentary sources and then give examples of old and new
temperature reconstructions for Europe.

2. Problems

One of the main problems with using documentary sources for climate
reconstruction is the quality of the basic data. With descriptive
material, data can be totally spurious, poorly dated, or from an entirely
different region from that stated in the source. Only data that are
demonstrably accurate, correctly dated and from a known location should be
used. Many secondary sources that give lists or compilations of climate
data fail to meet these criteria. Some such sources have been shown to
contain large amounts of unreliable material; see, for example, Bell and
Ogilvie (2). Unfortunately, some well known historically-based
reconstructions of European climate have been based on secondary sources
and so must be considered of doubtful reliability, especially prior to the
15th century.
Examples of the inconsistencies that can arise from the use of
historical data that have not been strictly verified as reliable are given
in Table 1 (from Ingram et al. (3)). This Table compares decadal indices

Table 1: Decadal climate indices for Europe. The figures shown are correlation coefficients with the sample size shown in brackets[*]. The Lamb and
Alexandre indices are for regions around latitude 52°N at the longitudes
shown, while Pfister's indices are for Switzerland, approximately 47°N,
8°E. Data from Lamb (4), Alexandre (5,6) and Pfister (7).

	Lamb (12°E)	Alexandre (5°E)	Pfister (8°E)
Summer Wetness			
Lamb (0°E)	0.51 (44)	0.41 (44)	0.27 (28)
Lamb (12°E)		0.28 (44)	0.16 (28)
Alexandre (5°E)			0.77 (7)
Winter Severity			
Lamb (0°E)	0.54 (24)	0.22 (24)	0.27 (28)
Lamb (12°E)		0.37 (24)	0.17 (28)
Alexandre (5°E)			–

[*] Sample size of 44 corresponds to the decades 1100s to 1610s (some
decades missed because of insufficient data), 24 corresponds to 1100s to
1390s (some decades missed because of insufficient data), 28 corresponds to
1550s to 1820s and 7 corresponds to 1550s to 1610s.

of summer wetness (precipitation) and winter severity (temperature) from
three sources, Lamb (4), Alexandre (5,6) and Pfister (7). Alexandre's and
Pfister's indices are based on strictly verified material, while Lamb's
indices are based on material of varying reliability including some sources
shown by Bell and Ogilvie (2) to be of poor quality. The correlation
between Lamb's 0°E winter severity data and Alexandre's 5°E winter severity
data is much lower than would be expected based on a comparison of recent
instrumental records (past 200 years), pointing to unreliability in the
former. For both winter severity and summer wetness, Lamb's 0°E and 12°E
indices show higher correlations than occur between Lamb's 0°E and
Alexandre's 5°E indices. This is probably because data from different
parts of Europe were amalgamated in some of the secondary sources used by
Lamb. In an attempt to improve the quality of Lamb's pioneering work in
this field (and noting that Lamb was unaware of the unreliability of some
of the sources he used) we have re-examined a large body of documentary
historical material. Further details are given in Farmer and Wigley (8),
and an example of the resulting revised indices is given below.

3. Winter (DJF) severity indices for western Europe

Decadal winter severity indices were constructed from strictly
verified, reliable documentary sources from England for the period 1220 to
1419 A.D. These are compared with the indices of Lamb, Alexandre and
Pfister (cited above) in Figure 1. Some of Alexandre's data are only given
as running 5-decade means in the original publication (6). The decadal
indices for Alexandre after 1399 were inferred from the 5-decade values in
the following way. If Y_i and Y_{i+1} are consecutive 5-decade means (with Y_i
$= (X_{i-2} + X_{i-1} + X_i + X_{i+1} + X_{i+2})/5$ where X_i are the decadal index

values), then $Y_{i+1} - Y_i = (X_{i+2} - X_{i-2})/5$. Thus, with Y_i and Y_{i+1} known, if X_{i-2} is known, X_{i+2} can be calculated. Because of the overlap in the time periods covered in Alexandre (5) and Alexandre (6) we were able to estimate decadal indices through to the 1610s in this fashion.

Figure 1 shows that, for data after 1419, Lamb's and Alexandre's indices are in close accord, but many differences occur prior to this date. The Lamb and Wigley et al. (i.e. the present work) indices also show a number of differences prior to 1419. The Wigley et al. data appear to agree reasonably well with Alexandre, but there are some frustrating gaps in Alexandre's record due to a lack of basic data. For decades earlier than the 1220s, we do not believe that, for England, enough data exist at present to be able to produce indices of any value.

A reasonable record of decade-by-decade changes in winter severity can be obtained by using the Wigley et al. data presented here up to the 1410s and Lamb's data subsequently. This composite record shows a cooling trend from the late 12th century (evidence for the earliest decades comes only from Alexandre) to around 1340, a long warming trend to around 1510 (but with possible large inter-decadal variability in the 15th century), and a subsequent cooling trend to at least 1600. This cooling trend climaxed in particularly cool decades in the 1590s and 1600s and we know from evidence for later times that cool conditions were maintained throughout the 17th century. A more detailed analysis of the data, year by year, shows that the first severe winter of this 16th/17th century cool period (the climax of the Little Ice Age) occurred in 1563/4 after at least a century of relatively mild winters (see ref. 8 for further details).

REFERENCES

1.   WIGLEY, T.M.L., INGRAM, M.J. and FARMER, G., Eds. (1981). Climate and History. Cambridge University Press.
2.   BELL, W.T. and OGILVIE, A.E.J. (1978). Weather compilations as a source of data for the reconstruction of European climate during the Medieval period. Climatic Change 1, 331-48.
3.   INGRAM, M.J., FARMER, G. and WIGLEY T.M.L. (1981). Past climates and their impact on Man: a review. Ref. 1, pp. 3-50.
4.   LAMB, H.H. (1977). Climate: Present, Past and Future, Vol. 2. Methuen, London.
5.   ALEXANDRE, P. (1976a). Le climat au moyen âge en Belgique et dans les régions voisines (Rhénanie, Nord de la France). Centre Belge d'Histoire Rurale Publication No. 50.
6.   ALEXANDRE, P. (1976b). Le climat dans le sud de la Belgique et en Rhénanie de 1400 à 1600. Prélude au 'Petite Age Glaciaire' de l'époque moderne. Annales du XLIV-ième Congrès, Huy, 18-22 Août 1976. Fedaration des Cercles d'Achaeologie et d'Histoire de Belgique, ASBL.
7.   PFISTER, C. (1981). An analysis of the Little Ice Age climate in Switzerland and its consequences for agricultural production. Ref. 1, pp. 214-248.
8.   FARMER, G. and WIGLEY, T.M.L. (1984). The Reconstruction of European Climate on Decadal and Shorter Time Scales. Final Report to the Commission of the European Communities under Contract No. CL-029-81-UK(H).

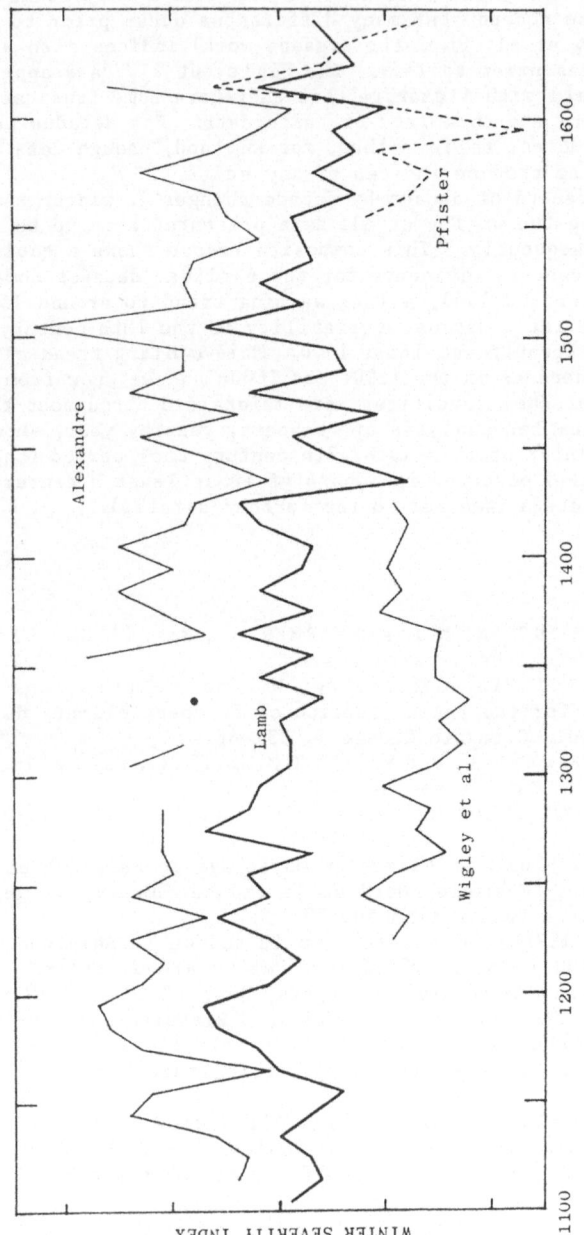

Fig. 1 : A comparison of various historical winter temperature indices

STAGES OF CLIMATIC CHANGE FROM FULL GLACIAL TO HOLOCENE
IN NORTHWEST SPAIN, SOUTHERN FRANCE AND ITALY:
A COMPARISON OF THE ATLANTIC COAST AND THE MEDITERRANEAN BASIN

W.A. WATTS
University of Dublin, Ireland

Summary

From 17,000 to 11,000 years before present polar water retreated
northward and westward in the North Atlantic. A brief readvance took
place between 11,000 and 10,000 yr. B.P. It might be anticipated
that the retreat of polar water would be reflected most strongly in
the vegetation record on the Atlantic Coast of northwest Spain and
less strongly in the Mediterranean region which is more distant from
the climatic influence of the Atlantic. At Sanabria Marsh and Lago
de Ajo in northwest Spain herb-dominated vegetation was present
before 13,000 years ago. Ice in the cirque of Lago de Ajo had melted
some time before 13,000 yr. B.P. and was not reconstituted.
Subsequent vegetation history suggests that invasion of tree birches
followed by pine and oak took place between 13,000 and 10,000 yr.
B.P. There is no clear evidence for a Younger Dryas event in
northwest Spain, but a regression of oak and temporary reappearance
of birch and pine at Sanabria Marsh is datable to the early Holocene.
Ocean cores from the Atlantic west of the Iberian Peninsula and in
the Bay of Biscay appear to record a Younger Dryas event which may
also be weakly recorded in the Pyrenees. More detailed comparison of
the land and ocean records is required to resolve a possible conflict
of evidence. In southern Italy at Lake Monticchio a full glacial
herb-dominated vegetation is recorded. The transition to the
Holocene is marked first by a peak of birch pollen and then by the
presence of deciduous trees. Pine is insignificant in the developing
vegetation at deglaciation. In southern Italy there is no evidence
for climatic fluctuations during deglaciation.

In a series of now classic papers, Ruddiman and McIntyre were able to
show that the maximum southward extension of polar water in the North
Atlantic took place about 17,000 years before present (yr. B.P.) when the
Laurentide and Scandinavian ice sheets reached their greatest size during
the last major glaciation. At that time polar water, defined as water
sedimenting low-carbonate marine sediment with only a single species of
polar Foraminifera, extended to about $40°N$, approximately the latitude of
Lisbon (1). After 17,000 polar water retreated northward and carbonate
ooze with a diverse flora and fauna was sedimented. The polar water had
withdrawn from northwest Europe by 13,000 yr. B.P. when the polar front
lay west of $25°W$ in the mid-North Atlantic to the west of Ireland. A
further retreat took place to the north and west by 11,000 yr. B.P. but a
notable readvance to the east dates to about 10,000 yr. B.P. (1). This is
correlated in time with the Younger Dryas cold phase on land. While the
cause of the readvance is somewhat speculative, it seems likely to have

been the direct cause of the climatic deterioration of the Younger Dryas.

Watts (2) has argued that there should be a regional correlation between the movement of the polar water front in the North Atlantic and climatic change expressed by vegetation changes recorded in pollen-diagrams. One would expect that the movements of polar water would have a particularly strong reflection in the pollen record of sites close to the Atlantic Coast and would be progressively weaker to the east in continental or Mediterranean Europe as one passed out of the range of climatic patterns determined largely by events in the Atlantic. This appears to be the case, for pollen-diagrams from the Alpine region, especially, for example, the slopes on the north side of the Po Valley (3), show little evidence of climatic fluctuation in response to the readvance of polar water at about 10,000 yr. B.P. One might, in fact, expect quite different types of vegetation record and climatic inference from the Atlantic Coast and the Mediterranean, suggesting that a comparison of the two regions may lead to a useful perception of climatic events during continental deglaciation. In western Europe the determination of many scientists to impose a common stratigraphic and chronological framework on data from the time of deglaciation has been a serious obstacle to progress because, in principle, climatic events should be time-transgressive in their expression in the fossil record from south to north (1) and should display regional characteristics. Because of these arguments, the author's research within the EEC Climatology Programme attempted to view Europe on a continental rather than a national scale. The research was structured so as to compare events on a north-south transect in Atlantic Coastal Europe with the climatic/vegetational sequence in the Mediterranean region.

Pollen-analysis of lake sediments permits the reconstruction of vegetational history, subject to certain technical difficulties which are well-known to palynologists such as the over-representation of certain genera, and the potential for pollen transport over long distances with the possibility that relatively abundant occurrence of a pollen type may not provide conclusive evidence for the local presence of the species. Ocean cores, from which pollen-diagrams may also be produced (4), are unsatisfactory in certain respects because of slow sedimentation rates and bioturbation, so that it is difficult or even impossible to obtain fine resolution in either pollen-analysis or, above all, radiocarbon dating. Pollen data permit much finer resolution in both dating and vegetation reconstruction because sedimentation may be an order of magnitude faster and bioturbation appears to be of lesser significance. However, as the land surface is so much more diverse than the ocean floor, factors such as geological substratum, elevation, local climate characteristics such as rainshadow effects, and the proximity or remoteness of potentially immigrant plant populations, as well as technical difficulties inherent in the method, make for much greater complexity in interpretation.

The first site discussed is referred to here as Sanabria Marsh (fig. 1). It is believed to be the same site as the 'Laguna de las Sanguijuelas' of Menéndez Amor and Florschütz (5). It was desired to revise the earlier work. The site is a marsh, an overgrown small lake, occupying what is probably an ice-block hollow in a moraine near a large natural lake, Lago de Sanabria. It is located in northwestern Spain, immediately north of the frontier with Portugal (42°6'N, 6°44'W, 1050 m elevation). This granitic region has high mountains rising to over 2000 m to the west. There is very clear geomorphological evidence for recent glaciation. The modern vegetation is dominated by deciduous oaks (Quercus

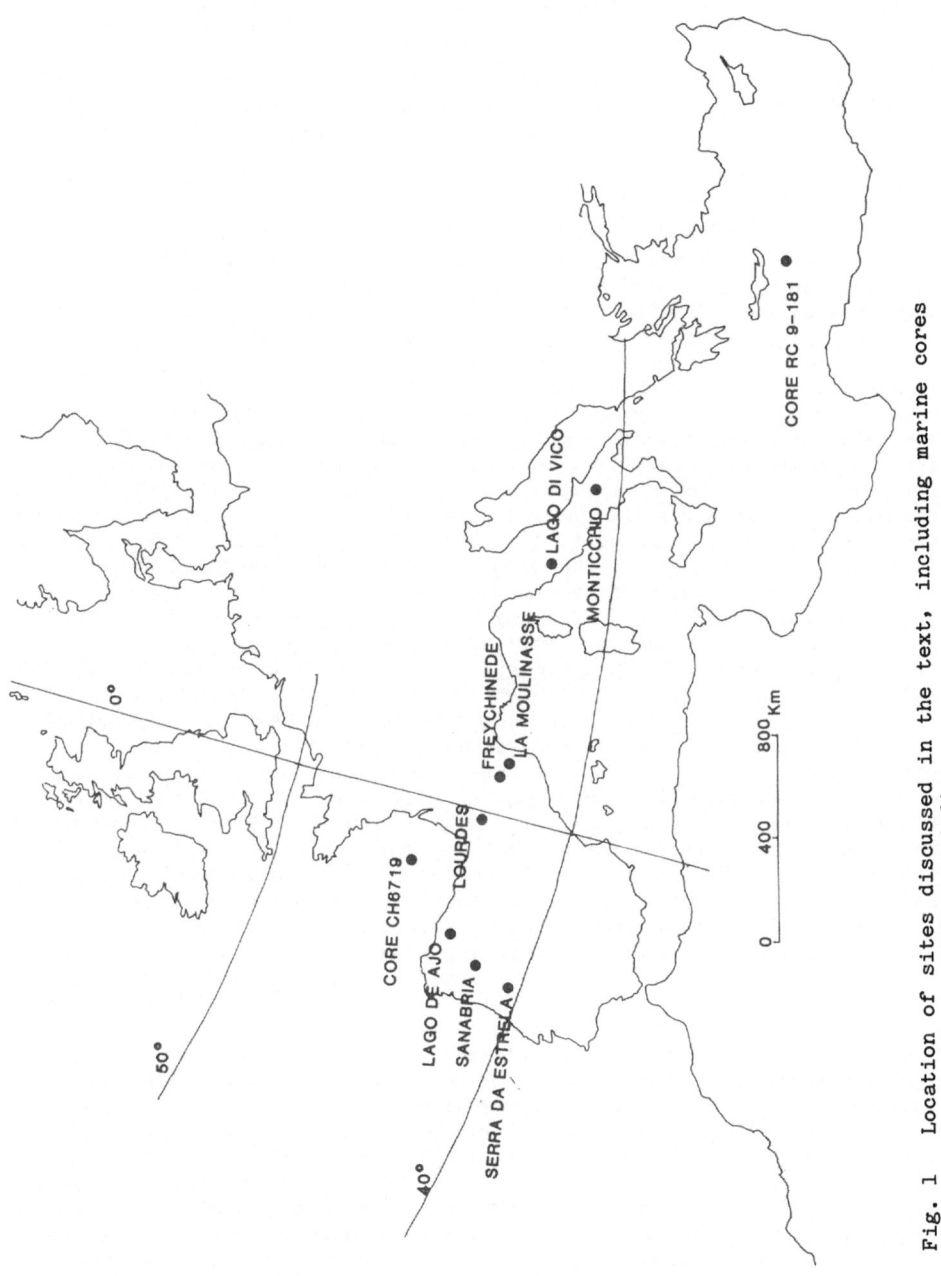

Fig. 1 Location of sites discussed in the text, including marine cores referred to in (4) and (18).

sp.) with virtually no other tree species. There are numerous heaths (Erica spp.) and diverse herbs. The granitic soils make the site very suitable for radiocarbon dating because the problem of 'old carbon' does not need to be considered.

At Sanabria, Menéndez Amor and Florschütz (5) recorded an early zone with high percentages of herbs, grass (Gramineae), sage (Artemisia) and pine (Pinus). This phase ended just before 11,585 ± 220 yr. B.P. Two dates, 12,830 ± 280 and 13,700 ± 300, confirm the age of the early phase which is described as 'Older Dryas'. There are minor fluctuations in the representation of birch (Betula) and pine in this zone which is otherwise rather homogeneous. New studies (6, fig. 2) confirm that grasses, sage, pine and herbs were the predominant pollen types before 12,585 ± 330 B.P. Juniper (Juniperus), not recorded in the older study, is also common. There are no macrofossils of trees. This seems to record an invasion of pioneer vegetation after a local upland glaciation. The character of the vegetation is difficult to determine. The abundance of Artemisia in deposits of full-glacial age from southern Europe has led many authors to refer to steppe-like vegetation. It may be safer to say herb- and grass-dominated vegetation with juniper and abundant Artemisia. Pine at this time is probably the result of long-distance transport from populations at lower elevations and perhaps considerably farther south in the Iberian Peninsula. At Sanabria the unusual presence of Calluna (fig. 2) and other Ericaceae (heathers) suggests a heath-like aspect to the vegetation. A subsequent peak of Betula with B. pubescens macrofossils dated to after 11,585 yr. B.P. (5) was succeeded by peaks of Pinus and of Quercus, the latter dated to 9490 ± 110 yr. B.P. in the present study. Pine is represented first by abundant macrofossils of P. mugo/uncinata type and, at a higher level, of P. sylvestris. Menéndez Amor and Florschütz referred the oak peak to the end of a late-glacial 'Allerød' phase. This is clearly in conflict with the new data. After 9490 yr. B.P. oak fell steeply and a rapid succession of birch and pine took place, dated in the new study to 8200 ± 90 yr. B.P. The subsequent oak rise which was to dominate the whole Holocene began, according to Menéndez Amor and Florschütz (5), at 8160 ± 190 yr. B.P. Frequent tree birch macrofossils show that forest trees were present throughout the period from 9490 and 8160 yr. B.P. although pine macrofossils do not occur at this level. The Sanabria data are puzzling because they suggest an early pioneer phase, followed by immigrant birch, two pine species and oak leading into the early Holocene without recording any climatic fluctuations. Then a revertence to a birch-dominated flora, perhaps with pine, took place briefly in what radiocarbon dates show to be the early Holocene. This indicates a local forest sequence inconsistent in timing with others known, or a local forest catastrophe such as a fire or disease episode which caused a successional process to be initiated once more. Further radiocarbon dates are required to confirm this event.

Sanabria Marsh is similar in some respects to Lagoa Comprida in the Serra da Estrela of central Portugal. The mountains are granitic, recently glaciated, and rise to nearly 2000 m. A group of cirque lakes lie at about 1600 m. The mountain tops have much bare rock with juniper, Erica species and herbs. It is probable that much deforestation has taken place in the recent past by burning and grazing (7). Lagoa Comprida has a pine peak dated to 9200 ± 270 yr. B.P. followed by an oak peak between 9080 ± 200 and 8310 ± 160. Oak then declines steeply in favour of birch with which it shares dominance throughout the Holocene. The decline of oak in the early Holocene repeats the history of Sanabria. The event is

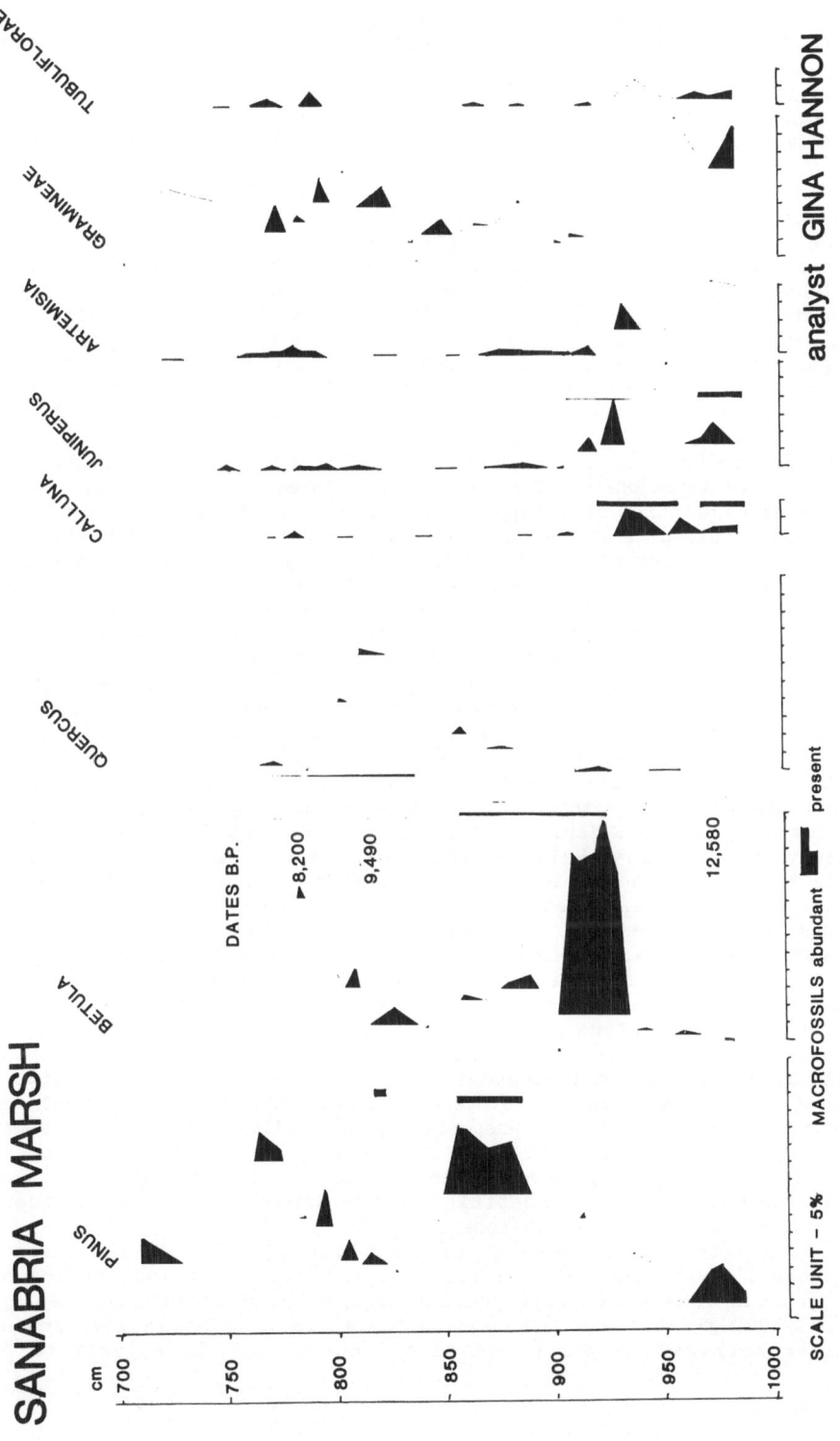

Fig. 2   Pollen-diagram (selected taxa only) from Sanabria Marsh.

clearly early Holocene rather than Late-glacial. Lagoa Comprida (7) does not contain a late-glacial record. Another adjoining site under current study as part of the EEC Climatology Programme shows early abundance of Artemisia.

Lago de Ajo lies at 1570 m in the northwest Cantabrian mountains of Spain (43°3'N, 6°9'W). Peaks reaching 2400 m occur nearby. The lake is a large cirque in dolomitic limestone with high surrounding cliffs. The vegetation is alpine and predominantly herbaceous with frequent juniper bushes. The tree-line, which lies only a little lower than the lake, is formed by beech (Fagus) forest, here near its western limit in Europe. Beech exists in nearby pure stands, with some Betula pendula, Sorbus aria (whitebeam), Corylus (hazel) and Acer (maple). Oaks and pines are absent from the near environment of the lake. The cirque is at the head of a long west-facing valley. Deciduous oaks occur on dry lower slopes. It is probable that considerable pollen transport from the lowlands takes place to this and to other high elevation sites in the region. In a pollen-diagram from Lago de Ajo (8, fig. 3), the vegetation before 14,270 is dominated by sage, grass, Ephedra (a xeromorphic shrub), diverse herbs and pine. Peaks of birch and pine between 14,270 and 12,610 yr. B.P. suggest an early tree invasion, but they are unaccompanied by macrofossils. There is a brief resurgence of Artemisia-rich pollen spectra after 12,610 yr. B.P. which hints at a climatic reversion. However, at the elevation of Lago de Ajo it seems more probable that virtually all of the early pollen spectra are heavily influenced by long-distance transport from lowland plant communities. There may have been only a sparse local vegetation cover. This conclusion is reinforced by the low pollen concentrations which characterise the oldest sediments (8). At about 12,610 birch rises to a second peak, followed by oak, while pine makes a lower percentage input. Macroscopic tree birches are continually present from after 12,610 until the Corylus rise at 9780 yr. B.P. Macroscopic pines are also present in the upper half of this stage. This indicates that birch alone at first, then with pine, formed woodland or forest beside the lake which is now above the tree-line. At this site the Corylus rise clearly marks the beginning of the Holocene. It is of interest that Ulmus (elm) is also present in the earlier Holocene and, as happens universally in Europe, falls suddenly to a low level in the mid-Holocene, presumably as the result of a disease. Beech, advancing from the east (9), formed its present forests after 3840 yr. B.P., somewhat earlier than the prediction of Huntley and Birks (9) which is based on rather few data from Spain.

Several points of interest are revealed by Lago de Ajo. There is no identifiable Younger Dryas event. In fact the cirque lake was open and sedimenting well before 14,000 yr. B.P. This suggests that the ice that formed or enlarged the cirque had disappeared decisively early during deglaciation and did not reappear. The tree-line may have been higher than at present early in the Holocene, although macrofossils of birch and pine might have been wind-transported up the valley. The tree-line may have been lowered during the later Holocene. Native pines do not occur obviously within the region at present, as is also true at Sanabria. They appear to have lost in competition to the densely-grown, shady beech forest, if not to earlier oak-forest.

At a site near Lourdes at an elevation of 430 m in the eastern Pyrenees (10) the published pollen-diagram records a long period of homogeneous vegetation with grasses, sedges (Cyperaceae) and Artemisia from 19,000 yr. B.P. to 13,500 yr. B.P. At this point an invasion of juniper followed by birch and pine took place. It appears to be

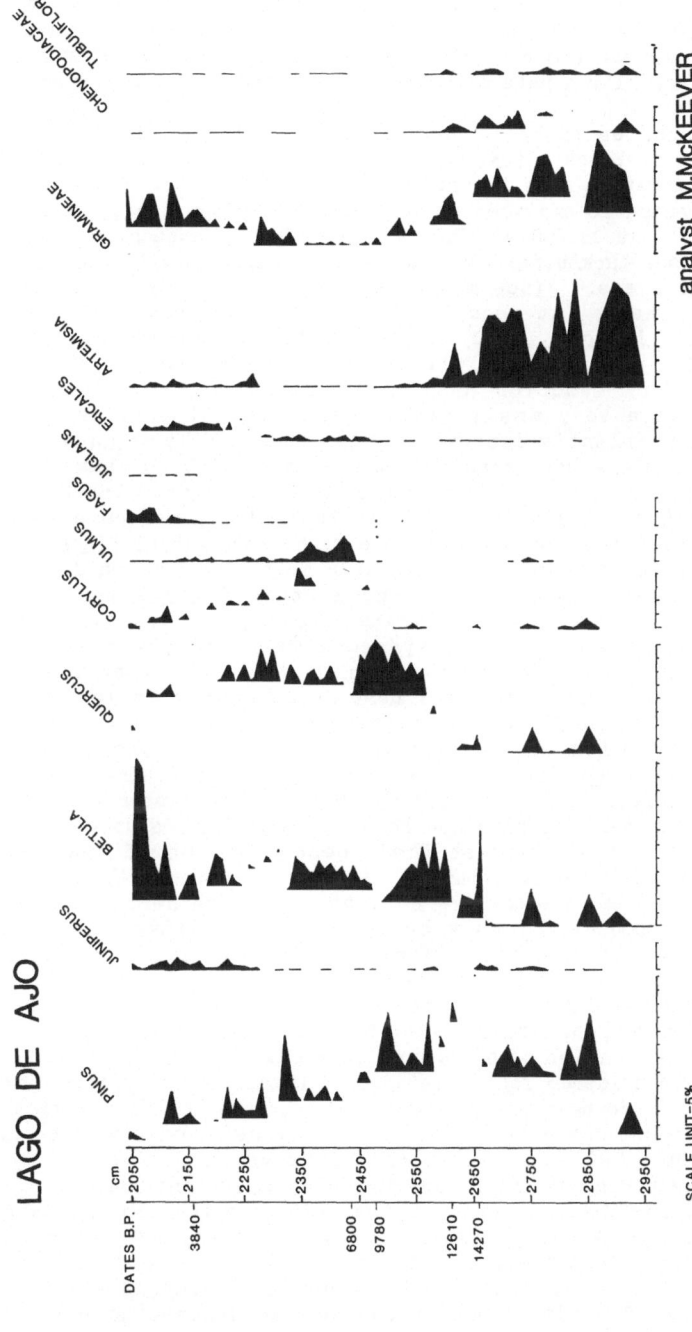

Fig. 3   Pollen-diagram (selected taxa only) from Lago de Ajo.

essentially successional with no record of reversion. A very similar record with similar dates is available from Poueyferré near Lourdes (11). At La Moulinasse farther east in the Pyrenees at 1330 m (12) a section is recorded which may show some climatic fluctuations. Unfortunately the late-glacial part of the core ('Younger Dryas and Allerød') has only 15 cm of sediment, too slow a rate of sedimentation and with too few samples for critical evaluation.

Recent studies in the Pyrenees claim to document a Younger Dryas event at a bog at Biscaye (13), close to Lourdes at an elevation of 400 m. At this low-elevation site the first evidence of vegetational change after the glacial period is provided by the expansion of Juniperus shortly after 14,820 yr. B.P. At 13,250 yr. B.P. a strong Betula expansion, accompanied by Hippophae (sea-buckthorn), marked the establishment of communities rich in shrubs and trees. Pinus may also have been an invader in the later part of this phase. At about 10,860 yr. B.P. a very brief and small resurgence of Juniperus and of Artemisia can be observed which is correlated with a short-term fall in pollen concentration. This is seen as evidence for reduction of forest cover and expansion of herbs. However, it is a very small feature and may not provide sufficient evidence for a climatic deterioration. It would be important to know whether Betula alone was present in the late-glacial woodland or whether Pinus was also present. If it could be assumed that the Betula fall before the 'Younger Dryas' which causes the fall in pollen concentration was partially replaced by long-distance rather than local Pinus pollen the interpretation would be more secure. At a more elevated and drier eastern site, Freychinede (14, 1350 m), no Younger Dryas episode could be identified in a very similar pioneering succession to that at Biscaye.

To compare Spain and the Pyrenees with Italy a site, Laghi di Monticchio, was cored. The two lakes which constitute Laghi di Monticchio lie in an ancient volcanic crater near Melfi east of Naples in southern Italy (40°56'N, 15°36'E) at an elevation of 530 m. It is surrounded by wooded hills to 1300 m which are dominated by beech, with a diversity of other deciduous forest trees. A preliminary pollen-diagram (15, fig. 4) shows a long vegetation record for the Holocene and later portion of the Last Glacial Period. The presumed Holocene occupies the upper 735 cm of a 2550 cm core. In the latest Pleistocene portion of the core, the vegetation is dominated by grasses, sage and chenopods (Chenopodiaceae), indicating a treeless steppe-like environment. The herbaceous flora of this time was specially diverse in species. The climatic interpretation is that the late Pleistocene was very dry, at least in the summer growing season and probably, though not necessarily, colder than today. In the Holocene greater availability of water in the growing season favoured the expansion of tree populations. The Pleistocene/Holocene transition is marked by the expansion of birch and then oak, followed by mesic trees. There is little evidence for climatic fluctuation in the pollen data which would hint at a 'Younger Dryas' event. The expansion and invasion of the tree populations took place without any reversion once it had begun. A date of 21,200 ± 500 B.P. is available from Dr. M. Stuiver's laboratory for the earliest part of the birch rise between 726 and 738 cm. A date of 31,800 ± 1200 for the sediments between 1204 and 1220 cm comes from the same laboratory. Although surprisingly old, they are consistent with still older dates from the lower part of the core which is not reproduced here. The possible explanations are a very early Betula expansion with the inference of early climatic change, an undetected hiatus, or a systematic cause for excessively old dates. The last seems unlikely, as

Laghi di Monticchio

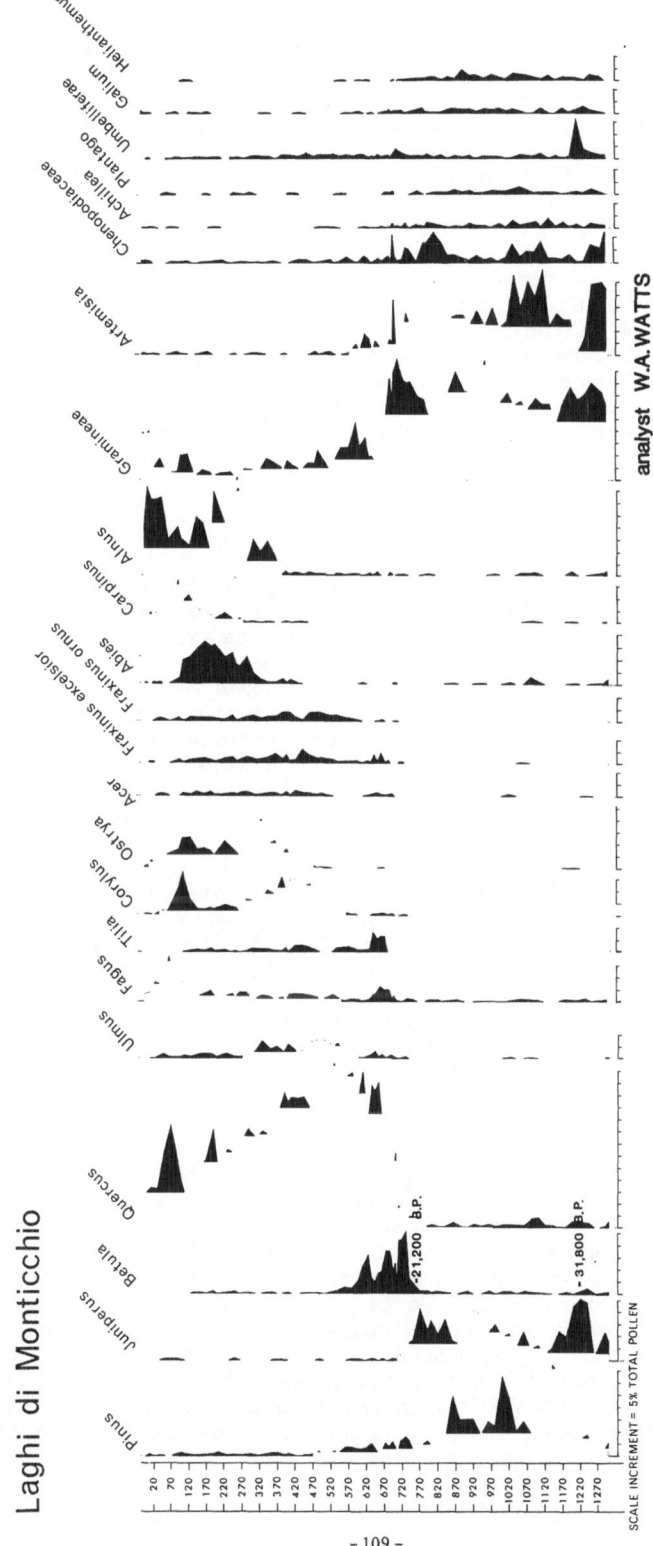

Fig. 4 Pollen-diagram (selected taxa only) from Laghi di Monticchio.

the sediments are carbonate-free and have no evidence for redeposited old organic material. Further dates in the 'Holocene' part of the profile are required to clarify the problem. The extreme dryness of the Mediterranean basin during the Last Glacial is already well established (16, 17). The assumed early Holocene at Monticchio is characterised by oak, beech, lime (Tilia) and ash (Fraxinus excelsior). This is followed by a mid-Holocene dominated by oak alone with a regression of the more mesic trees, indicating a probably drier and warmer summer climate. The early 'mesic' period may correlate with the early Holocene northward expansion of the African monsoon, now documented by the presence of sapropel horizons in east Mediterranean marine cores (18) as well as by an accumulating array of evidence from palynology. It is noteworthy also that trees such as Abies (fir) which were present in southern Italy during the middle part of the glaciation then disappeared, and did not reappear until late in the Holocene, apparently by migration from northern Italy. Southern Italy cannot have been a 'refugium' for many tree species during the Pleistocene.

Conclusion

The objective of the research was to establish when vegetation was established on the Atlantic Coast of the Iberian Peninsula at deglaciation and how this compared with the record from the Pyrenees and from Italy. It was anticipated that a series of late-glacial events recognisably similar to those known from Ireland (2, 19) would be found with a somewhat different expression and timing in Spain, assuming that the migration of vegetation in response to deglaciation would be time-transgressive. Italy, remote from the movement of North Atlantic polar water, was predicted to be less affected by climatic fluctuations than Spain.

In the event it appears that deglaciation took place early in Spain and in the Pyrenees and that local ice caps or cirque glaciers had retreated or disappeared before 14,000-13,000 yr. B.P. There is no evidence for their later rejuvenation. The first vegetation was rich in grass, Artemisia and herbs and was probably treeless. It persisted for a considerable period, succeeding the similar or identical vegetation which had been present in the Mediterranean countries throughout the later Pleistocene. At all sites in Spain vegetation with tree birches, pine species and oaks invaded the herbaceous vegetation after about 12,000 yr. B.P. and continued into the beginning of the Holocene, when after about 9,500 yr. B.P. oaks finally became dominant, sometimes in association with hazel and other deciduous species. There is no clear evidence for a Younger Dryas cold period, nor for any lesser cold episodes in the Older Dryas. These events seem confined in their full expression to northern and northwest Europe which appear to have been nearer to and more responsive in their vegetation cover to fluctuations in the position of the polar water front. It is remarkable that the ocean record shows a clear Younger Dryas event in the Bay of Biscay (4) and, in unpublished data, west of the Iberian Peninsula. The apparent conflict of data requires further examination. It is possible that the marine record, for local reasons, provides a stronger signal of this climatic event. A more exacting examination of the data and their chronology may prove desirable. The regression of oak at Sanabria and in Serra da Estrela is a remarkable phenomenon, but it lies early in the Holocene, about 9000 yr. B.P., and may be an unidentified climatic episode or, more probably, the expression of a significant ecological event such as disease, fire or an unrecognised successional process involving a greater diversity of tree species than are found in northwest Europe. Italy, as predicted, passed quickly from a 'glacial' to an 'interglacial' environment. These events, seen for too long through the distorting telescope of northwest European perceptions, are of great inherent interest. Considerable research is still required to establish the chronology and nature of the events objectively and to assess their significance in relation to events elsewhere in continental Europe.

REFERENCES

1. RUDDIMAN, W.F. and McINTYRE, A. (1973). Time-transgressive deglacial retreat of polar waters from the North Atlantic. Quaternary Research 3, 117-30.

2. WATTS, W.A. (1979). Regional variation in the response of vegetation to Lateglacial climatic events in Europe. p. 1-21 in Studies in the Lateglacial of northwest Europe (ed. J.J. Lowe, J.M. Gray and J.E. Robinson). Pergamon Press.

3. SCHNEIDER, E.R. (1978). Pollenanalytische Untersuchungen zur Kenntnis der spätund postglazialen Vegetationsgeschichte am Südrand der Alpen zwischen Turin und Varese (Italien). Bot. Jahrb. Syst. 100, 26-109.

4. DUPLESSY, J.C., DELIBRIAS, G., TURON, J.L. and DUPRAT, J.C. (1981). Deglacial warming of the northeastern Atlantic Ocean: correlation with the palaeoclimatic evolution of the European continent. Palaeogeography, Palaeoclimatology, Palaeoecology 35, 121-144.

5. MENÉNDEZ AMOR, J. and FLORSCHÜTZ, F. (1961). Contribución al conocimiento de la historia de la vegetación en España durante el Cuaternario. Estudios Geologicos 17, 83-99.

6. HANNON, G.E. (1984). Late Quaternary vegetation of Sanabria Marsh, northwest Spain. Master's Thesis, Trinity College, Dublin.

7. JANSSEN, C.R. and WOLDRINGH, R.E. (1981). A preliminary radiocarbon dated pollen sequence from the Serra da Estrela, Portugal. Finisterra 16, 299-307.

8. McKEEVER, M. (1984). Comparative palynological studies of two lake sites in western Ireland and northwestern Spain. Master's Thesis, Trinity College, Dublin.

9. HUNTLEY, B. and BIRKS, H.J.B. (1983). An Atlas of past and present pollen maps for Europe: 0-13,000 years ago. Cambridge University Press.

10. KOLSTRUP, E. (1980). Climate and stratigraphy in northwestern Europe between 30,000 B.P. and 13,000 B.P. with special reference to the Netherlands. Meded. Rijks Geol. Dienst 32-15, p. 181-253.

11. DE VRIES, H., FLORSCHÜTZ, F. and MENÉNDEZ AMOR, J. (1960). Un diagramme pollinique simplifié d'une couche de 'gyttja', située a Poueyferré près de Lourdes. Koninkl. Nederl. Akad. Wetensch. 63B, 498-500.

12. JALUT, G. (1973). Analyse Pollinique de la Toubière de la Moulinasse: nord oriental des Pyrénées. Pollen et Spores 15, 472-509.

13. MARDONES, M. and JALUT, G. (1983). La Tourbière de Biscaye (Alt. 409 m, Hautes Pyrénées): approche paléoécologique des 45,000 dernières années. Pollen et Spores 25, 163-212.

14. JALUT, G., DELIBRIAS, G., DAGNAC, J., MARDONES, M. and BOUHOURS, M. (1982). A palaeoecological approach to the last 21,000 years in the Pyrenees: the peat bog of Freychinede (alt. 1350 m, Ariège, South France). Palaeogeography, Palaeoclimatology, Palaeoecology 40, 321-359.

15. WATTS, W.A. (1984). A long pollen record from Laghi di Monticchio, southern Italy: preliminary account. Geological Society of London - in press.

16. ZEIST, W. VAN and BOTTEMA, S. (1982). Vegetational history of the eastern Mediterranean and the near east during the last 20,000 years. p. 277-321 in Palaeoclimates, Palaeoenvironments and Human Communities in the Eastern Mediterranean Region in Later Prehistory (ed. J.L. Bintliff and W. van Zeist). BAR International Series 133.

17. FRANK, A.H.E. (1969). Pollen stratigraphy of the Lake of Vico (Central Italy). Palaeogeography, Palaeoclimatology, Palaeoecology 6, 67-85.

18. ROSSIGNOL-STRICK, M. (1983). African monsoons, an immediate climate response to orbital insolation. Nature 303, 46-49.

19. WATTS, W.A. (1982). Abrupt climatic changes: the terrestrial record. p. 74-80 in Palaeoclimatic Research and Models (ed. A. Ghazi). Publ. D. Reidel, Dordrecht.

PALEOCLIMATIC RECONSTRUCTION IN BELGIUM AND IN GREECE

BASED ON QUATERNARY LITHOSTRATIGRAPHIC SEQUENCES.

by R. PAEPE and I. MARIOLAKOS
Vrije Universiteit Brussel   University of Athens
(Belgium)   (Greece)

with collaboration of

A. METTOS, V. SABOT, J. THOREZ, G. LIVADITIS, E. VAN OVERLOOP,
M.E. HATZJOTIS, J. HUS, LIN JINGXING, R. VANHOORNE

Summary

Continental soil-sedimentological lithostratigraphical records from
the Eastern Mediterranean and the Southern North Sea prove to be comparable and reveal long term periodicities which allow subdivision of
the Pleistocene into four Maxi Cycles of each 600.000 yrs. The latter
may be split up into Cryomeres of 250.000 yrs and Thermomeres of
350.000 yrs each comprising a number of cold-warm cycles at 100.000
yrs interval. So far the 413.000 yrs is not established. Periglacial
conditions prevailed since at least 1.6 Ma yrs ago in latitudes as
far south as Crete.
The last so called Upper Pleistocene loess stratigraphy as known for
the middle European belt appears to extend as far south as the
Southern Peloponesos and yields perfect litho-stratigraphic correlation and periodicities (23.000 and 41.000 yrs).
The Holocene Stage shows 19 humid-dry cycles with a treshold at 2.700
YBP subdividing this Stage into two parts of prior coarser sedimentation and finer thereafter. A 2.500 yrs periodicity is controlling the
archaeological-geological evidence in Greece.

In comparing the North Sea basin with the Eastern Mediterranean it
was firstly aimed at the comparison of boreal versus subtropical climatic
sequences during the Quaternary, secondly whether or not such sequences
differ in stable and unstable regions. If the latter statement proves negative, it is assumed that the observed sequences are of global climatic
origin and hence a tool for the study of climatic changes during the
Quaternary.
Secondly, it may be questioned which are the most suitable parameters
for setting up correlation between deposits of entirely differing facies.
For a long time Quaternary cooling in marine deposits was solely indicated by the first occurrence of arctic molluscs : in the Amstelian
Stage (F.W. HARMER, 1896) of the North Sea basin and in the Calabrian
Stage (M. GIGNOUX, 1910) of the Mediterranean.
It is only since in the late sixties drastic evolution in bio-
stratigraphical methods took place that more evidence about climatical
variations became available. Moreover methods of absolute dating enhanced greater precision about the appearance and extinction of species
and the duration of the Stages involved.
Very soon pollenanalysis lead to the first systematic Subdivision
of Quaternary continental sediments and establishment of the Plio-

Pleistocene boundary in the Netherlands (W.H. ZAGWIJN, 1974), in Italy
(F. LONA, 1971) and in Greece (T.A. WIJMSTRA, 1969). On the other hand
foraminifera studies greatly contributed to the subdivision of marine
sediments not only in coastal and off-shore regions but by and large in
abyssal deepsea sediments (N.J. SHACKLETON & N.D. OPDYKE, 1977, N.J.
SHACKLETON & M.B. CITA, 1979) as well. Finally studies appeared in which
marine deposits were studied from both micropaleontological and palynological point of view (R. BERTOLDI, 1977, G. PASINI & M.L. COLALONGO,
1982, W.H. ZAGWIJN & J.W. Chr. DOPPERT, 1978). It made possible to in-
tercorrelate both marine and continental deposits between northwest
Europe and northwest Mediterranean (J.P. SUC & W.H. ZAGWIJN, 1983).
    Despite of the fact that all paleontological, paleobotanical and
dating material is evidently derived from lithostratigraphical sections,
very little if any consideration is usually given to the climatic sedimentological evolution of the very deposits themselves. Thus the lithostratigraphic framework is lacking causing great difficulties as to the
comparison of the stratigraphic position of the observed phenomena or
of the obtained results. Most stratigraphic problems result from this
lack of precision with regard to the lithostratigraphic column.
    Since the 1960's other authors like K. BRUNNACKER is Germany, G.J.
KUKLA in Czechoslovakia, J. FINK in Austria, R. PAEPE in Belgium and
J. SOMME and J.P. LAUTRIDOU in Northern France and Normandy built up
lithostratigraphies exclusively based on sedimentary sequences alone.
Detailed study of periglacial loess and coversand proved hereby quite
promising for systematic reconstruction of the entire paleoclimatic
evolution most often from Neogene till Present. Especially relict soil
horizons offer a wide variety of investigations which subsequently are
used as climatic indexes : from sediment analysis through thin section
study and clay mineralogical X-ray diffraction, and not at least the
very macroscopic study of soil horizons in the field indicating type of
soil intensity of development, frequency of recurrence in the sedimentary sequence and especially the geomorphological surface on which they
have developed (R.V. RUHE, 1975).
    Furthermore, sediment structures such as frost wedges and cryotur-
bations yield immediate evidence of cold climate, incidently Polar Desert
conditions at the level of which they occur in the lithostratigraphic
column.
    In intermingling evidence from sediment structures and pollenana-
lysis, W.H. ZAGWIJN and R. PAEPE (1968), came to the first complete
paleoclimatic curve for the Last Glacial Stage.
    With regard to the Pleistocene as a whole, G.J. KUKLA (1972) was
the first to set up cyclicities in the soil stratigraphic sequences of
Cerveni Kopec (Czechoslovakia), completed hereafter with late J. FINK's
section of Krems (Austria) (J. FINK and G.J. KUKLA, 1977). At least 21
glacial/interglacial cycles indicated A through V occurred within the
timespan of approximately 2,4 Ma. These continental stages easily correlated with N.J. SHACKLETON's and N.D. OPDYKE (1973) deepsea record up
to oxygen isotope Stage 24 (G.J. KUKLA, 1978).
    It was thus proved that lithostratigraphy combined with pedostra-
tigraphy of continental deposits such as loess and related deposits of-
fered complete sequences of warm/cold cycles in precisely the same way
as in deep sea records.
    Sediment series unlike paleontological species offer an immediate
response to the prevailing conditions. Therefore they are considered to
serve as good indicators for climatic variance and periodicities.
    Furthermore they reflect all landscapes that passed through time at

a given spot (J. BUDEL, 1974), unlike plants and animals which are subject to wandering themselves.

THE QUATERNARY CONTINENTAL STAGES OF GREECE

Recently J. THOREZ (1982) developed a new method in claygeology by which it became possible to establish the paleoclimatic indexes of soils and sediments in many quaternary sections of Greece, in particular in those along the East coast of Attica studied by R. PAEPE since 1976.

In applying the clayweathering results to the lithostratigraphic sections of Meltemi, Kokkino Limanaki and Spiliazesa, the paleoclimatic curve was established for the last 10 Ma (R. PAEPE, 1982).

In addition, a specific palaeoclimatic curve was elaborated for the last 10.000 years (Holocene) in collaboration with archaeologists and palynologists who reviewed a great number of old and new excavations first in the coastal area from Marathon to Sounion, second in the Academias Platonos of Western Athens (R. PAEPE, M.E. HATZIOTIS, J. THOREZ, E. VAN OVERLOOP & G. DEMAREE, 1982).

Although emphasis was laid on the continental stages for both long term and short term curves, ample possibility was given in sections along the Attica East coast to locate marine transgression phases within Pleistocene and Holocene Series as well.

THE EAST MEDITERRANEAN PLEISTOCENE RECORD

As stated before by one of the authors (R. PAEPE, 1982) the Pleistocene record of Attica compares well with : the sequences obtained from the Drama-Tenaghi-Philippon area (Macedonia) studied by P. ANASTASIADIS; from the Pinios valley (Tessalia) studied by G. LIVADITIS; from Peloponnesos and Kreta studied by I. MARIOLAKOS; from the island of Naxos (by V. SABOT), Skyros and Lesbos (by R. PAEPE).

Taking the central position of Greece into account, one may easily extrapolate the results obtained in Greece to the northeast Mediterranean as a whole. Actually after first inspection of the neighbouring area's as southern Yougoslavia, Turkey, Libanon and Syria there is quite some paralellism to be observed between the pleistocene sediments of all these regions.

The Plio-Pleistocene Boundary

M. MITZOPOULOS (1948) and G. CHRISTODOULOU (1957) were the first to describe the macro- and micro fauna of the deposits of the coast of Rafina, located at 25 km distance from Athens on the East Coast of the Attica peninsula. Both authors are of the opinion one is dealing with Lower Pliocene Subseries.

G. CHRISTODOULOU, however, mentions above an Ostrea Zone in layer 8 between 11 and 19 m. a series of yellow marls in which he finds three Assemblage Zones from bottom to top namely :
- *Bulimina elongata d'Orbigny, Bulimina gibba Fornasini*
- *Elphidium crispum Linneus*
- *Nonion boueanum d'Orbigny*

The lithostratigraphic section is so poorly recorded that it is difficult to compare it with the newly elaborated sections of Meltemi and Kokkino Limanaki by R. PAEPE in the period of April 1974 through February 1982 the ideal composite profile of which is used in Table 1.

LIN JINGXING succeeded nevertheless to recognise three similar Assemblages Zones : *Nonion* and *Elphidium* Assemblage Zone in a Yellow Marl deposit overlying the *Pecten/Bulimina* Crag at the base of the Meltemi section. The top of the Yellow Marl, in which appears *Nonion boueanum* bears a Red Soil which R. PAEPE originally considered totally as of Pliocene in Age. In the very upper part of this soil another *Elphidium crispum/ Elphidium macellum* Assemblage Zone is recognized.

In conclusion, the Plio-Pleistocene transition is located at the boundary between a series of Yellow Marls overlying the Pecten Crag the latter containing the *Bulimina elongata/gibba Assemblage Zone*, wherein and below which also occur *Uvigerina sp., Cassidulina sp., Globigerina falconensis Blow, Elphidiella sp.* indicating subtropical to cold temperate climatic conditions.

It is the first time that part of the uppermost Yellow Marl deposits in Greece are considered to be of Pleistocene Age with the appearance of cold guests such as *Elphidium crispum* and *macellum*.

None of the guide microfossils of the western Mediterranean especially of the Calabrian Stage (such as *Globorotalia crassaformis* and *Globorotalia inflata*, nor *Hyalinea balthica*) were mentioned.

The First Warm Stage

In a new series of Yellow Marls above the Red Soil the *Ammonia beccarii* and *A. Limbatobeccarii* Assemblage Zone occurs. Sometimes in between the Red Soil and the Yellow Marl an *Ostrea* Crag interferes, however, containing still great amounts of *Elphidium crispum*. A disturbed zone of convoluted bedding structures may in some places occur at the lower boundary of the *Ostrea* Crag.

After a new level of marls, another *Ostrea* Crag occurs in which foraminifera occurrence is drastically reduced. Hereupon follow sterile marls with PKI (pedokomplex 1) at the top.

Paleomagnetical analysis points to at least two normal and one reversed component in the deposits in between the Red Soil and PKI (J. HUS, 1982). This may infer a position beyond the Jaramillo event, or even within the Olduvai event if one of the foregoing signals was to weak. From the palaeontological point of view, the latter interpretation seems most plausible. The very pronounced reversed magnetisation in sediments below the Red Soil may point to an early Matuyama position.

In this light, the first series of Yellow Marls with the Red Soil in top of it is considered to indicate the first cooling called hereafter RAFINAN Stage.

It most likely corresponds to the PIII steppe phase of the northwest Mediterranean (2.3 - 2.1 Ma) for which clay-weathering indexes revealed temperate dry-wet conditions (J. THOREZ, 1982).

The series of Yellow Marls with Ostrea banks extending hereupon till PKI then correspond to the first generally long term warm phase of the Pleistocene, called hereafter MELTEMIAN Stage. This Stage, however, splits up in three parts of Yellow Marls separated by two marine crag deposits occurring respectively at 1.9 Ma and 1.8 Ma approximately. Finally the whole Stage is comprised within a soil building at the beginning (Red Soil about 1.95 Ma) and another one (PKI) developing at about 1.65 Ma. Both are outspoken interglacial soil developments.

Clay weathering indexes reveal warm wet conditions again, which may in fact only apply to both of the marine transgression phases but not to the Yellow Marls which are believed to be also warm and most probably dryer. The Meltemian Stage is the first "warm phase" corresponding most likely to the PIV/Pl 1 Stage of J.P. SUC (1984) but which in fact comprises three cool periods and four warm ones i.e. the soils at the end and at the beginning and both marine crags in the middle. Judging from the presence of convoluted bedding at the base of the first marine crag (1.9 Ma), and at the top of the Stage just above PKI (1.65 Ma) in entirely different deposits such as red pediments, periglacial circumstances may be put forward for their origin.

The occurrence of similar cryoturbatic phenomena within and at the top of the pediment sands was also attested in other places by the present authors : at Messini nearby Kalamata (Peloponesos) and at Rethymnon (Kreta). Not only do they occur in an extreme low latitudinal position (35 - 36°N) but they appear in the stratigraphic column at a moment where deposition of Yellow Marls came to an end. Thus they seem to indicate an important climato-sedimentological break as well. Furthermore it must be stated that no forams occur in the profile after the last marine crag.

The upper part of the Lower Pleistocene

Above the already mentioned convoluted periglacial bedding a series of calcrete horizons interfering with marly horizons follow. Number of the layers are affected by convoluted bedding. A general temperate dry or wet climate is inferred from the clay weathering indexes although the convoluted bedding may point at more severe extremes. No doubt one is dealing here with a so-called cold phase where milder interstadials resulted in the building up of calcrete layers. It is proposed to name it the SPATAN Stage, lasting in time from 1.65 through 1.4 Ma and comprising three cold/warm cycles, however, of weaker intensity than the foregoing.

Marine invasions interrupt these series of recurrences of which two separated by a lake or lagoonal phase seem to characterize another long term warm cycle. As for the previous warm cycle, its series ends up with a paleosoil of the latosolic type namely PKII. For the whole the name KOKKINO LIMANAKIAN STAGE is introduced. It covers the timespan of 1.4 to 1.05 Ma subdivided again into three outspoken cold/warm cycles respectively attesting temperate and warm wet climate.

Another series of marls and calcretes follow hereupon covering the period 1.05 through 0.81 Ma attesting three new but weaker cold/warm cycles. It builds up the Loutsan Stage which may recall the Spatan Stage, however, with a remarkable increase in the gravel compound. Incidently frost wedges may occur at the upper boundary of this Stage in the section of Meltemi East.

The general conclusion is that there exists within the timespan of 2.25 to 0.81 Ma (conventionally the Lower Pleistocene), approximately within 1.44 million years, three major long term colder phases and two major long term warm phases totalling 13 short term cold/warm cycles as a whole.

The Red Soils of the latosolic type represent together with the marine transgressions the warm interglacials; the calcrete horizons the temperate interglacials. Three cold peaks are attested at respectively about 1.95 and 1.60 Ma and 810 K yrs.

The major subdivision correlates well with the pollenanalytical subdivision of the northwestern Mediterranean, the minor subdivision into 100 K cycles coinciding well with the oxigen-isotope stages of the deep-sea record and with the cycles obtained in the loess area of Central Europe.

The Middle Pleistocene

Most drastically is the change in sedimentation pattern with the beginning of the Draman Stage. It starts with an important fluviatile erosion immediately followed by the deposition of a series of coarse fluviatile gravels. The latter are periodically interrupted by red brown lessivé type of paleosoils connotated as PK IIIa through PK IIId. PK IIIa which is the most steady in the profile is paleomagnetically dated as of before the Brunhes normal polarity Zone (J. HUS et al., 1982).

The soils no longer reflect latosol and hence savanna but a lessivé type of soil and hence deciduous forest. The conglomerate is attesting of short intermittent fluviatile phases during which huge piedmont deposits were built up. Pediplanation has become an all-round process now, so that it is reasonable to say that semi-arid conditions prevailed. Unlike the foregoing Loutsan Stage which remains totally temperate as to climate, the gravel deposits only attest of temperate dry-wet conditions whereas the PKIII soils first indicate warm dry wet (PK IIIa and b) later warm wet conditions (PK IIIc and PK IIId). It is quite evident that the warm pedocomplexes represent the interglacial stages and the conglomerates the semi-arid temperate stages.

Another climato-sedimentological revolution took place hereafter i.e. after 450 K yrs. Cold peaks occur at 370 and 200 K yrs.

The whole is labelled as Spiliazesan Stage (R. PAEPE, 1973) according to the work carried out in Laurium.

The next Stage is characterised by the strongly developed Potami paleosoil dating most probably from about 100 K yrs followed by wind born deposits of the Last Glacial Stage and the Holocene Series as of 10.000 yrs ago (see hereafter).

PERIODICITIES OF THE PLEISTOCENE SERIES

Some major periodicities also result from the foregoing. Considering the foram-defined Plio-Pleistocene boundary (LIN JINGXING), the Red (Latosolic) Soil which contains this boundary is dated about 1.98 Ma. On the other hand as PK IIIa is dated before the Brunhes-Matuyama boundary but of the interglacial non latosolic soil type it must be located at the beginning of Kukla's cycle with termination 810 K yrs. It means that the 9 soil cycles and 4 marine transgressions which cover the time-span of 1.44 Ma years since the beginning of the Pleistocene show approximatively a 110.000 y periodicity which encompasses the calculated eccentricity period e of 104.000 yr. (A. BERGER, 1973). It also means that soils and marine phases tally with KUKLA's cycles V through K.

In rearranging all lower Pleistocene soils and marine transgression phases in the light of this very periodicity, three major cooler phases called hereafter cryomeres appear : Rafina 2.25 - 2.00 Ma, Spata 1.65 - 1.40 Ma, Loutsa 1.05 Ma - 810 K each lasting 250.000 years; they all alternating with two major generally warm phases, called hereafter thermomeres : Meltemi 2.0 - 1.65 and Kokkino Limanaki 1.4 - 1.05 Ma each lasting about 350.000 years.

A couple of one cyromere and one thermomere is labelled here Maxi Cycle totalling up to 600.000 years. The Maxi Cycles starting each with a cool phase namely at 2.25 Ma, 1.65 and 1.05 Ma year, undubiously go along with profound climato-sedimentological changes.

Whereas Yellow Marls are still the dominant facies of the Rafina-Meltemi Maxi Cycle I; these facies diminuish rapidly in the next Spata-Kokkino Limanaki Maxi Cycle II, where especially in the cool Spata cryomere, they interfere with calcrete horizons.

It also occurs that at the start of Maxi Cycle II about 1.65 Ma extreme cold conditions reigned leading to the existence of permafrost

conditions as far South as 35° North (Crete) as it attested by the presence of the frost wedges and cryoturbations. Since all observed frost
wedges are less than 20 cm in width extreme cold conditions were only
of short duration. Therefore, it may perhaps become difficult to track
such cold peaks in deepsea cores.

As to the soils, it clearly appears that latosolic soils such as
PKO, PKI and PKII are located within the thermomeres whereas the calcrete-
horizons (or their red brown soil equivalent) appear to exist only within
the Spata and Loutsa cryomeres. Finally the marine transgressions occupy
the middle of both thermomeres, the end and the beginning being occupied
by the paleolatosols.

This infers that certainly the beginning and the end of the thermomeres (i.e. at 2.00, 1.65, 1.40 and 1.05 Ma) were characterised by open
savanna of the peritropical type of landscape evolution leading to flat
latosolic surfaces amidst which inselbergs occurred (J. BUDEL, 1974).
In average every 350.000 years such savanna landscape initiated most
probably at a new geomorphological level. The temperature maxima of the
middle of the thermomeres are occupied by marine transgressive-regressive
phases respectively dated at 1.9, 1.8, 1.35 and 1.25 Ma, i.e. at 100.000
years intervals from each other and from the next paleolatosol. The relative distance in time between the two groups of marine layers evaluates
evidently at 600.000 years.

During the cryomeres it may be questioned whether real steppe conditions prevailed or not as was recently advocated by J.P. SUC and W.H.
ZAGWIJN (1982). Certainly the calcretes, however very thin (20 cm), in
between marine and limnic layers indicate dry conditions. When calcretes
are lacking red soils are found instead but nothing of the type of a
chernozem or brunizem. Moreover as calcrete horizons only occur as of
1.6 Ma ago within the Spata and Loutsa cryomeres the equivalent of the
PIII steppe (sensu Suc & Zagwijn) is not found back in the Rafina
cryomere.

Maxi Cycle III is composed of the Loutsa cryomere ending up with
termination J at 810 K yrs, and of the Drama thermomere ending up at approximately 450 K yrs. It means that cryomere IIIa belongs to the Lower,
whereas thermomere IIIb already to the Middle Pleistocene Subseries.

The Drama thermomere, however, correlates at least in time with
WIJMSTRA's Boz Dagh and Phalakron Interglacials combined.

The Loutsa cryomere is lasting again 250 K yrs while the Drama
thermomere 350 K yrs. Especially the soils of PKIII seem greatly involved with following terminations : 810 K yrs, 710 K yrs, 570 K yrs and
480 K yrs. Certainly not did these soils develop at the beginning of the
(cold/warm) cycles, but towards the middle of them when temperate dry/
wet conditions changed to warm wet conditions.

Thus the Loutsa-Drama Maxi Cycle III very much reflects the Spata-
Kokkino Limanaki Maxi Cycle II separating an initial cold to temperate
from a warm cycle. This is in sharp contrast with the foregoing Rafina-
Meltemi Maxi Cycle I which mainly remains in the warm column all over
600 K yrs of time and again with the next Spiliazesa-Attika Maxi Cycle IV
of which climatic periods are equally distributed over all three columns.
It means that both amplitude and frequency of climatic changes reached a
maximum during Maxi Cycle IV. From the afore it is known to correspond
to the period of maximum deposition of eolianites.

According to the spacing in time of soils which formed at respectively 330 K for PK IV and 240 K yrs for PK V and which generally are of
weaker development than the immediate foregoing soils and the one next
after (Potami S.) the Spiliazesa cryomere lasted for about 250 K yrs. This

implies that the last thermomere must have started about 190 K yrs when extreme cold conditions were prevailing. Taking the standard deviation into account in the geological sequence it is quite reasonable to correlate this cold phase with the one indicated by Ruddiman and Mc Intyre (1971) at 225 K yrs when the first cold invaded the North Atlantic Ocean. From this it immediately appears that the Potami Soil of the Last Interglacial Age (circa 125 K yrs) appears slightly after the start of the Attica Stage.

The Holocene Modern Soil appears at about 80 - 100 K yrs later than the Last Interglacial one depending on the position of the beginning of the Holocene. Its position reminds to the situation of PK IIIb revealed in the Drama thermomere IIIb if one assumes at least that the very last thermomere is of the same duration as the previous ones and consequently not finished yet. Yet perhaps two other soils, certainly one, are still to be developed according to the model of the Drama Stage. This should occur within the next 160 K yrs AP (After Present), which is exactly the timespan during which PK IIIa through PK IIIb developed i.e. from 810 K yrs till 650 K yrs. Even when the timespan of one cycle may reach its maximum at 90 K yrs it is certainly sufficient to leave 70 K yrs for the other one to develop.

Anyhow it shows the position of the Potami Soil (Last Interglacial, Stage 5e) at somewhat the same position as PK IIIa i.e. 50.000 years after the start of the Drama thermomere. It also shows the position of the Holocene (Recent) Soil as just another interglacial soil in between all other soils of the same thermomere. As will be pointed out further in this paper the pedostratigraphy of the Upper Pleistocene contains more soils than just one Last Interglacial (Potami) soil and evenso for the Holocene which is indicated by the best developed soil. So it may be questioned which Holocene Soil is actually dealt with.

Anyhow it may be advocated that from now till 115 K yrs AP generally cooler climatic conditions will prevail; after 115 K yrs warm conditions will prevail again during the same thermomere IVb. Thereafter, it is most likely to enter a new Maxi Cycle V beginning with cryomere Va lasting until 390 K yrs AP.

Conclusions for the Long term Pleistocene Eastern Mediterranean Cycles

So far it was possible to produce only simple geological deductions based on the anology and number of paleosoil occurrences in the loess deposits of Central Europe (G.J. KUKLA and J. FINK, 1977) on the one side, their correlation with the deep sea oxygen-isotope stages and 100 K cycles based on the astronomical Milankovitch periodicity of 104.000 yrs (A. BERGER, 1978) on the other hand.

Paleontological and paleomagnetical dating allowed fixation of the Plio-Pleistocene boundary as well as the Brunhes-Matuyama boundary thus fixating a timescale for the long term Pleistocene record.

Besides the 100 K distribution of warm and cold periods it became evident that a superposed grouping of warmer interglacial cycles and cooler glacial cycles was feasible which led to the recognition of longer term cycles labelled thermomeres and cryomeres as previously suggested by G. LUTTIG (1965).

During the thermomeres strong interglacial soil development occurs and also marine transgressions attain a high mean sea level (M.S.L.); cold phases are not so cold.

During the cryomeres interglacial soil development is different and generally weaker (calcrete of horizons or brown soils) than during the thermomeres; glacials are extreme cold.

Thermomeres and cryomeres ever since the beginning of the Pleistocene show an evolution through time with regard to soil development (latosol) sirosim, occurrence of marine transgressions, nature of deposits (lahars, lagoons, piedmont colluvium, eolianite). Despite this change in lithofacies, the timespan of duration for both thermomeres and cryomeres may be considered constantly the same throughout the Pleistocene : 350.000 yrs for the thermomeres and 250.000 yrs for the cryomeres.

This evaluation is a direct consequence of the 100 K distribution of the interglacial features. It follows that the warm periods as PIV/P1 corresponding with the Tiglian Interglacial is about 150 K yrs shorter i.e. the Meltemi Stage or thermomere Ib which starts at 2.0 Ma and ends at 1.65 Ma. It covers the timespan from Reunion till end Olduvai event.

Also is there a shift of Pl III with regard to the Kokkino Limanaki Stage and equally Pl II with regard to Spata (see Table 1).

The shifts in time do not occur as to be of major importance. Far more important is the fact that the Pleistocene in the Eastern Mediterranean may be subdivided into four Maxi Cycles of 600 K yrs totalling 21 interglacial/glacial cycles which perfectly encompass G.J. KUKLA's cycles of Central Europe.

In the light of the foregoing, especially taking the position of PK IIIa into account and the paleomagnetic results with regard to the distribution of the paleosoils, the start of Pleistocene is estimated at 2.25 Ma.

The classical subdivision into an Upper, Middle and Lower Pleistocene and finally a Holocene Series is a finite one which in the light of the foregoing is an absolute absurdity. The Holocene is not a series but just one phase of thermomere IVb. The Upper Pleistocene starts later than the beginning of thermomere IVb which will moreover, continue until 160 K yrs AP.

LONG TERM CYCLES IN THE SOUTHERN NORTH SEA BASIN

Many studies have been carried out in various countries of Northern Europe on the (litho)-stratigraphy of the North Sea Basin. As pointed out by G.J. KUKLA, 1978, the systematic correlation with the loess cycles as well as with O^{18}-Stages is still problematic. One or three Holsteinian Stages, one or four Cromerian Stages ?

It is quite questionable why so far only the Cromerian Stage has been considered as a fourfold complex. Why the Waalian "interglacial" and the Tiglian "interglacial" have never been considered as such. In the same way Eemian and Holsteinian may be considered as complexer, and not longer as a single valued interglacial.

The problem is that in Northern Europe unlike Greece no continuous section, even not in a borehole is available. Second, the investigation is clearly separated into continental and coastal marine geology. However, in Belgium and Northern France, in combining loess and coastal area sections good insight was gained with regard to the relationship of soils, sediments and marine transgressions (R. PAEPE and J. SOMME, 1975).

A compilation of the present knowledge is produced here on basis of the Belgium evidence in the marine and loess sectors.

For the classical Upper- and Middle Pleistocene area's like the coastal plains, the Flemish Valley, the IJzer estuary or Gulf of Lo have been taken into consideration, especially with regard to marine transgressions. As for the soil sequences the loess and the transitional area's have been taken mainly into account, the coversand area producing little if any of the soil levels.

For the Lower Pleistocene Subseries the Campine Formation of Northern Belgium has been chosen as a reference (J. HUS, et al., 1976).

Soil sequences and related marine transgressionlevels are indicated in Table I under the stratigraphic column Belgium. Age and lithostratigraphic position of these deposits have been sufficiently discussed in previous publications. The column may immediately been correlated with the Greek sequences appearing at the left side of Table I.

The Lower-Pleistocene and Plio-Pleistocene Boundary

The Pre-Tiglian Stage which has been correlated to the PIII Stage in the northwestern Mediterranean (J.P. SUC and W.H. ZAGWIJN, 1983) corresponds to the Upper Member of the Mol (Kieseloölith Sand) Formation (R. PAEPE, 1976). It is characterised by the occurrence of the first periglacial features frost wedges, etc.) as stated as early as 1943 by R. TAVERNIER and A. HACQUAERT. Thus it is greatly contrasting to the Lower Member of the Mol Formation occurring under the Arendonk lignite Member containing a Pliocene flora (R. VANHOORNE, 1970).

The marine Plio-Pleistocene boundary has been established in the Merksem Sand Formation by P. LAGA (1973) on basis of the foraminifera. Earlier J.H. VAN VOORTHUYSEN (1957) pointed out that the marine Plio-Pleistocene boundary was indicated by the foraminifera appearance of *Elphidium oregonense* at the boundary of *Ammonia-Quinqueloculina* and *Bucella-Cassidulina* Subzones within the *Cribononion excavatum* and *Elphidiella Hannai* Zone. This allowed him to distinguish Harmer's Amstelian from Icenian Stages, which W.H. ZAGWIJN subsequently related to the above mentioned continental Stages as well.

As a similar foraminifera transition was observed in the Rafina Ia Stage (see above) and the Yellow Marls are thought of cooler temperate climate an every long distance lithostratigraphic correlation is feasible.

The correspondance between the Rijkevorsel Member of Tiglian Age (2.05 - 1.65 Ma) and Meltemi Stage of Pl I - PIV Age (2.1 - 1.6 Ma) is even more striking. With respectively PK 0 and PK I correspond the lower peat bog (St. Lenaerts) and upper peat bog (Rijkevorsel) the latter containing *Azolla tigliensis*. In between the presence of two high bioturbated beach deposits at De Toekomst seem to indicate comparable sealevel changes as to the two marine crag deposits at Rafina - Meltemi. Perhaps is it possible that at the beginning of the Tiglian Stage the Meerle Sands indicate another marine influence. Anyhow the whole of the Rijkevorsel Member is the first estuarine appearance within the Campine Formation.

The next Eburonian Stage (1.65 - 1.4 Ma) testifies of cold conditions again. Generally three (Sint-Franciscus) sometimes five cold peat levels interfering with periglacial coversands, cryoturbations and frost wedges interfere. In both lithostratigraphic position and number of cold levels and milder soil forming levels they reflect the Spata calcretes and marls with occasionally frost wedges and cryoturbations.

The Turnhout Member of Waalian Age shows again a series of sands (1.4 - 1.05 Ma), bioturbated estuarine deposits and warm peat bogs with *Azolla filiculoides* very similar to the sequence of the Tiglian Stage and fairly corresponding to the Kokkino Limanaki IIb Stage of Greece.

Hereupon follow the deep cryoturbatic structures of the next cold phase called the Menapian Stage (1.05 - 0.81 Ma). No soils have been hitherto reported from the Belgium territory but in Greece three calcrete horizons split up the corresponding Loutsa IIIa Stage. Pollenanalytically a threefold subdivision has been established by W.H. ZAGWIJN (1975). The

Menapian Stage forms the upper boundary of the classical Lower Pleistocene Subseries.

The Middle and Upper Pleistocene

Despite numerous investigations carried out in this part of the Pleistocene many doubts remain in the North Sea Basin.

As stated before, erosion gaps are too numerous as to produce a continuous series of sediments and soils as in Greece. It is therefore only possible to dispose of a composite profile.

From the Netherlands'palynological sequence the existence of four warm and three cold periods is known for the "Cromerian" Stage (0.81 - 0.45 Ma or 810 K - 450 K yrs). This at first sight perfectly tallies with the Draman IIIb Stage in which the PK IIIa, b, c and d soil levels are interfering with conglomerates. As each of the soils in Greece perfectly corroborate the terminations of G.J. KUKLA's cycles, their position in time was thought to be really corresponding to the terminations established elsewhere.

Applying the same way of reasoning to the North Sea, the combined sections of Melle (Flemish Valley), of Herzeele (Gulf of Lo - Ijzer) and of Han-sur-Lesse (paleosoils) become important.

As for the soils, at least four interglacial Middle Pleistocene Soils are known to exist from the Han-sur-Lesse terrace, which according to its morphostratigraphic position is early mid-pleistocene in age. It is known that the Melle complex consists of two peat bogs which are separated from each other by a marine deposit that may be considered of (Upper) Cromerian certainly of Holsteinian Age. It also infers some possible correlation with the two uppermost soils of Han-sur-Lesse. The Herzeele marine deposit supposedly of Cromerian Age must then occupy an intermediate position in between the lowermost soils of the Han-sur-Lesse Soil series.

The Belgian "Cromerian" Stage tallies with the Draman IIIb Stage as to the number of soils. However, the presence of marine transgressions is unknown so far on the Greek side. Nevertheless, the marine incursions during the "Cromerian" Stage of the North Sea are not entirely surprising. Just as the Tiglian and the Waalian one can see that it is a warm thermomere, which besides warm interglacial paleosol phases may contain some relicts of marine invasions. Anyhow, the fourfold subdivision of the Draman Stage encompasses perfectly the fourfold subdivision of ZAGWIJN's Cromerian Complex.

Climatically, there is perhaps a greater differentiation than in foregoing thermomere periods. Whereas the previous Menapian period was entirely cold/dry in the North Sea Basin against temperate dry and wet in Attica, it becomes dominantly temperate in both regions but with warm peaks during the interglacials in Greece and extreme cold peaks during the glacials in the North Sea Basin.

Upon the "Cromerian Complex" i.e. the Cromer Cryomere follows a series of extremely complicated glacial and interglacial stages in northwestern and northern Europe as a whole.

The upper Melle (Peat) II may correlate with the Frimmersdorf Interglacial Soil underneath the Glacial Elsterian Stage, the soil being dated at 440 K yrs (STREMME, 1982) and thus corresponding to PK IIId. Hereupon follow the so called Schelde gravels and coversands of Elsterian Age corresponding to the eolianites of Greece. The latter are in turn overlain by the Zeebrugge Marine Stage deposits of Holsteinian Age (R. PAEPE, 1983).

This transgression is followed by cold deposits in which top occurs a textural-B-horizon designated as Tubize Soil in the Loess Region (R.

PAEPE, 1966). The latter may correlate in time with the Wacken interglacial dated about 330 K yrs. From hereon, extreme cold conditions are attested by the presence of several frost wedge levels in deposits of the Saalian Stage in between the soil horizons.

The next marine transgression is the Izenberge Crag (A. RUTOT, 1885, J. SOMME & R. PAEPE, 1975) of about 290 K yrs overlain by glacial deposits again of Saalian Age. It is this very crag that was first used to determine the Flandrian Stage (A. RUTOT, 1895) which at that time was labelled Lower Pleistocene.

Cold deposits follow again in which sporadically peat horizons such as the Daussoulx Soil developed. The latter may represent the TREENIAN Interglacial Stage dated about 230 K yrs. This soil is overlain by the "old loess" deposits which then should correspond to the cold Warthe Stage. Finally the whole is flooded again by marine waters of the Oostende transgression with *Tapes senescens var. eemiensis* and *Corbicula fluminalis* and immediately covered by the Eeklo Soil of last interglacial age. Hereupon are resting periglacial sediments of the Weichselian Stage with the Holocene soil in top of it.

It is quite obvious that the coldest glacial phases in Europe such as Elsterian, Saalian, Warthian and Weichselian, correspond to the phases of eolianite building in Greece. Moreover, the Wacken interglacial soil corresponds to PK IV whereas PK V to the Treenian Soil, the whole followed by the last interglacial Eeklo Soil of Eemian Age (128 K yrs) corresponding to the Potami Soil of Pangaion Age in Greece (see postscriptum).

Sea level changes and periodicities during the Pleistocene

Middle and Upper Pleistocene sea level changes seem complicated not only by their cyclicity in time but by the topographic variation of occurrence of the marine deposits as well : transgressions of Herzeele at + 10 m, Melle at 0 m, Zeebrugge at - 20 m, Izenberge at + 15 m and Oostende at + 2 m M.S.L. Investigations in view of an explanation of such differences are now under study. Next to a possible climatic cause for the change in elevation (W.H. ZAGWIJN), ice surgeries (J. HOLLIN & G. MILLER) and neotectonic movements (I. MARIOLAKOS, R. PAEPE) are taken into account.

Whatever their height, they seem to occur at 200 K, 150 K and 100 K with a maximum possible periodicity of 150 K for all transgressions within the Brunhes normal polarity phase.

Lower Pleistocene sea level changes seem to be grouped at 600 K in the thermomere phases within which they occur at 100 K intervals, both in Greece (Eastern Mediterranean) and Belgium (North Sea Basin).

Between the first "Cromer Complex" transgression and the last Waalian transgression a 500-600 K time interval may be inferred.

MIDDLE TERM AND SHORT TERM CYCLES

The classical Upper Pleistocene periodicities are known since long from continental deposits of the Boreal periglacial regions thanks to numerous radiocarbon datings of peatbogs and other ones as the coral reef dating of Barbados and elsewhere.

Nevertheless, it remains most striking that morphostratigraphic resemblance exists between continental loess deposits (W.H. ZAGWIJN and R. PAEPE, 1968; R. PAEPE, 1966; R. PAEPE, J. SOMME and J.P. LAUTRIDOU, 1974) especially as with regard to the occurrence of the Last Interglacial Soil (Eeklo Soil; Potami Soil), the Early Last Glacial Warneton Soil Complex encompassing the three classical Amersfoort (68.000 YBP), Brörup (65.000 YBP) and Odderade (58.000 YBP) interstadials, the Middle Last Glacial interstadials Poperinge-Moershoofd (45.000 YBP), Hoboken-Hengelo (39.000 YBP) and the Zelzate-Denekamp (27.000 YBP) and finally the Late Glacial Bölling-Alleröd (12.000 YBP) interstadials.

Recent studies in Southern Peloponnesos at the loess section of Koroni investigated by I. MARIOLAKOS and R. PAEPE (1984, unpublished) reveal lithostratigraphic sequences quite similar to Amersfoort (W.H. ZAGWIJN, 1963), to Warneton and Tongrinne (R. PAEPE, 1966) and still-fried (J. FINK, 1965). All soils of the periglacial area appear in this section which infers, as frost wedges also are attesting, periglacial conditions in the most southern part of Europe.

From a simple geological extrapolution into the future BERGER's, 1979 natural coolings predictions at 60.000 YAP, roughly 100.000 yrs after the 50.000 YBP cooling and at 24.000 YAP, 40.000 yrs after the 18.000 YBP cooling are predictable.

The Holocene subdivision with 19 pedostratigraphic phases (R. PAEPE et al. 1983; R. PAEPE et al. 1984) also reveals important short term periodicities especially in Greece.

The major periodicity derived from the combined archeological and geological cycles is 2.500 yrs attesting the 6.000 YBP humid peak and the 2.700 YBP drought (Kallikleios Soil). This periodicity of 2.500 yrs clearly appears in W. DANSGAARD's et al. (1984) curve of the Greenland ice cores as the 2.550 yrs cycle. Before and after 2.700 YBP sedimentation patterns are entirely different : generally coarser before and finer hereafter. Also after 2.700 YBP periods of drought occur more frequently with a 1.000 yrs time intervals at 200 AD, 1.000 AD and to-day. It encompasses the statement of BERGER (1974) about decreasing insolation during the summer months since 3.000 YBP, which the same author assumes not to become significantly larger before the 24.000 YAP.

. Keeping in mind that 2.700 YBP is a treshold in the Holocene evolution, it may be concluded as to the existence of an even larger cycle roughly 7.500 yrs if the beginning of the postglacial warming up is considered to have started with the Bölling-Alleröd interstadial some 11.000 years ago and the Holocene at 10.300 YBP. It means that the next change is to occur about 4.800 YAP which is near to BERGER's prediction of the next cold peak (or should one read the next dry peak) at 4.000 YAP.

Postscriptum : At the time this paper was submitted for printing, the authors became aware of the recent publication on "The Pre-Weichselian Glaciations of north-west Europe" by J. EHLERS et al. (1984). In this paper the geographical occurrence of the Warthe and Drenthe Stages are clearly established from one region to another whereas the Treenian Stage is abolished and simply mentioned as an interval.

REFERENCES

1. BERGER, A. (1978). Théorie astronomique des paléoclimats, une nouvelle approche, *Bull. Soc. belge Géologie, 87/1*, 9-25
2. BERTOLDI, R. (1977). Studio Palinologico della serie di Le Castella
   (Calabria). *Accad. Naz. dei Lincei VIII:62*, 547-555
3. BUDEL, J. (1974). Klima Geomorphologie. Gebr. Borntraeger. 304 p.
   Berlin
4. CHRISTODOULOU, G. (1957). Peri tinon Pliocenikon trimatophoron tis
   Raphinis; *en Athinais* 24-29
5. CHRISTODOULOU, G. (1961). Die Foraminiferen des marinen Neogens
   (Astien) von Attika. *Inst. Geol. Subs. Res. 8/1*, 47 p.
6. DANSGAARD, W., JOHNSEN, S.J., CLAUSEN, H.B., DAHL-JENSEN, D.,
   GUNDESTRUP, N. and HAMMER, C.U. (1984). North Atlantic Climatic
   oscilliations revealed by deep Greenland ice cores, *Climate processes and Climate Sensitivity, Geophysical Monograph 29, Maurice
   Ewing volume 5*, 288-298
7. EHLERS, J., MEYER, K.D. and STEPHAN, H.-J. (1984). The Pre-Weichselian glaciations of North-West Europe, *Quat. Sc. Rev. 3*, 1-40
8. FINK, J. (1965). The Pleistocene in Eastern Austria, *Geolog. Soc.
   Amer., Special paper*, 84 p.
9. FINK, J. and KUKLA, G.J. (1977). Pleistocene climates in central
   Europe : at least 17 interglacials after the Olduvai Event. *Quaternary Research 7*, 363-371
10. GIGNOUX, M. (1910). Sur la classification du Pliocène et du Quater-
    naire dans l'Italie du Sud. *C.R. Acad. Scien. 150, 13*, 841-844
11. HARMER, F.W. (1896). On the Pliocene deposits of Holland and their
    relation to the English and Belgian Crags with a suggestion for the
    establishment of a new zone "Amstelian" and some remarks on the
    geographical conditions of the Pliocene Epoch in Northern Europe.
    Q.J. Geol. Soc. Lond. 52, 748-782
12. HUS, J., PAEPE, R., GEERAERTS, R., SOMME, J. and VANHOORNE, R. (1976).
    Preliminary magnetostratigraphical results of Pleistocene sequences
    in Belgium and northwest France, *Quat. Glac. Northern Hemisphere,
    IGCP 3*, 99-128
13. KUKLA, G.J. and KOCI, A. (1972). End of the Last Interglacial in the
    loess record. *Quaternary Research 2*, 374-383
14. KUKLA, G.J. (1978). The classical European glacial stages : corre-
    lation with deep-sea sediments. *Transactions of the Nebraska Aca-
    demy of Sciences VI*, 57-93
15. LAGA, P. (1978). Stratigrafie van de mariene Plio-Pleistocene afzet-
    tingen uit de omgeving van Antwerpen met één bijzondere studie van
    de foraminiferen; Thesis 299 p., Leuven
16. LIN JINGXING (1983). Preliminary Report on Foraminifera in Greece,
    Inst. Geol. Chinese Academy of Geol. Sciences, 8 p.
17. LUTTIG, G. (1965). Interglacial and Interstadial Periods. *The Jour-
    nal of Geology, 73/4*, 579-591
18. MITZOPOULOS, M. (1948). Das Pliozän von Raphina (Attika). *Prakt. Ak.
    Ath. 23*, 295-301
19. PAEPE, R. et MORTELMANS, G. (1969). Sur la présence de sols fossiles
    pléistocènes pré-eémiens entre Hal et Tournai. *Bull. Soc. belge Géol.
    78*, 57-68
20. PAEPE, R., SOMME, J. and LAUTRIDOU, J.P. (1980). Principes, méthodes

et système de la stratigraphie du Quaternaire dans le Nord-Ouest de la France et la Belgique; Problèmes de Stratigraphie quaternaire en France et dans les pays limitrophes. *Suppl. Bull. Ass. Fr. Et. Quat.*, *N.5.*, *1*, 148-162

21. PAEPE, R. and VANHOORNE, R. (1970). Stratigraphical position of periglacial phenomena in the Campine Clay of Belgium, based on palaeobotanical analysis and palaeomagnetic dating, *Bull. Soc. belge Géologie*, *79*, *3-4*, 201-211

22. PAEPE, R., HATZIOTIS, M.E., THOREZ, J., VAN OVERLOOP, E. and DEMAREE, G. (1982). Climatic indexes on the basis of sedimentation parameters in geological and archaeological sections. *Palaeoclimatic Research and Models*, ed. A. GHAZI, EEC, 129-138, Reidel, Dordrecht

23. PAEPE, R. and DERAYMAEKER, D. (1973). Geomorphological and Quaternary Mapping of the Adami-Potami area (S.-E. Attica), *Thorikos 1969*, 79-99

24. PAEPE, R. and SOMME, J. (1975). Marine Pleistocene transgressions along the Flemish Coast (Belgium and France), *Quat. Glaciations in the Northern Hemisphere*, *2*, IGCP 73/1/24, 108-116

25. PAEPE, R. and VANHOORNE, R. (1976). The Quaternary of Belgium in its relationship to the stratigraphical legend of the geological map. *Verh. Geol. Kaart en Mijnkaart Belgie*, *18*, 38 p.

26. PAEPE, R. (1971). Quaternary marine formations of Belgium. *Quaternaria*, *XV*, 99-104

27. PAEPE, R., BAETEMAN, C., MORTIER, R., VANHOORNE, R. (1981). The marine Pleistocene sediments in the Flandrian Area. *Geol. Mijnbouw*, *60;* 321-330

28. PAEPE, R. (1971). Dating and position of Fossil Soils in the Belgian Pleistocene Stratigraphy. *Paleopedology*, 261-269

29. PASINI, G. and COLALONGO, M.L. (1982). Status of research on the Vrica Section (Calabria, Italy), the proposed Neogene/Quaternary Boundary-Stratotype section, in 1982. *Report* presented at XI INQUA Congress, Moscow. *Ist. Geol. Marina del C.N.R.*, 1-75. Bologna

30. RUDDIMAN, W.F. and McINTYRE, A. (1981). The North Atlantic Ocean during the last deglaciation. *Palaeogeogr.*, *Palaeoclimatol.*, *35*, 145-241

31. RUHE, R.V. (1975). Geomorphology. Geomorphic Processes and Surficial Geology. Houghton Mifflin Company, 246 p. Boston

32. RUTOT, A. et VAN DEN BROECK, M.E. (1885). Note sur la nouvelle classification du terrain Quaternaire de la Basse et de la Moyenne Belgique, *Bull. Soc. R. Malac.*, *20.*

33. SHACKLETON, N.J. and OPDYKE, N.D. (1977). Oxygen isotope and palaeomagnetic evidence for early Northern Hemisphere glaciation. *Nature* *270*, 216-219. London

34. SHACKLETON, N.J. and CITA, M.B. (1979). Initial Report, *D.S.P.D. 47*, 433-445

35. SHACKLETON, N.J. and OPDYKE, N.D. (1973). Oxygen isotope and palaeomagnetic stratigraphy of equatorial Pacific cores V28-238 : oxygen isotope temperatures and ice volumes on a 10^5 year - 10^6 year scale. *Quaternary Research 3*, 39-55

36. SOMME, J. et PAEPE, R. (1978). La Formation d'Herzeele : un nouveau stratotype du Pléistocène Moyen marin de la Mer du Nord. *Bull. Ass. franç. Etude Quaternaire*, *1*, *2*, *3*. 81-149

37. STREMME, H.E., FELIX-HENNINGSEN, P., WEINHOLD, H. and CHRISTENSEN, S. (1982). Paläoböden in Schleswig-Holstein. *Geologisches Jahrbuch*, *F14*, 311-361

38. SUC, J.P. (1984). Origin and evolution of the Mediterranean vegetation and climate in Europe, *Nature 307*, *No 5950*, 429-432

39. SUC, J.P. and ZAGWIJN, W.H. (1983). Plio-Pleistocene correlations between the northwestern Mediterranean region and northwestern Europe according to recent biostratigraphic and palaeoclimatic data. *Boreas 12*, 153-166. Oslo

40. SYMENIODIS, N., BACHMAYER, F. and ZAPPE, H. (1979). Pikermi, *Field Guide to the Neogene of Attica*, Univ. of Athens, Series A, 33, 1-11

41. TAVERNIER, R. and HACQUART, A. (1946). Compte rendu de l'excursion conduite par A. HACQUART et R. TAVERNIER, Géologie des terrains récents; *Bull. Soc. belge Géol.*, 452-478

42. THOREZ, J. (1982). Clay geology, a palaeoclimatic tool for Quaternary Series. *Striolae, INQUA Newsletter 4*, 10-16, Uppsala

43. VAN VOORTHUYSEN, J.H. (1957). The Plio-Pleistocene boundary in the North Sea Basin; *Geologie en Mijnbouw, n.s. 19, 7*, 263-266

44. WIJMSTRA, T.A. (1969). Palynology of the first 30 metres of a 120 m deep section in Northern Greece. *Acta Bot. Neerl. 18*, 511-527

45. ZAGWIJN, W.H. (1974). The Plio-Pleistocene boundary in western and northern Europe. *Boreas 3*, 75-97

46. ZAGWIJN, W.H. (1975). Variations in climate as shown by pollen analysis, especially in the Lower Pleistocene of Europe. In Wright, A.E. & Moseley, F. (eds.) : *Ice Ages : Ancient and Modern*, 137-152, Liverpool

47. ZAGWIJN, W.H. and DOPPERT, J.W. Chr. (1978). Upper Cenozoic of the southern North Sea Basin : Palaeoclimatic and Palaeogeographic evolution. *Geologie en Mijnbouw 57*, 577-588

48. ZAGWIJN, W.H. (1963). Pleistocene Stratigraphy in the Netherlands based on changes in Vegetation and Climate. *Verhand. Kon. Ned. Geol. Mijnb. Gen., N.5, 24*, 139-156

49. ZAGWIJN, W.H. und PAEPE, R (1968). Die Stratigraphie der Weichselzeitlighen Ablagerungen der Niederlande und Belgiens. *Eisz. und Gegenwart, 19*, 129-146.

50. HUS, J., MORTIER, R. and PAEPE, R. (1982). Lithostratigraphical, palaeomagnetical and sedimentological study of Pleistocene outcrops in the Campine Area (N. Belgium). *XI INQUA CONGRESS; Abstracts, Vol. II*, 203, Moscow.

PLEISTOCENE
LITHO-AND PEDOSTRATIGRAPHIC SEQUENCES OF GREECE
(EASTERN MEDITERRANEAN) and BELGIUM (SOUTHERN NORTH SEA)

MEDITERRANEAN CHRONOSTRATIGRAPHIC STAGES		GREECE (ATTICA)		AGE Ka BP or Ma BP	CYCLES	TERMINATIONS	PALAEOMAGNETIC SCALE	CLIMATIC INDEXES & CLIMATIC CURVES	BELGIUM		NORTH SEA STAGES
POLLEN W.P. WIJMSTRA (1969) / J.P. SUC (1982)	SEDIMENTOLOGY	SOILS & LITHO-FACIES	FORMATION MEMBER (SOIL)					C T W C T W	FORMATION MEMBER (SOIL)	SOILS & LITHO-FACIES	W.H. ZAGWIJN (1975)
RECENT	RECENT		RECENT / Eolianite gravel	11 Ka	B						Weichsel
PANGAION	ATTIKA IVb		POTAMI SOIL	100 / 128	II				OOSTENDE		EEMIAN
			Eolianite	200	C / III				LOESS		Warthe
SYMVOLON				240					DAUSSOULX		TREENIAN
	SPILIAZESA		PK V / gravel	300	D / IV	B			IZENBERGE		Saale
LEKANIS	IVa		PK IV / Eolianite gravel	330 / 400	E / V				TUBIZE / ZEEBRUGGE		WACKEN / Saale / HOLSTEIN
				480	F / VI						Elster
BOZ DAGH &			PK IIId conglomerate	500 / 570	G / VII				H A N S. L E S S E C A M P I N E F O R M A T I O N	MELLE / MELLE / MELLE	FRIMMERSDORF
	DRAMA IIIb		PK IIIc conglomerate	600 / 630	H VIII				LOESS		CROMER
PHALA-KRON			PK IIIb conglomerate	700 / 710	I / IX				HERZEELE		COMPLEX
			PK IIIa	800 / 810	J / X	M					
	LOUTSA IIIa		CALCRETE	900	K / XI	Q90					MENAPIEN
			CALCRETE		L / XII	Q99					
			CALCRETE	1.0 Ma							
PL. III	KOKKINO		PK II	1.1 / 1.2	M / XIII / N						
	LIMANAK IIb		MARINE / LAGOONAL / MARINE	1.3	XIV / O / XV	M			MARINE / MARINE		WAALIAN
PL. II	SPATA IIa		CALCRETE marl / CALCRETE marl / CALCRETE marl / CALCRETE cryoturbation	1.4 / 1.5 / 1.6 / 1.7	P / XVI / R / XVII / S / XVIII	167			EOLIANITE / BEERSE MEMBER / RIJKEVORDSEL / EOLIANITE		EBURONIAN
PL. I	MELTEMI Ib		PK I / yellow marl / MARINE / yellow marl / MARINE	1.8 / 1.9	T / XIX / U / XX	O / Q89			MARINE / FORMATION / MARINE		TIGLIAN
P. IV			PK 0	2.0	V / XXI	Q01 / Q04			SINT LENAERTS / MEERLE		
P. III	RAFINA Ia		LATO SOIL (RED SOIL) / marl	2.1 / 2.2		Q12 / Q14 / M			UPPER MOL FORMATION		PRE-TIGLIAN
P. II			PECTEN GRAG	2.3 / 2.4					ARENDONK LIGNITE / LOWER MOL FORMATION		REUVER
				2.5 / 2.6 / 2.7 / 2.8 / 2.9		248 / G / 292 / K					

PROJECT OF THE COMMISSION OF THE EUROPIAN COMMUNITIES
CLIMATOLOGY RESEARCH PROGRAMME 05/002

Centre for Quaternary Stratigraphy
R. PAEPE & I. MARIOLAKOS , 1984

CHAPTER II : CLIMATE MODELLING

- Evidence of bimodality in the statistics of northern hemisphere circulation

- Simulation of European climate with an atmospheric general circulation model

- Modelling the impact of soil properties on European climate

- Interactive ocean-atmosphere models for climate forecasting

- Climate modelling activities at the Max-Planck-Institute of meteorology, Hamburg

- Ocean modelling in Cambridge

- Subpolar circulation, deep water formation and air-sea-ice interactions in Labrador and Greenland seas

- Development of an economical soil model for climate simulation

EVIDENCE OF BIMODALITY IN THE STATISTICS OF NORTHERN HEMISPHERE CIRCULATION

A. SPERANZA
CNR-FISBAT c/o Dip. Fisica
Bologna-Italy

Summary

After a critical re-analysis of Charney-Devore (CDV) type of minimal theories of general circulation, non-linear self-interaction of planetary waves is identified as a key mechanism to take into account in order to produce better agreement with observations. A new improved version of CDV is studied and its theoretical predictions concerning statistical properties of general circulation are verified by means of data analysis. Particular attention is devoted to the problem of multiple equilibria and associated bimodality of occupation frequency in phase-space. Bimodality is found on a statistical sample of 14 years of 500 mb height observations.
Extensions of the theory, suggested by energetic considerations, to include baroclinic effects are finally discussed.

1 Introduction

In dealing with the complex structure of atmospheric fields operations that should be based on a knowledge of the general statistical properties of the system (in particular space and/or time averages) are performed all the time. However the problem of describing and physically interpreting such properties has not been the object of general interest until recently. As a typical example we may mention the fact that the time-average (usually over a season), zonally non-symmetric disturbances of the atmospheric circulation are traditionally called by meteorologist "stationary", with obvious reference to a fixed point of Navier-Stokes equations for atmospheric flow, although we can hardly find in literature a specific mention of the problem of calculating such a stationary circulation (see (1) for a discussion of such a problem).

On the other hand the accumulation of experience in modeling the general circulation, both for operational and scientific purposes, has made clear that we are dealing with a system with global (in phase-space) properties that are well marked and potentially explanable. For example, the variation in time of forecasting skill (see, for example (2) for a discussion of this problem) and the distribution in space of the so called systematic error (see (3)) seem to be characterized by general properties that are relatively independent on the specific model used.

It is clear that if we want to be able to enter the domain of long range forecast, trying to predict the evolution of some statistical parameters beyond the deterministic predictability of the system, we have to be able to define rigorously, at least from a qualitative point of view, the "slow" variables of our model. By doing this we have entered the domain of "climatic" studies.

As in the problem of climate modeling, we have little a priori hope of deducing the correct laws of slow evolution of the system by the ingegneristic construction of more and more complicated systems in an ordered hyrarchy. We have rather to resort, as it has happened very often in physics, to the inductive construction of prototype laws that possess the essential symmetrics of the observed system and are able to fit some macroscopic properties that are considered central in the slow dynamics.

This is what we have tried to do in dealing with the so called "low-frequency variability" of the atmosphere (time scales approximately from 10 days to 2 month) and the associated meteorological phenomenology of "blocked" and "zonal" states of the middle-latitude circulation.

Our work started from re-analysis the classical model of Charney and DeVore (4) (from now on CDV) in the hight of some observed properties of atmospheric statistics and evolved into the construction of a more articulated, although not much more complex, minimal model of general circulation which desplays much more realistic statistical properties.

The essential observational (bimodality of ultra-long wave amplitude statistics) and theoretical (resonance folding) phenomenologies can be find in (5).

Some improvements of theory and data analysis are described in papers mentioned throughout the following text.

2 Statistical properties of CDV system.

The simplest formulation of the barotropic CDV system can be obtained assuming that topography is sinusoidal:

$$h(x) = h_0 \cos K_m x \tag{1}$$

and the streamfunction is determined by the superposition of a uniform zonal flow and a wave of the same wavelegth as topography:

$$\psi(x,t) = -V(t)y + A(t)\cos K_m x + B(t)\sin K_m x. \tag{2}$$

If we also assume that vorticity is dissipated by a Laplacian friction $-\nu\nabla^2\psi$ we obtain from the barotropic equations:

$$\dot{A} + \frac{\nu}{K_m} A + (V - \frac{\beta}{K_m^2})B = 0$$

$$\dot{B} - (V - \frac{\beta}{K_m^2})A + \frac{\nu}{K_m} B + \frac{f_0 h_0}{K_m^2 H} V = 0. \tag{3}$$

The equations for time evolution of zonal flow neeed to close the system (3) can be obtained in different ways (for example taking the limit for vanishing latitudinal shear of the zonal flow as in (6)). For a linear frictional dissipation and a constant forcing of zonal momentum the resulting equation is:

$$\dot{V} = \frac{f_0 h_0 K_m}{4H} B - \nu(V - V^*). \tag{4}$$

The phase space of the system of eq. (3-4) is characterized by global properties which can be deduced to a large extent by the local properties of critical points. The stationary solutions of (3-4) are easily found as the intersections between the linear solution for B(V) from the system (3) and the other linear relation (4). Fig. 1 shows a graphycal desplay of such relationships for sensible meteorological values of parameters, together with an ensemble of "instantaneous" (every twelve hours) values of and deduced from observational data of two winters.

For a certain range of values of momentum forcing there are three different equilibria (i.e. there is a bifurcation in the V* direction in parameter space).

Stability analysis of the different equilibria shows that only the one
characterized by intermediate values of V and B is unstable; the others are
stable. The statistical properties of such a system (see (7) and (8)) are
expected to be non-homogeneous: the system should spend most of its time in
the proximity of the equilibria.

Consequently, the probability density in the parameter space should
show a pronounced bimodality.

3 Comparison of CDV theory with statistics deduced from observations.

The "instantaneous" (every twelve hours) values of V and B (for zonal
wavebumber 3) calculated from the observed northern hemisphere 500 mb data
of two winters are also shown in Fig. 1. From comparison with CDV solutions
it is immediately seen that while the predicted variation of the wave amplitude is realistic, the observed dispersion of the zonal wind is by far less
them theoretically predicted (notice that the slope of our "form-drag" relationship (4) are already a factor of four larger then the one of the CDV paper).

This impression is confirmed by statistical analysis. Fig. 2 shows the
frequency of occurrence of different values of V and B (zonal wavenumbers
2-4) calculated from 14 years of data. The distribution is clearly bimodal
in B, but not in V.

4 Modifical unidimensional theory.

A good fitting of the observational statistics can be obteined introducing non-linearity (associated with wave selfinteraction) into the theory. To
this purpouse we introduce latitudinal variation assuming:

$$h(x,y)=h(x)g(y)$$
$$\psi(x,y,t)=-V(t)y + A(x,t)g(y) \tag{5}$$

and project the barotropic equation with Laplacian dissipation on to the $g(y)$
latitudinal structure, obtaining:

$$\partial_t(A_{xx}+\alpha A)+V\partial_x(A_{xx}+\alpha A)+\frac{f_0}{H} V\ h_x+\beta A_x+\frac{3}{2}\delta AA_x +$$
$$-\nu(A_{xx}+\alpha A)=\gamma A(A_{xx}+\frac{f_0}{H}h)-A(A_{xx}+\frac{f_0}{H}h_x) \tag{6}$$

where the non-linear selfinteraction coefficients:

$$\alpha \equiv <gg_{yy}>$$

$$\gamma \equiv <g^2 g_{yy}>$$

$$\delta \equiv <gg_y g_{yy}> \tag{7}$$

have been defined $(<\cdot>=\int \cdot dy$ is the scalar product in y).

Perturbation analysis in near-resonant regime (9) of eq. (6) shows that
the resonant peak of Fig. 1 is "bent" by nonlinearity as in Fig. 2. Now it
is possible to stabilize wave-amplitude to different values while maintaining
the variability of zonal flow in a limited range.

5 Further improvements of the theory.

Consideration of other observational facts (energetics, spectral correlations, etc...) suggests that the real process operating at the ultra-long
(zonal wavenumbers 1-5) is essentially baroclinic and possibly involves several wavenumbers. Improvements of the theory in these directions are under

study. The basic features of the baroclinic theory, in particular as far
concerns resonance bending and energetics have already been outlined (see
(10)).

6 Conclusions.

The meteorological implications of the above described research are made
clear by an analysis of the superposition of atmospheric circulations in the
proximity of the different equilibria. Fig. 3 shows some examples of these
superposition: the similarity with "blocked" and non-blocked states is obvious
The amplitude of zonal wavenumbers 2-4 proves to be an excellent "global"
blocking indicator: only very localized cases of atlantic blocking are missed
by such an index (see (11) and (12) for further details concerning data analysis on bimodality).
In conclusion we believe we are beginning to capture some essential properties of low-frequency variability.

REFERENCES

1. SHNEIDER, E.K.and LINDZEN, R.S. (1977). Axially symmetric steady-state
    models of the basic state for instability and climate studies. Part I.
    Linearized calculations.
2. NAVARRA, A. and SPERANZA, A. (1984). Coexistence of stochasticity and
    determinism in simple truncated models of barotropic flow in -planes.
    In "Predictability of Fluid motions" by B.J.West and G. Holloway. Ameri-
    can Institute of Physics.
3. WALLACE,J.M., TIBALDI, S. and SIMMONS, A.J. (1983) Reduction of systematic
    forecast errors in the ECMWF model through the introduction of an envelope
    orography. Q.J.R.M.S. Vol. 109, 683-717.
4. CHARNEY,J.G. and DEVORE, J. (1979). Multiple flow equilibria in the at-
    mosphere and blocking. J. Atmos. Sci., Vol. 36, 1205-1216.
5. BENZI, R., MALGUZZI, P., SPERANZA, A. and SUTERA, A., (1984). On the the-
    ory of stationary waves and blocking. Submitted to Q.J.R.M.S.
6. HART,J.E. (1979). Barotropic, quasi-geostrophic flow over anisotropic
    mountains. J.Atmos.Sci., Vol. 36, 1736-1746.
7. EGGER, J. (1981). Stochastically driven large scale circulation with mul-
    tiple equilibria. J.Atmos.Sci., Vol. 38, 2606-2618.
8. BENZI, R., HAUSEN,A.R. and SUTERA, A. (1984). On stochastic perturbation
    of simple blocking models. Q.J.R.M.S., Vol. 110, 393-409.
9. MALGUZZI, P., SPERANZA, A. (1981). Local multiple equilibria and regional
    atmospheric blocking. J.Atmos.Sci., Vol. 38, 1939-1948.
10. SPERANZA, A. (1984). The statistical properties of general circulations.
    In "Large scale anomalies and blocking". Edited by Benzi, R., Saltzamn,
    B. and Wiin-Nielsen, A.. Academic Press.
11. SUTERA, A. (1984). Probabilistic distributions of ultra-long planetary
    waves in winter circulations. In "Large scale anomalies and blocking".
    Edited by Benzi, R., Saltzman, B. and Wiin-Nielsen, A.. Academic Press.
12. HANSEN, (1984). Observational characteristics of atmospheric planetary
    waves with bimodal amplitude distributions. In "Large scale anomalies and
    blocking". Edited by Benzi, R., Saltzman, B. and Wiin-Nielsen, A.. Aca-
    demic Press.

Figure Captions

Fig. 1. CDV parameter space (zonal wind V in abceisse and wave-amplitude in ordinate). The dotted line represents the solution B(V) of eqs (3) and the continuous stright lines are form-drag relationships corresponding to different values of V* in eq (4). The dots are values of zonal V and amplitude B of zonal wavenumber 3 computed from observed 500 mb height over two winters.

Fig. 2. Amplitude of the resonant wave for equations (6) as a function of the zonal wind. The effect of nonlinearity produces the folding of the resonance: for the zonal wind more than one equilibrium solution exists.

Fig. 3. Two-dimensional histogram in CDV parameter space calculated from 14 years of observations of 500 mb height in the Northern hemisphere. The horizontal coordinates are V (the shortest axis) and amplitude of zonal wavenumbers 2-4 (the longer axis).
The vertical axis marks the number of cases in which the observed values were in intervals of V and wave-amplitude.
The numbers do not represent physical values, but only channels of the statistical distributions.

Fig. 4. Composite maps of the two maxima of the probalitiy distribution a) the low amplitude mode; b) the high amplitude mode; c) difference between a and b.

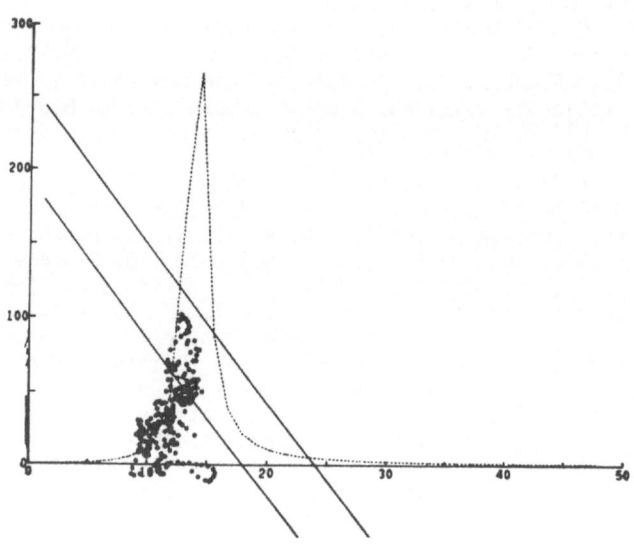

Fig. 1.

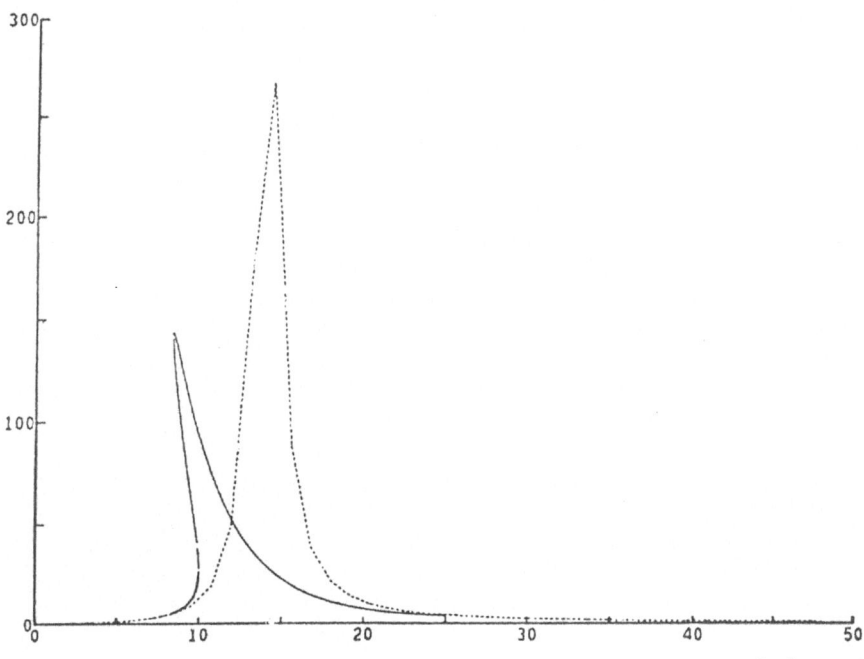

Fig. 2.

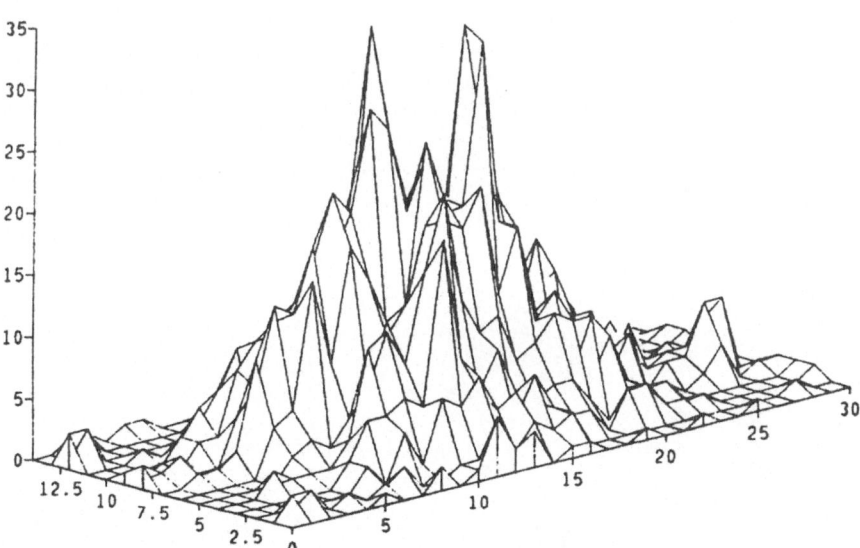

Fig. 3.

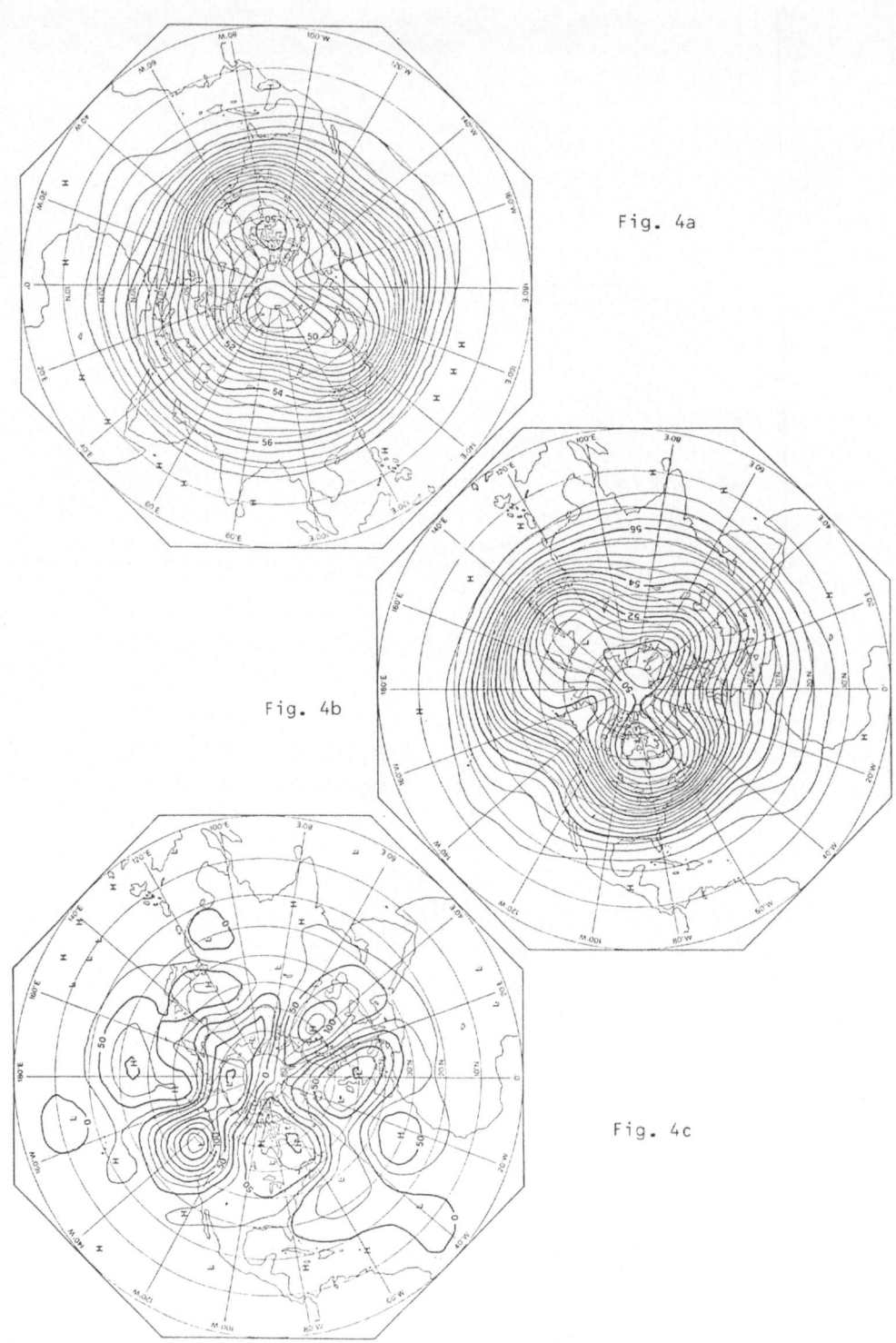

Fig. 4a

Fig. 4b

Fig. 4c

SIMULATION OF EUROPEAN CLIMATE WITH AN ATMOSPHERIC
GENERAL CIRCULATION MODEL

A. SLINGO and H. WILSON
Dynamical Climatology Branch, Meteorological Office, London Road,
Bracknell, Berkshire, RG12 2SZ, England

Summary

The Meteorological Office 11-Layer Atmospheric General
Circulation Model is under intensive development as part of a coupled
ocean-atmosphere model of the earth's climate system. This is being
used in studies of the physical basis of climate and of its response
to natural or man-induced perturbations. Such a model is potentially
a powerful tool for climate research, but for it to be applicable to
the real world it must be able to simulate the present-day climate
accurately. No model in existence is entirely satisfactory in this
respect and the study of systematic errors in the simulations forms an
important part of such research. The most serious problem with the
11-layer model, which it shares with many other high-resolution
models, is a tendency for excessive westerly flow in the northern
mid-latitudes in winter. In this presentation the problem is
discussed and results presented from various attempts to remove it by
incorporating improvements to the model's formulation.

1. Introduction
   This presentation describes work which follows on from that carried
out for contract CLI-031-UK(N), entitled "European climate produced by
an atmospheric general circulation model". For that contract,
integrations of the Meteorological Office 11-layer Atmospheric General
Circulation Model (AGCM) were used to study the characteristics of the
simulated European climate, with particular emphasis on surface and
near-surface variables, which were compared with climatological data.
The sensitivity of the simulations to changes in the methods employed to
model various physical processes, such as the transfer of radiation and
turbulent processes within the atmospheric boundary layer, was also
examined.
   In general, the model produces a good simulation of the observed
European climate in summer, but in winter it maintains too strong a
westerly flow over the area. This is illustrated in Figure 1, which
compares the global distribution of the pressure at mean sea level
(PMSL) for the northern winter season from the third year of a recent
8-year annual-cycle integration of the model with the observed January
mean for 1931-1960 (1). In the southern hemisphere, the strengths and
positions of the sub-tropical high pressure cells and of the circumpolar
depression belt are well simulated. In the northern hemisphere,
however, there is excessive westerly flow over the mid-latitudes, not
only over Europe but at most longitudes. This is associated with
pressures which are about 20 mb too low at high latitudes, the Icelandic

low pressure area being about 25 mb too deep and displaced
north-eastwards from its climatological position. The strong westerly
flow over Europe is exacerbated by an over-intense sub-tropical high
pressure cell which is positioned over the western Mediterranean.

The observed PMSL distribution shows considerable differences
between the two hemispheres. In the southern hemisphere there is strong
mid-latitude flow which in the mean is predominantly westerly with
little longitudinal variation. In the northern hemisphere, the
substantial land area and presence of several major mountain ranges
produces a weaker mean flow with considerable longitudinal variations in
the meridional component. The asymmetric land-sea distribution is
clearly the cause of the differences between the hemispheres, but the
precise reason has been debated for many years (2). Two factors which
have received the most attention are "orographic" forcing of the flow by
the major mountain ranges and "thermal" forcing brought about by the
different thermal characteristics of land and sea. Other differences
include the surface roughness which is generally higher over land and
shows large geographical variations (3), while over the sea it is
dependent on the wind speed (3,4).

In the model results, the northern hemisphere winter circulation is
more similar to that in the southern hemisphere than it is to the
observed circulation. This suggests that the forcing by the land/sea
distribution is too weak and was the reason for carrying out preliminary
studies of the sensitivity of the model to enhanced land surface
elevations and roughness as the final part of the contract work. More
recently, the effects of including an "envelope" orography, a
geographical distribution of land surface roughness and exchange
coefficients over the ocean which depend on the wind speed have been
assessed. Results from other experiments in which improvements have
been made in the representation of radiative processes and with a new
parametrization of the effects of gravity wave drag will also be shown.
Only the global PMSL distribution is illustrated, although it will be
appreciated that many more diagnostics are studied in assessing the
effects of these changes.

2. The 11-Layer Model

A complete description of the 11-layer model has not yet appeared
in the open literature, but a summary taken from (5) is reproduced here.
The results shown in Figure 1 were obtained with this version.

The 11-Layer Model is a global finite-difference AGCM which is
similar in many respects to the 5-Layer Model (6). The vertical
co-ordinate is sigma (pressure divided by its surface value) and the
11-layers are irregularly distributed in the vertical to provide
enhanced resolution near the surface and in the upper troposphere (see
Figure 4 of (7)). A limited area version was used in the GATE
experiment (8) and for experiments with a cloud parametrization scheme
(7). The model was re-programmed for efficient execution on a Cyber 205
computer and is now usually integrated on a regular 2.5 x 3.75 degree
latitude-longitude grid (72 x 96 points) with a 10 minute time-step.
The land-sea mask was chosen to enable the model to be coupled to
dynamical ocean and sea-ice models which are also being developed. The
land surface elevations in each grid box were taken as the mean of the
values from a $1° \times 1°$ dataset (9). Multipoint filtering in the E-W
direction of the increments for the 17 rows nearest to each pole
maintains computational stability at high latitudes. Non-linear

diffusion (to control grid-point noise) is applied to a linear combination of temperature and potential temperature, designed to have a small variation in each sigma surface for a horizontally homogeneous atmosphere, rather than to potential temperature as previously. This produces a better formulation near steep topography and removes a part of the global-mean cooling of the model when integrated from real data. The dynamical routines now use virtual temperatures in order to include the effect of the moisture content on the air density.

Physical processes are modelled in four sub-routines, which deal with boundary layer and surface exchanges, convection, large scale precipitation and finally radiation and clouds. The boundary layer is assumed to occupy the lowest three model layers in all conditions and transfers of heat, moisture and momentum between the layers are modelled by an eddy diffusivity approach (10). The treatment of the surface is similar to that used in the 5-layer model (11), with interactive soil moisture and snow depth. Evaporation is limited for soil moisture less than 5 cm of water and runoff occurs to prevent the soil moisture from exceeding 15 cm. The convection scheme treats both unsaturated and saturated convection as it penetrates the vertical grid of the model (12). Large-scale precipitation, in the form of rain or snow, is formed whenever the relative humidity exceeds 100%. The radiation scheme is similar to that described in (11) and in most of the integrations shown here imposed, zonally-meaned climatological cloud cover was used.

One novel feature of the model is the use of a 360-day year, made up of twelve months each of 30 days, which leads to a considerable simplification of the logic required to run the model and produce diagnostics. The small error thus introduced into the incoming solar radiation is minimised by a slight adjustment to the date of perihelion. The diagnostics are passed through a high-speed link to the front-end IBM computers for off-line processing.

3. Effect of introducing an envelope orography

The problem of excessive westerly winds over the northern mid-latitudes in winter is not unique to the 11-layer model but is a common feature of high resolution AGCMs. Wallace and others (13) attributed such a problem in the ECMWF medium-range weather forecasting model to insufficient orographic forcing by the major mountain ranges. They argued that the common practice of using a simple area-mean for the land surface elevation in a grid box underestimates the blocking effect of the highest peaks and ridges on the flow. They therefore increased the orography by adding an increment proportional to the standard deviation of the sub-grid-scale variability in the elevations. Such an orography follows the "envelope" containing the highest peaks and thus enhances the forcing of the flow.

The effect of various envelope orographies on the 11-layer model have been studied and some results are shown in Figure 2. All are 90-day means taken from annual-cycle integrations run in parallel with that in Figure 1. The only change to the model was to use the same high resolution orography dataset as in (13) to generate the mean orography (top), and envelope orographies with the mean plus $\sqrt{2}$ times (middle) and twice (bottom) the standard deviation, the latter being that used in (13). As expected, the two integrations using mean orographies are very similar, but the envelope orographies produce substantial changes in the northern hemisphere. With the higher envelope, there are pressure rises of up to about 20 mb in polar regions, the north Atlantic low being 17

mb less deep. The westerly surface flow over north America has given
way to a much slacker gradient and the Siberian high pressure area has
moved northwards. All these changes bring the model closer to the
climatological PMSL. However, pressures are generally too low in the
tropics and the Mediterranean anticyclone has moved north-eastwards, so
that there is still an excessive gradient to the north, although weaker
than before. In terms of the effect on the simulated European climate,
the excessively westerly flow using the mean heights has been replaced
by generally blocked conditions, pressure now being much too high over
the area. There has been little change at middle and high latitudes in
the southern hemisphere. The envelope orography produces a substantial
reduction (by about 20 per cent) of the globally-averaged zonal kinetic
energy, which is an improvement, but the ratio of transient to standing
eddy kinetic energy, which was already too low, is further reduced.
Generally speaking, it will be seen that the lower envelope orography
produces results which are intermediate between the other two
simulations.

4. Effect of changes in surface roughness
     In the standard version of the model described earlier, the surface
roughness length Z_o takes the value 0.1m for land points and 0·1mm for
sea points, so there is no geographical dependence of the land surface
roughness and no wind-speed dependence over the sea. Results from an
experiment in which Z_o over land was made a function of the sub-gridscale
variability in the surface elevations (from the same dataset used to
generate the envelope orographies) through a formulation similar to that
described in (14), which is in agreement with the limited observational
data, are shown at the top of Figure 3. Comparison with Figure 1 shows
improvements in certain areas; for example the surface flow is weaker
over north America and polar pressures are slightly higher, although
changes are generally small and indicate that the major systematic
errors are insensitive to changes in the specification of land surface
roughness.
     A somewhat greater response is seen in the results at the bottom of
Figure 3, which come from an integration in which the surface exchange
coefficients for momentum, heat and moisture over the sea were made a
function of the near-surface wind speed, using a simple formulation
based on a large body of observational data (4). This has produced
larger pressure rises at north polar latitudes and a weaker surface flow
over north America. However, the surface flow around the north Pacific
low pressure area is now too weak and the southern hemisphere depression
belt has also weakened, due to the increased surface drag. In addition,
there are considerably enhanced surface sensible and latent heat fluxes
where cold continental air is being advected over warm sea, eg near
Newfoundland. This is due to a strong wind-speed dependence of the
exchange coefficients in unstable conditions. This aspect of the
formulation may not be realistic and needs further study.

5. Effect of radiation and other changes
     Results from the first northern winter of an integration with a new
version of the model are shown at the top of Figure 4. Changes compared
with that described in section 2 are the use of less diffusion (which
increases and thus improves the modelled transient eddy kinetic energy),
a stomatal resistance to evaporation of 60 sm^{-1} for all land points
(which improves the partitioning of the surface turbulent flux into the

latent and sensible components) and the new longwave radiation scheme
and revised cloud and surface radiative properties described in (5)
(which produce a more realistic radiation budget and remove some
systematic errors). In terms of the PMSL distribution the effect of
these changes in the northern winter is quite small. North polar
pressures are generally higher than shown in Figure 1 but a much longer
integration is needed to determine the statistical significance of the
changes.

In all the integrations shown so far, the cloud amounts have been
prescribed from zonally-meaned climatologies. Considerable work has
gone into developing a cloud-prediction scheme, based on that in (7) but
with some changes (5) and recent results from the new version of the
model with this scheme are shown at the bottom of Figure 4. Interactive
cloud has led to a number of improvements, for example pressures are
higher near the north pole so that the Icelandic low is less deep and
the Siberian high extends further north. One intriguing change is that
the high pressure area over the western Mediterranean, which in the
earlier integrations showed no tendency to move towards its correct
location in the Atlantic, has moved westwards although its central
pressure is still too high. The pressures in the southern hemisphere
depression belt are slightly higher than without interactive cloud, but
only by a few mb. The cloud distributions are in general realistic, but
there is too much low cloud over the winter continents and over the
sub-tropical oceans the low cloud tends to form beneath the high
pressure cells rather than in the correct locations over the cooler sea
surface to the east. While in this respect the cloud fields are not
significantly worse than those from other models, this is nevertheless a
potentially serious error in a coupled model as the predicted sea
surface temperature distribution in these areas depends critically on
the low cloud field. This is an active research area, not only in
tuning the present model but also in investigating more physically-based
methods for predicting cloud and representing their effects on the
thermodynamic and radiative fields.

6. Effect of a parametrization of gravity-wave drag

In the studies summarised in sections 3 and 4, efforts were made to
reduce the westerly flow by enhancing the direct effect of surface
irregularities, through increasing either the surface friction or the
surface elevation. However, it is also known that gravity waves may be
generated by air flowing over such irregularities. As these propagate
upwards, they may lead to shear or convective instabilities which
dissipate the wave energy and act as a drag on the flow. The drag is
thus applied within the atmosphere as opposed to at the surface. A
simple parametrization of gravity wave drag has been developed and
preliminary integrations carried out (15). This parametrization uses
the high resolution orography dataset to relate the magnitude of the
gravity wave stress to the sub-gridscale variability in the surface
elevations. The scheme has recently been applied to the model described
in section 2, with the exception that a simplified "climatological"
radiation scheme and real initial data were used. Preliminary results
are shown in Figure 5 for constant January integrations without (top)
and with (bottom) the parametrization and were kindly made available by
T.N. Palmer and R. Swinbank of the Synoptic Climatology Branch. The
PMSL distribution without gravity wave drag shows the characteristic
features described earlier, although the southern hemisphere depression

belt is less deep than in the annual-cycle integrations. The integration with gravity wave drag shows substantial improvements in many areas. There are large pressure rises in northern high latitudes, with a much stronger Siberian high pressure area, weak surface flow over north America and a less deep and less extensive Icelandic low. In addition, the high pressure area over the western Mediterranean has weakened and moved westwards. Overall, the simulation is better than with envelope orography, although it must be stressed that these are preliminary results and much more work is needed to refine the parametrization and study its effect through a large number of diagnostics in extended annual cycle integrations.

REFERENCES
1. SCHUTZ C. and GATES W. L. (1971). Global climatic data for surface, 800 mb, 400 mb: January. R-915-ARPA, Rand Corporation, Santa Monica, California.

2. HOSKINS B.J. and KAROLY D.J. (1981). The steady linear response of a spherical atmosphere to thermal and orographic forcing. J. Atmos. Sci., 38, 1179-1196.

3. GARRATT J.R. (1977). Review of drag coefficients over oceans and continents. Mon. Wea. Rev., 105, 915-929.

4. WU J. (1982). Wind stress coefficients over sea surface from breeze to hurricane. J. Geophys. Res., 87, 9704-9706.

5. SLINGO A. and WILDERSPIN R.C. (1984). Development of an improved GCM radiation scheme. Submitted to the Proceedings of the International Radiation Symposium, Perugia, Italy, 21-29 August 1984.

6. CORBY G.A., GILCHRIST, A. and ROWNTREE P.R. (1977). United Kingdom Meteorological Office five-level general circulation model. Methods in Computational Physics, 17, 67-110, Academic Press, New York.

7. SLINGO Julia M. (1980). A cloud parametrization scheme derived from GATE data for use with a numerical model. Quart. J.R. Met. Soc., 106, 747-770.

8. LYNE W.H., ROWNTREE P.R., TEMPERTON C. and WALKER Julia M. (1976). Numerical modelling using GATE data. Met. Mag., 105, 261-271.

9. GATES W.L. and NELSON A.B. (1973). A new tabulation of the Scripps topography on a 1° global grid. Part 1; Terrain heights. R-1276-ARPA, Rand Corporation, Santa Monica, California.

10. CARSON D.J. (1982). Current parametrizations of land surface processes in atmospheric general circulation models. Land surface processes in atmospheric general circulation models, ed. P.S. Eagleson, Cambridge University Press.

11. SLINGO Julia M. (1982). A study of the earth's radiation budget using a general circulation model. Quart. J.R. Met. Soc., 108, 379-405.

12. ROWNTREE P.R. (1984). Sensitivity of global simulations to the formulation of the evaporation of convective condensates. Proceedings of the Workshop on Convection in large-scale numerical models, 28 November to 1 December 1983. ECMWF, Shinfield Park, Berkshire, England.

13. WALLACE J.M., TIBALDI S. and SIMMONS A.J. (1983). Reduction of systematic forecast errors in the ECMWF model through the introduction of an envelope orography. Quart. J.R. Met. Soc., 109, 683-718.

14. HANSEN, J., RUSSELL G., RIND D., STONE P., LACIS A., LEBEDEFF S., RUEDY R. and TRAVIS L. (1983). Efficient three-dimensional global models for climate studies: Models I and II. Mon. Wea. Rev., 111, 609-662.

15. PALMER T.N. and SHUTTS G.J. (1984). Preliminary results of the effect of a parametrization of gravity wave drag in the Meteorological Office 11-layer operational model. Met O 13 Branch Memorandum No. 149.

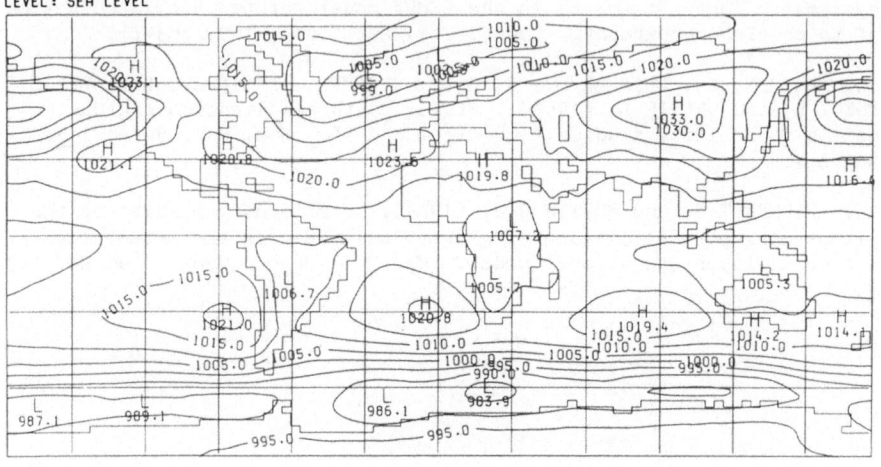

PMSL - JANUARY CLIMATOLOGY - SCHUTZ AND GATES
MEAN OF   1 SAMPLES: PMSL
AVERAGE FROM 0Z ON 1/1 DAY 1 TO 0Z ON 31/1 DAY 31
LEVEL: SEA LEVEL

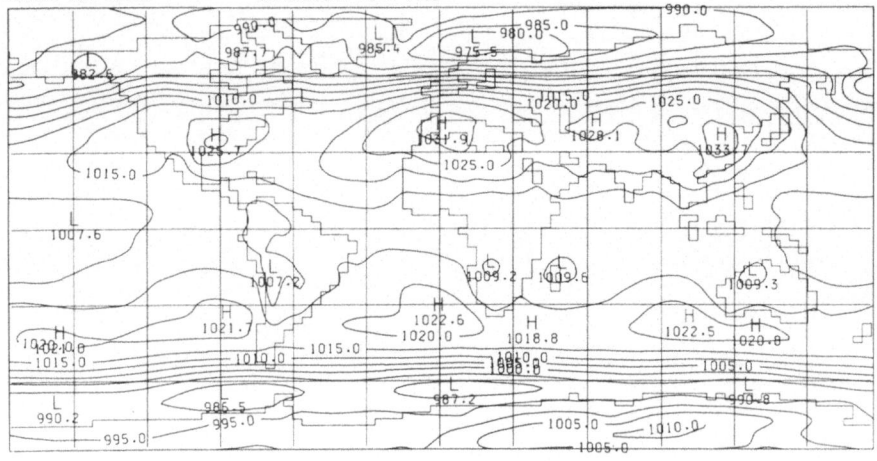

2ND ANNUAL CYCLE, 3RD WINTER - ANCN0190
MEAN OF   1 SAMPLES: PMSL
AVERAGE FROM 0Z ON 2/12/2 DAY 1 TO 0Z ON 1/3/3 DAY 90
LEVEL: SEA LEVEL   EXPERIMENT NO.: 1122

Figure 1   Comparison of observed mean PMSL distribution for January
            (top) with that for a northern winter season from the model
            (bottom).

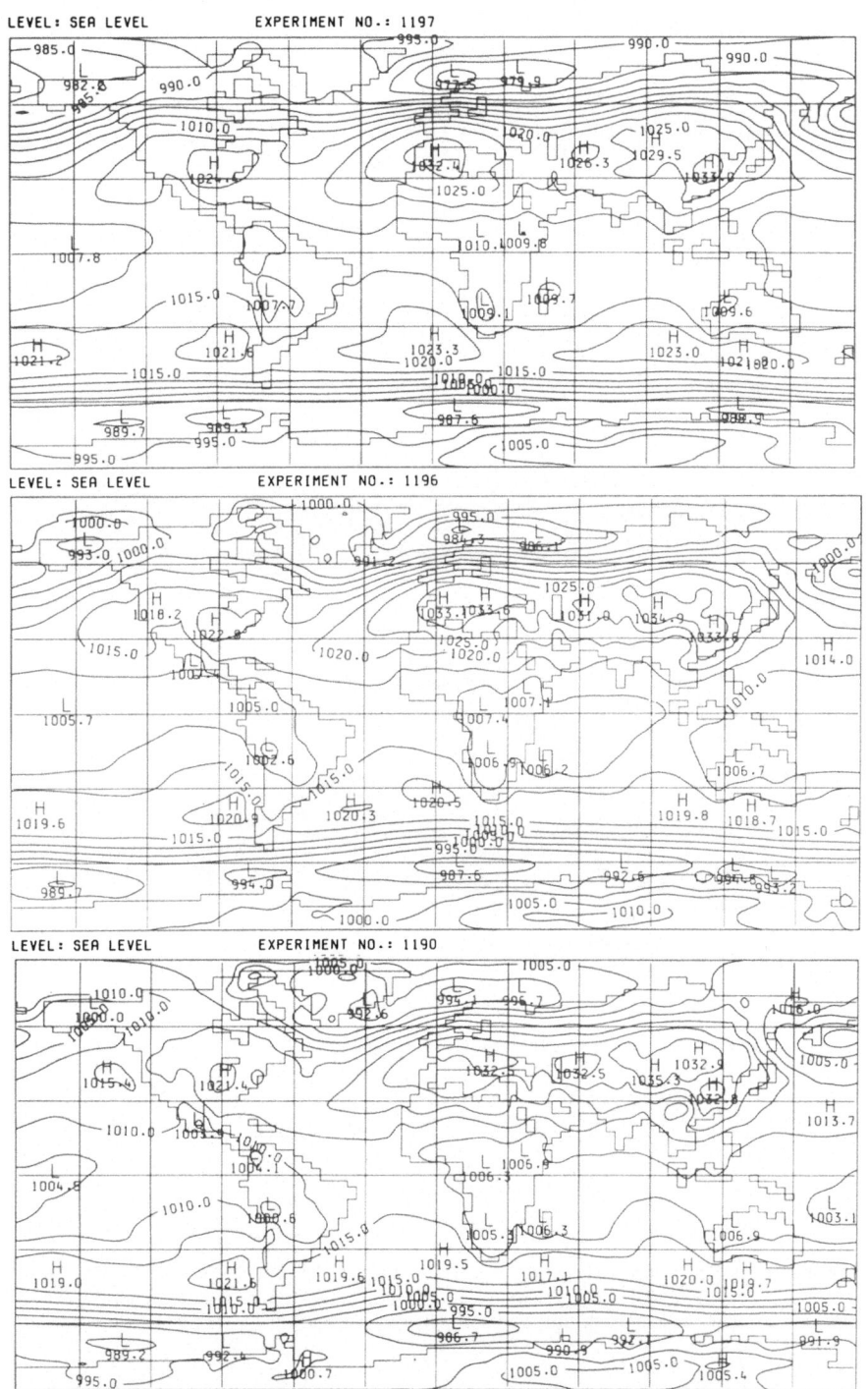

Figure 2   PMSL from control (top) and two envelope orography integrations.

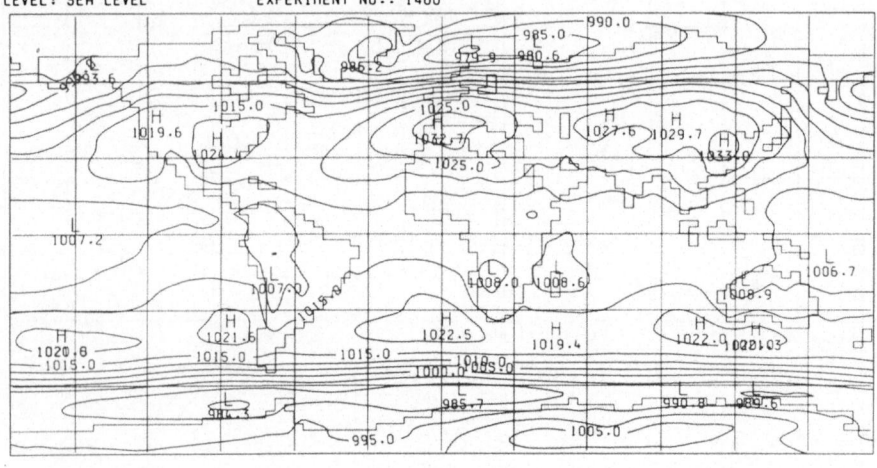

VARIABLE SURFACE ROUGHNESS OVER LAND - EXPT. YZOAO190
MEAN OF   1 SAMPLES: PMSL
AVERAGE FROM 0Z ON 2/12/2 DAY 1 TO 0Z ON 1/3/3 DAY 90
LEVEL: SEA LEVEL   EXPERIMENT NO.: 1400

EXPT. WAD90190
MEAN OF   1 SAMPLES: PMSL
AVERAGE FROM 0Z ON 2/12 DAY 1 TO 0Z ON 1/3/1 DAY 90
LEVEL: SEA LEVEL   EXPERIMENT NO.: 1401

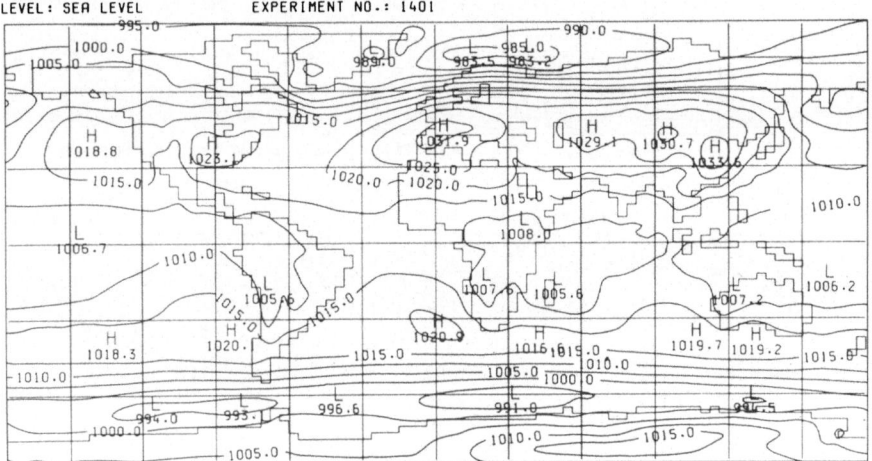

Figure 3 PMSL from integrations with variable surface roughness length
over land (top) and with surface exchange coefficients over
the sea which depend on the wind speed (bottom).

- 148 -

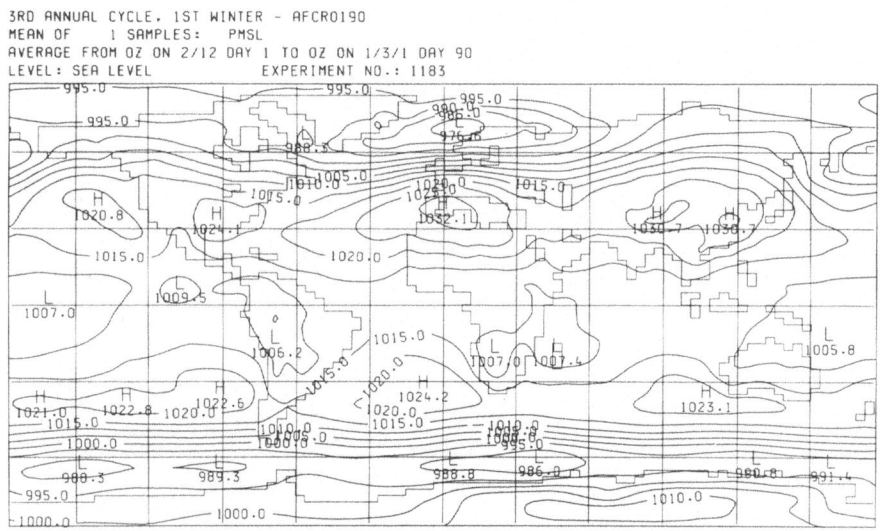

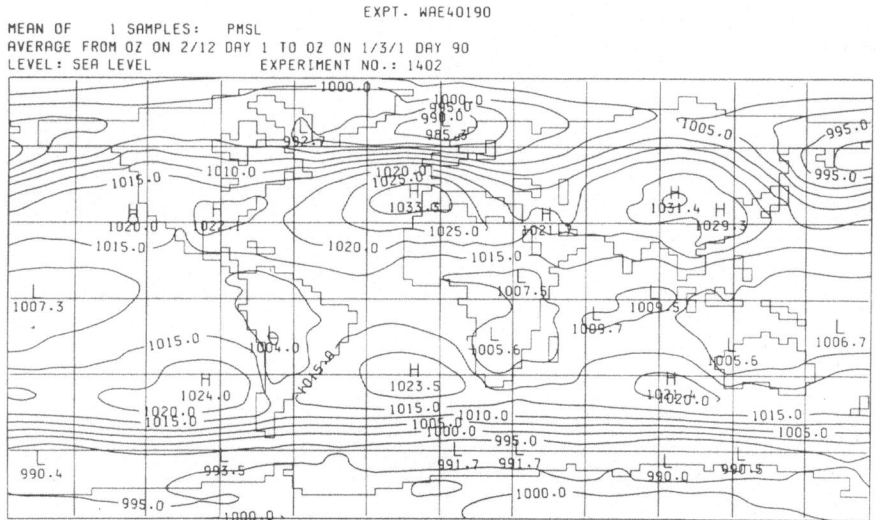

Figure 4   PMSL from an integration with the new version of the model,
          described in section 5 (top), and from a parallel integration
          with interactive model-predicted cloud (bottom).

-149-

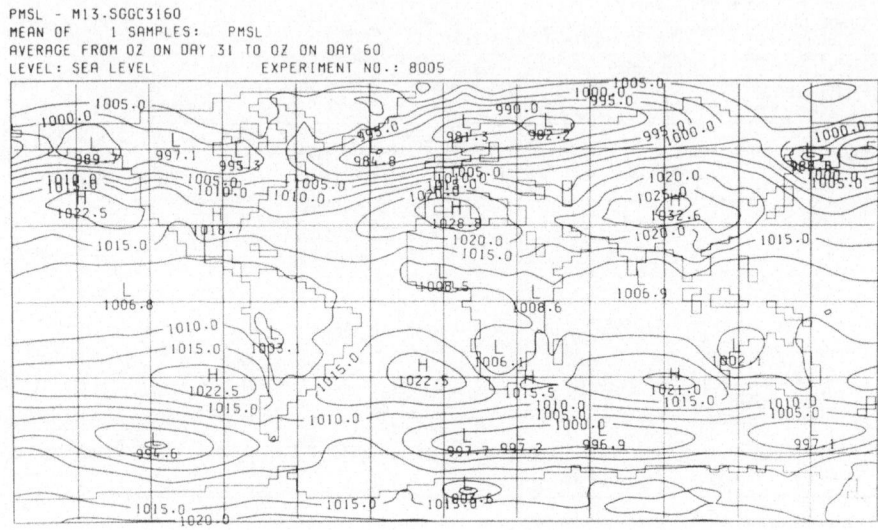

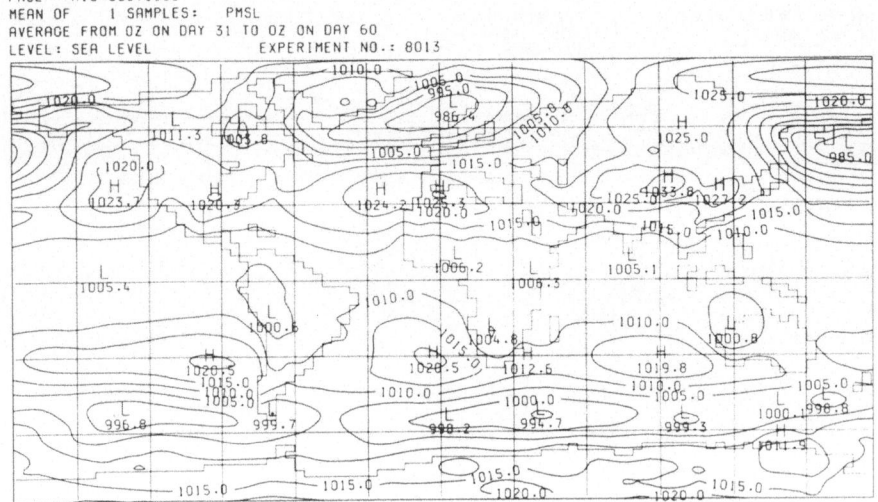

Figure 5   PMSL from control (top) and from an integration with a
parametrization of gravity-wave drag (bottom).

MODELLING THE IMPACT OF SOIL PROPERTIES

ON EUROPEAN CLIMATE

K. LAVAL

Laboratoire de Météorologie Dynamique, Paris, France.

Summary

A change of the state of the atmosphere can be due to variations of the
exchanges between earth and atmosphere. We analyse different simulations
of July climate obtained with our GCM. The experiments are achieved
with different parameterizations of evaporation over land or with
different cloudiness. The decrease of evaporation has always for
consequence a decrease of precipitation and of humidity of the
atmosphere.

   The variation of the release of latent heat due to evaporation
affects heat balance of the soil. Although radiative flux varies
slightly when evaporation changes, the largest variation occurs
in the sensible heat flux term. The ratio of latent heat released
by evaporation to the turbulent fluxes between earth and atmosphere
depends essentially on the parameterization of evaporation.

1. Introduction

   Land surface processes can affect regional climate. General Circulation
Models (GCM) have shown that a variation of evaporation over land leads
to a variation of precipitation and to changes in the circulation of the
atmosphere (1). Rowntree and Bolton (2) have analysed the effect of an
anomaly of soil moisture over Europe for the month of July. They show
that a local decrease of soil moisture induces a decrease of precipitation
which affects not only the region where the variation of soil moisture is
prescribed but also adjacent land areas. Yeh et al (3), using a model with
idealized topography which simulates the annual cycle, show that a wet
anomaly can persist several months in middle latitudes. In this numerical
experiment, the precipitation is increased where soil is wet during

approximately 2 months.

With the General Circulation Model of LMD, we have performed experiments which simulate the climate of July and have studied the effect of a variation of evaporation on climate.

In these simulations, over Europe, the grid is about 5° in longitude and latitude, and 11 levels are prescribed. The model includes parameterization of boundary layer, radiative fluxes, precipitation and evaporation.

The experiments start with initial conditions of 11 June 1979 and are carried out for 50 days. The mean state of July is defined as an average over the last 30 days.

2. European climate simulated by the control experiment

Let us show first the mean state of July obtained by the model compared to reality. In the control experiment, hereafter called F1, the pressure field, shown in figure 1, is quite realistic with the Atlantic anticyclone well located and extending towards northern Europe. This pattern induces westerlies in the north of Europe and easterlies in the south.

The rate of evaporation is shown in figure 2. The value of 3 mm.day^{-1} compares well with the results given by the Atlas of the World Water Balance published by Unesco (4). The rate of precipitation (figure 2) is about 1 mm.day^{-1} with larger values over Great Britain, Sweden and Germany. Although the maxima are well located, the rate is generally lower than reality when we compare it with the distribution of rain given by Jaeger (5)

3. Parameterizations of evaporation

We have analysed the effect of a change of the parameterization of evaporation over land with our GCM.

The evaporation over land can be evaluated as the product of potential evaporation EP and of an aridity coefficient β. The usual method (6) is to write evaporation E with a formula of the bulk aerodynamic type :

$$E = \beta \ EP$$

$$EP = \rho \ C_D V(q_s(T_o) - q(z))$$

where ρ is density, C_D is a drag coefficient, $q_s(T_o)$ is the saturation mixing ratio of water vapor at the temperature of the ground T_o, and $q(z)$ is the mixing ratio of water vapor in the air, at a standard level z.

The parameterisation of the coefficient β depends on soil moisture W as follows

$$\beta = 1 \ \text{if} \ W > W_K$$

$$\beta = \frac{W}{W_K} \ \text{if} \ W \leq W_K$$

W_K is a critical value of W, often taken as 0.75 W_{MAX} where W_{MAX} is moisture at field capacity.

This method was used in the control experiment that I have shown previously.

Another method to compute potential evaporation can be based on Priestley and Taylor's equation (7). The potential evaporation reads :

$$LE = \alpha \, \frac{s}{s+\gamma} \, (R_n - F_o)$$

where L is the latent heat of evaporation, $s = \frac{\partial q_s}{\partial T}$, $\gamma = \frac{C_p}{L}$, R_n is the
net radiative flux, F_o is the heat flux into the ground.
The value of the coefficient α has been evaluated by (7) as 1.26 but
Perrier (8) obtained the value of 1.4.
    We have parameterized the evaporation over land with the Priestley
and Taylor's equation, determining α as a function of soil moisture
given by

$$\alpha = 0.4 \, \frac{W}{W_K} + 0.6 \qquad \text{for } W > W_K$$

$$\alpha = \frac{W}{W_K} \qquad \text{for } W \leq W_K$$

with the critical value of soil moisture $W_K = \frac{W_{MAX}}{2}$ and $W_{MAX} = 15 \text{ g cm}^{-2}$.
    This parameterization gives generally lower values of evaporation
when we compare it to the previous one (9). We have performed a second
simulation of July with the same conditions than F1 except for the
parameterization over land : this experiment, called F2, uses the
parameterization of Priestley and Taylor just defined. We compare the
mean state obtained with the two simulations F1 and F2.

4. Effect of a change of evaporation rate on climate

4.1. Clouds are fixed.
    The evaporation rate simulated by the experiment F2 is lower than
the one obtained by F1 case (figure 3). Differences higher than
1 mm day^{-1} occur in the north of Europe. We have averaged the results
over an area E shown in figure 4, and we get in figure 5, the variation
of 10-day mean of the evaporation and precipitation rates. Evaporation
decreases when soil moisture dries up. This induces a decrease of
precipitation (figure 3 and 5) which, also, is maintained all along the
integration of the model.
    The difference of evaporation and precipitation, averaged over
30 days (Table 1) is for the two experiments equal to 1.1 mm day^{-1}
which corresponds to a moisture flux divergence over this area. The
dependence of the evaporation rate with soil moisture can be approximated
by a linear function for the F2 case but in the F1 experiment, no simple
relation has been found. One can approximate the relation as $E \, \alpha \, W^r$ with
$r = 0.89$. So, we do not find the value $r = 0.5$ obtained by (2).
Figure 5 shows also that soil moisture is slightly lower in the F1 case
than in the F2 case at the end of the experiment. This is totally different
of the result obtained by (2). In their experiment, an anomaly of soil
moisture induces a variation of evaporation. In our experiment, initial
state is the same but the evaporation is reduced because of the scheme
chosen to compute evaporation.
    The mean precipitable water contained in the atmosphere is lower
in F2 experiment than in F1 case. The decrease does not extend on the
whole region and Germany for example is more humid. In the model, the
decrease of humidity of the air occurs in the low levels, under 850 mb
(Table 1).
    The decrease of evaporation is associated with an increase of
temperature and an increase of surface pressure (figure 6). Although

the decrease of surface pressure is consistent with the increase of
temperature, it is difficult to assess if these variations are not due
to the model variability. Let us note that this change induces weaker
westerlies or stronger easterlies over Europe.

It is interesting to compare the energy balance over the European
area for the two experiments F1 and F2. In these simulations, the scheme
which computes radiative fluxes uses distribution of cloudiness which is
constant with time. It is the reason why the solar flux is not changed.
The variation of infrared flux is essentially due to the increase of
temperature. This variation of 7.7 Wm^{-2} is lower than the variation of
the release of latent heat due to evaporation (24.6 Wm^{-2}) which is
balanced, for a large part, by an increase of sensible heat flux.

The ratio $\frac{LE}{LE+H}$ which is approximately equal to the ratio of latent
heat to net radiative flux is reduced from 0.57 to 0.39 in the dry case.

4.2. Experiment with variable clouds

If evaporation influences precipitation, it has certainly an effect
on clouds. The experiments F1 and F2 prescribed a fixed distribution
of clouds to compute radiative fluxes. This condition certainly induces
a bias in the results of the effect of a change of evaporation on
hydrological cycle.

We have performed two other experiments IR1 and IR2 where clouds
are not constant but are evaluated each time radiative fluxes are
computed. One must be aware that it is very difficult to obtain cloudiness
on a grid of about 300 km when we just know the mean values of vertical
profile of temperature, humidity, and vertical velocity. Nevertheless,
we have defined a scheme to compute the surface of clouds based on the
parameterization of condensation (10).

The initial state and all other conditions are the same than the
F1 and F2 experiments : IR1 uses evaporation scheme defined in F1 and
IR2 uses the one defined by F2. The mean state of July is defined as
an average over the last 30 days.

The pressure field in IR1 is shown in figure 7. The pattern is quite
realistic with westerlies over the north of Europe.

The evaporation rate (figure 8) simulated in IR1 experiment is larger
than in the F1 case : this is probably due to a larger soil heating
by solar flux in IR1 case as we shall see later. The maxima of 5 mm.day^{-1}
are certainly too high and one can expect a more realistic simulation in
the IR2 experiment which uses the Priestley and Taylor's equation
to compute evaporation. The precipitation rate also is amplified in
IR1 case compared to F1 run but the locations of the maxima are the same.

The change obtained between IR1 and IR2 for the evaporation rate
is 1 mm day^{-1} and the decrease of precipitation is 0.6 mm day^{-1}. The
moisture flux divergence has then changed from 1.9 to 1.5 mm day^{-1}
whereas it was the same in the F1 and F2 experiments (Table 1). The
differences between IR1 and IR2 of the evaporation and precipitation
rates are small at the beginning of the simulation and increases during
the integration (figure 9).

The decrease of the evaporation rate induces a decrease of temperature,
and a decrease of pressure. This lead to a variation of winds with
stronger easterlies over Europe. Let us note, however, that the
variability of surface pressure is higher in these experiments with
variable clouds and we cannot be sure that this variation is not only
due to the variability of the model.

The variation of heat balance between IR1 and IR2 is influenced by the variation of cloudiness. Solar radiative flux has increased from 292 Wm^{-2} to 305 Wm^{-2} because cloudiness is reduced. On the opposite, infrared infrared radiative flux is enhanced and, in the model the two effects balance. Then the decrease of the release of latent heat is compensated by an increase of sensible heat flux. In these simulations, the ratio of latent heat to the sum of turbulent fluxes varies from 0.58 to 0.44, values which are rather closed to the ones obtained previously.

5. Conclusion

We have performed different simulations of July climate to study the effect of a variation of evaporation on climate. The changes of evaporation were induced by the use of two different schemes defined to compute evaporation over land.

We got different values of evaporation over Europe from 1.6 mm day^{-1}, corresponding to dry conditions to 4.2 mm day^{-1} in the most humid case. The variation of evaporation goes with a variation of precipitation of the same sign.

The variation of temperature depends on the radiative fluxes. In the case of constant solar heating, the decrease of evaporation induces a warming but if cloudiness are variable, solar flux is changed and warming is lower.

Decrease of evaporation corresponds to a decrease of humidity in the low atmosphere and an increase of sensible heat flux.

In our model, the ratio of latent heat released by evaporation to the sum of turbulent fluxes is essentially a function of the parameterization of evaporation.

REFERENCES

1.   MINTZ, Y. (1981) : The influence of soil moisture on rainfall and circulation : a review of simulation experiments. Paper presented at the GARP Study Conference, Greenbelt, Md., USA.
2.   ROWNTREE, P. and BOLTON, J. (1983) : Simulation of the atmospheric response to soil moisture anomalies over Europe. Quart. J. R. Met. Soc., 501-526.
3.   YEH, T., WETHERALD, R. and MANABE, S. (1984) : The effect of soil moisture on the short-term climate and hydrology change. A numerical experiment. Monthly Weather Rev., 474-490.
4.   Atlas of the World Water Balance (1977) published by the Unesco Press, Paris.
5.   JAEGER, L. (1983) : Monthly and areal patterns of mean global precipitation. In "Variations in the Global Water Budget" published by D. Reidel Publishing Company.
6.   MANABE, S. (1969) : The atmospheric circulation and the hydrology of the Earth's surface. Monthly Weath. Rev., 739-774.
7.   PRIESTLEY, C. and R. TAYLOR (1972) : On the Assessment of the surface heat flux and evaporation using larger scale parameters. Monthly Weath. Rev., 81-92.
8.   PERRIER, A. (1978) : Importance des définitions de l'évapotranspiration dans le domaine pratique de la mesure de l'estimation et de la notion des coefficients culturaux. Société hydrotechnique de France, Toulouse.
9.   LAVAL, K. and OTTLE, C. : A new parameterization of evaporation over land surfaces defined in a climate model. Submitted to Monthly Weath. Rev.

10. LE TREUT, H., and K. LAVAL (1983) : The importance of cloud
    radiation interaction for the simulation of climate. In
    New Perspectives in Climate Modelling published by Elsevier
    Publishing Company.

TABLE 1

	F1	F2		IR1	IR2
$E_{mmday^{-1}}$	2.5	1.6		4.2	3.2
$P_{mmday^{-1}}$	1.4	0.6		2.3	1.7
E-P	1.1	1.0		1.9	1.5
Humidity soil → 850 mb g cm^{-2}	1.5	1.4		1.6	1.4
Energy KJ cm^{-2}	43.4	44.0		43.7	43.9
Solar F Wm^{-2}	189	189		292	305
LW Radiative Flux	62	69		76	90
LE Wm^{-2}	71	46		120	91
H_S	53	72		89	118
$\frac{LE}{LE+H_S}$	0.57	0.39		0.58	0.44

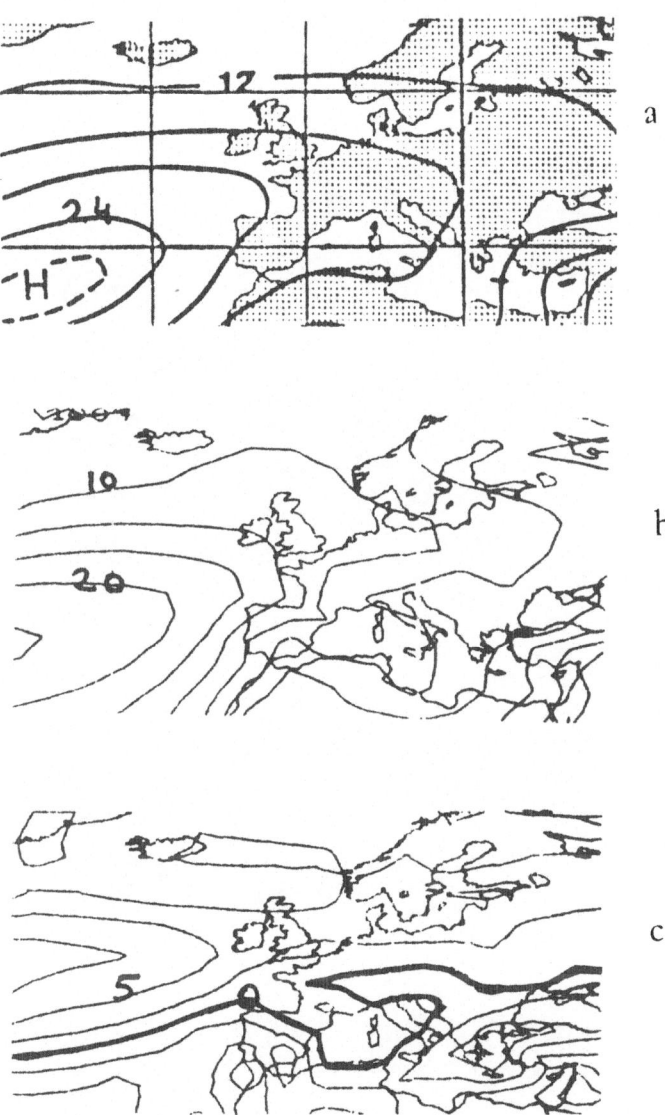

Figure 1 Sea level pressure
a) observed (Godbole and Shukla),
b) averaged for days 20-50
   of experiment F1
c) Zonal wind at 900 MB

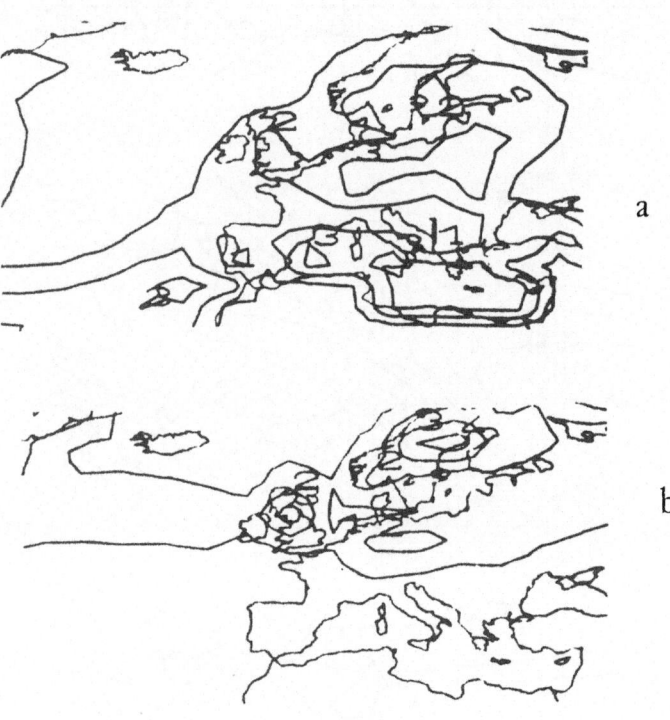

a

b

Figure 2   Evaporation (a) and
            precipitation (b) rates
            simulated by F1 experiment.

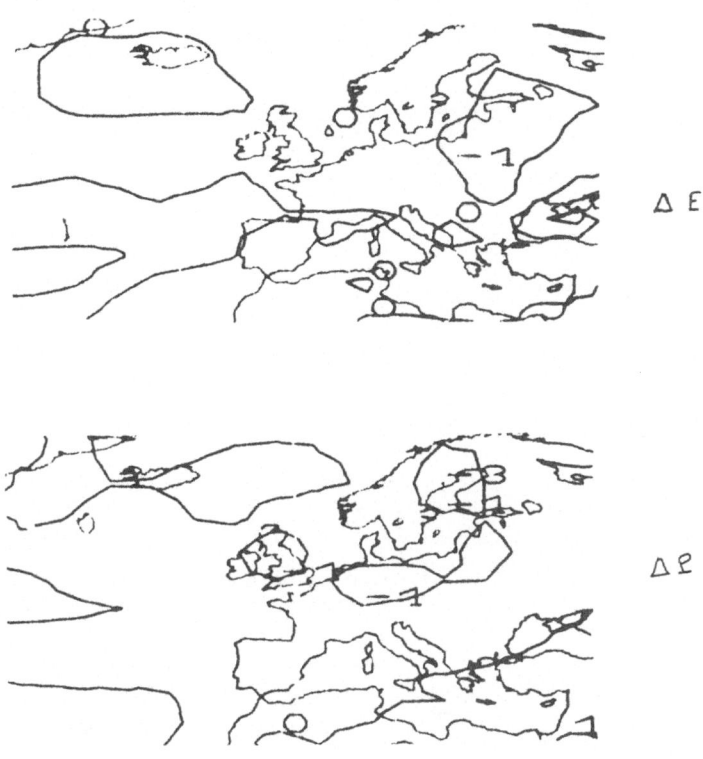

ΔE

$\Delta \mathcal{P}$

Figure 3 Variations of evaporation and
precipitation rates between
experiments : F2 - F1

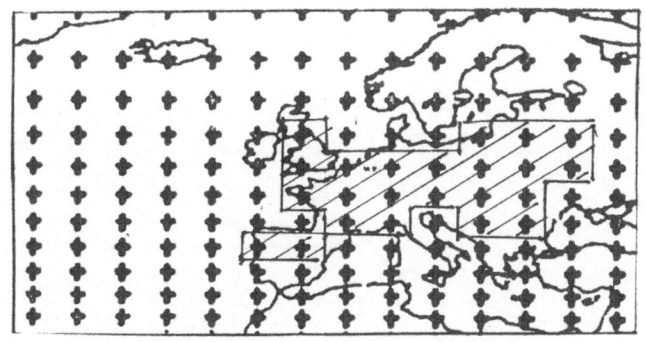

Figure 4 Area stippled is the region
where the averages shown in
Table 1 are computed

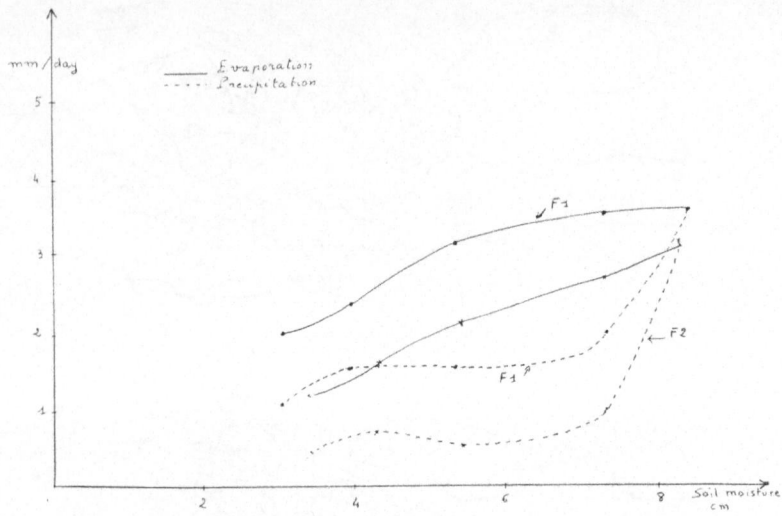

Figure 5   The variations of evaporation and rainfall plotted against soil
moisture averaged for each of the five 10-day periods of experiments F1 and F2

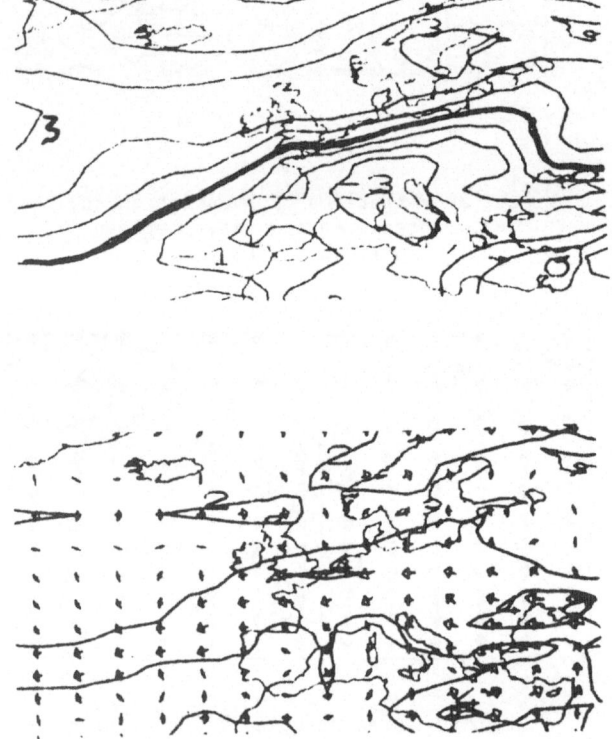

Figure 6   The differences between experiments F2 - F1 for (a) sea level
pressure and (b) zonal wind at 900 mb

Figure 7 Sea level pressure and zonal wind at
900 mb obtained by IRI experiment

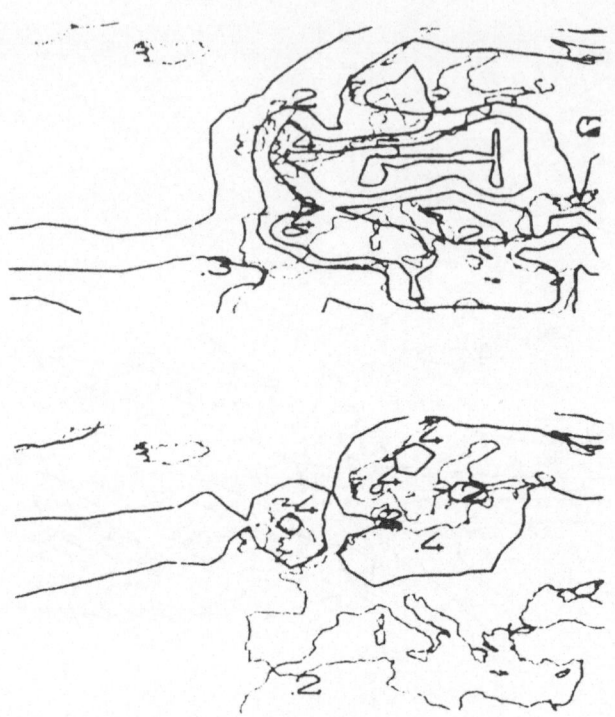

Figure 8 Evaporation (a) and Precipitation (b) rates simulated by
IR1 experiment.

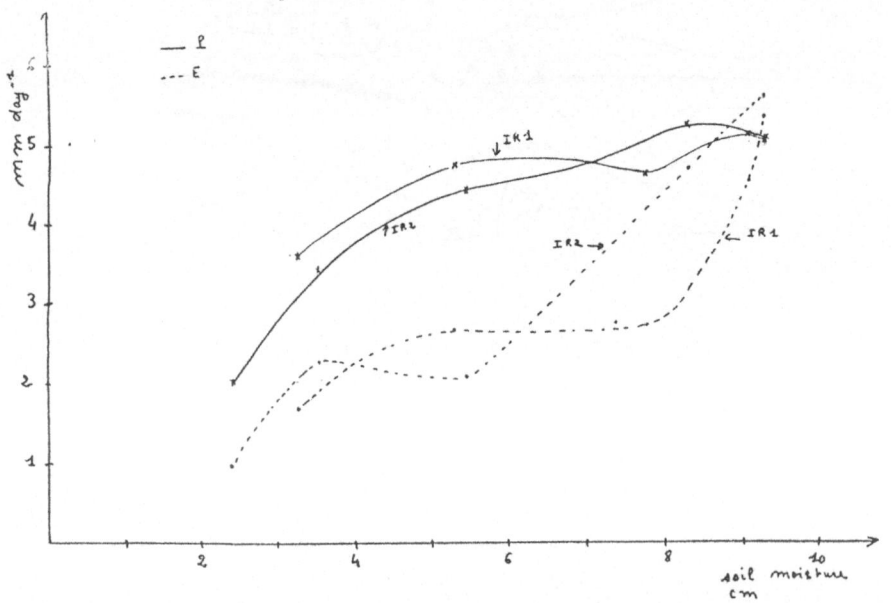

Figure 9 The variations of evaporation and rainfall plotted against soil
moisture averaged for each of the five 10 day periods of
experiments IR1 and IR2.

INTERACTIVE OCEAN-ATMOSPHERE MODELS FOR CLIMATE FORECASTING

Jacques C.J. NIHOUL
GeoHydrodynamics and Environment Research, Liège University, Belgium

Summary

One emphasizes the cogent effect of the ocean on climates and the need
of interactive ocean-atmosphere models for climate forecasting.
The main features of the present generation of ocean-atmosphere climate
models are discussed stressing the problems which subsist and the fields
where further improvements are required.
The GHER interactive ocean-atmosphere climate model is presented in
illustration.

1. Introduction

Central to atmospheric variations on climatic scales are the interactions with the ocean.
Fig. 1, compounded from recent review papers (e.g. 8, 10), shows schematically the nature and complexity of the interaction for different forecast intervals and different latitudes : while the atmosphere is largely
independent of the condition of the sea for less than a few days, - except
perhaps in the tropics where the diurnal reaction of the ocean may play a
role -, for longer forecast intervals, a sufficient knowledge of sea-surface temperature and sea ice distributions is required.
From a few days to a few months, however, one may be content with
"static" boundary conditions, climatological averages or observed anomalies. For forecast intervals exceeding a few months, dynamic informations
are required and must be provided by an appropriate ocean model interacti-
vely coupled with the atmospheric model.
It will be shown in the following that time averages over periods
of the order of one month are the most adequate state variables for interactive ocean-atmosphere climate modelling. Taking into account that one
is mainly interested here in climate variation on time scales less than
a few decades which correspond to typical human planning times and which
received the highest priority from the World Climate Research Programme
(WRCP 1980-2000) (10), one can see that forecast intervals from a few
months to a few decades will be the main concern of ocean-atmosphere climate models and that these will have to include practically the whole ocean
dynamics.

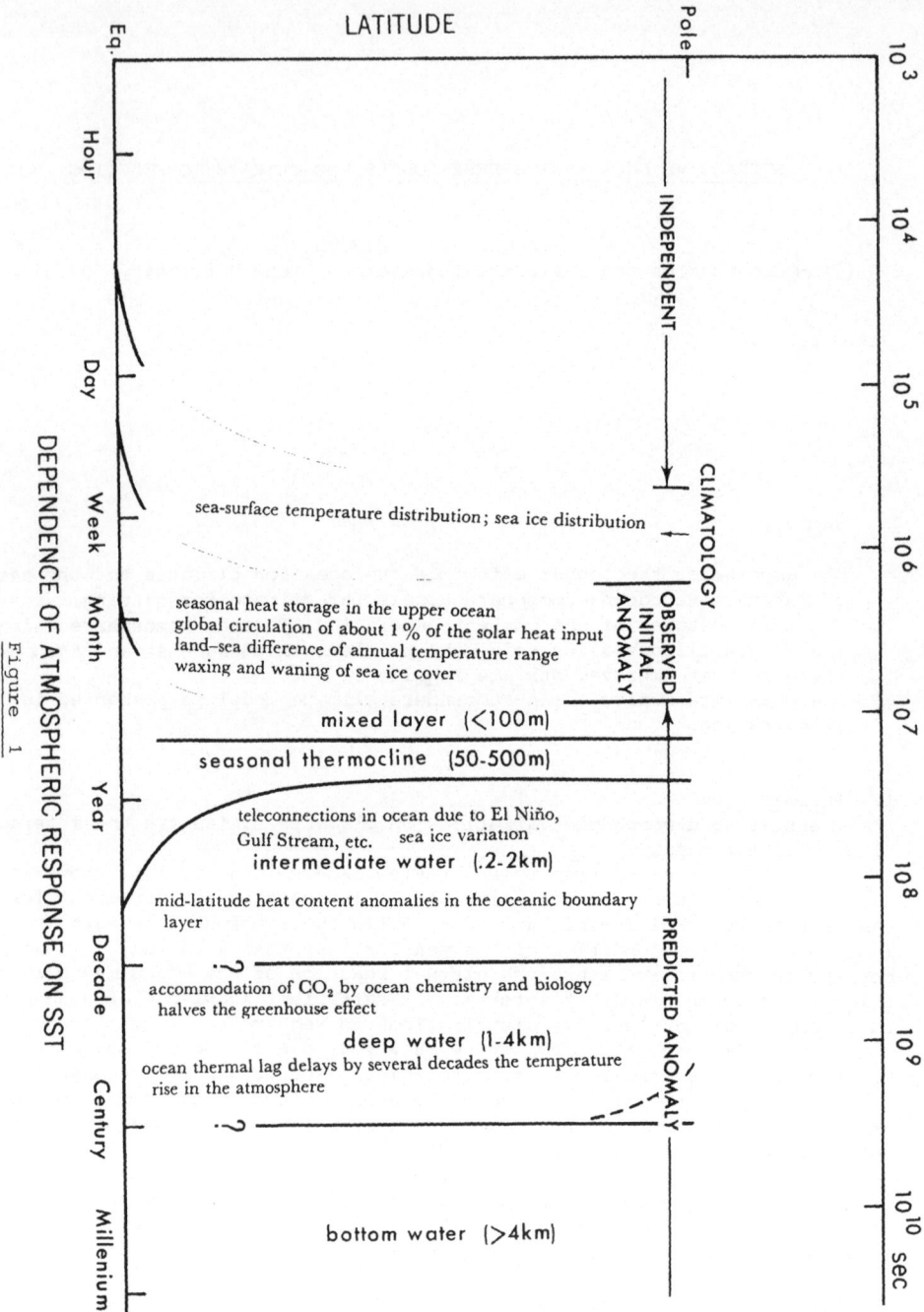

DEPENDENCE OF ATMOSPHERIC RESPONSE ON SST

Figure 1

LATITUDE

Eq.   Pole

Hour

Day   INDEPENDENT

Week   CLIMATOLOGY

Month   sea-surface temperature distribution; sea ice distribution

             seasonal heat storage in the upper ocean
             global circulation of about 1 % of the solar heat input   OBSERVED INITIAL ANOMALY
             land–sea difference of annual temperature range
             waxing and waning of sea ice cover

                        mixed layer (<100m)

             seasonal thermocline (50-500m)

Year   teleconnections in ocean due to El Niño,
                 Gulf Stream, etc.   sea ice variation
                    intermediate water (.2-2km)

             mid-latitude heat content anomalies in the oceanic boundary
                 layer

Decade   ·~?

             accommodation of CO_2 by ocean chemistry and biology   PREDICTED ANOMALY
                 halves the greenhouse effect

                        deep water (1-4km)
             ocean thermal lag delays by several decades the temperature
                 rise in the atmosphere

Century   ·~?

Millenium

                        bottom water (>4km)

10^3 10^4 10^5 10^6 10^7 10^8 10^9 10^{10} sec

2. Basic equations of ocean-atmosphere climate models

The models are based on the equations of Geophysical Fluid Dynamics applicable to the atmosphere and the ocean (e.g. 6). In this general form, all models are essentially the same but there are nevertheless some discrepancies in the representation of physical mechanisms, such as changes of phase and entropy production, which are often concealed behind different formulations and notations, originating in distinctive traditions in oceanography and meteorology.

Such fundamental differences may be, in the present generation of models, totally watered down in subsequent simplifications, averagings and parameterizations but they are worth keeping in mind in prevision of future refinements of the models.

In general, the basic geophysical fluid dynamics equations are transformed to simpler form by means of the "geohydrodynamic" (or "Boussinesq's") approximation.

In the geohydrodynamic approximation, the state of the atmosphere and the ocean is treated as a first order deviation from a known state of reference and one restricts to the leading terms in the series expansions of the state variables around the reference state.

Again, there are differences, from one model to another, both in the definition of the state of reference and in the way the series expansions are conducted but these differences are hopefully largely smoothed away in subsequent approximations.

By way of illustration, Table I gives the system of geohydrodynamic equations on which the GHER ocean-atmosphere climate model is based. The justification and a detailed discussion of these equations can be found in reference (6).

Table I

$$\frac{\partial v_1}{\partial x_1} + \frac{\partial v_2}{\partial x_2} + \frac{\partial w}{\partial z} = 0 \quad ; \quad w = r v_3 \quad ; \quad z = \int_0^{x_3} r \, dx_3 \quad ; \quad r = \frac{\rho_0}{\rho_\odot}$$

$$\frac{\partial v_1}{\partial t} + \underset{\sim}{v} . \nabla v_1 - f v_2 + m w = - \frac{\partial \pi}{\partial x_1} + \nu \, \nabla^2 v_1 \quad ; \quad \pi = \frac{p - p_0}{\rho_\odot}$$

$$\frac{\partial v_2}{\partial t} + \underset{\sim}{v} . \nabla v_2 + f v_1 = - \frac{\partial \pi}{\partial x_2} + \nu \, \nabla^2 v_2 \quad ; \quad \underset{\sim}{v} . \nabla = v_1 \frac{\partial}{\partial x_1} + v_2 \frac{\partial}{\partial x_2} + w \frac{\partial}{\partial z}$$

$$\frac{\partial w}{\partial t} + \underset{\sim}{v} . \nabla w - m v_1 = - \frac{\partial \pi}{\partial x_3} + b + \nu \, \nabla^2 w \quad ; \quad b = - \frac{\rho - \rho_0}{\rho_0} g$$

$$\frac{\partial b}{\partial t} + \underset{\sim}{v} . \nabla b = B + \lambda \, \nabla^2 b \quad ; \quad B = \frac{\beta_0 g}{\rho_0 c_p} R^T \equiv \frac{\beta_0 g}{\rho_0 c_p} (R + \underset{\sim}{T_v} : \nabla \underset{\sim}{v} - \Sigma \, Q^a \tilde{\chi}_0^a)$$

$$\frac{\partial \rho^a}{\partial t} + \underset{\sim}{v} . \nabla \rho^a = Q^a + \lambda^a \, \nabla^2 \rho^a \qquad a = 1, 2, \ldots$$

In these equations, the subscript $_0$ refers to the state of reference, $\underline{v}$ is the velocity vector, ρ_0 is a constant reference value,
$f = 2\,\Omega\,\sin\,\phi$ and $m = 2\,\Omega\,\cos\,\phi$ are respectively twice the vertical and
horizontal components of the earth's rotation vector.

The buoyancy b is used as a state variable but the equation for
buoyancy can easily be transformed into an equation for (potential) temperature if preferred ; taking into account the contribution of the constituents' concentrations and the corresponding mass balance equations at
the bottom of Table I (ρ^α denotes the mass per unit volume of the α cons-
tituent, viz salts,in the ocean, water vapor, condensed water ...,in the
atmosphere.

In the buoyancy production rate B, the second term represents heat
production by viscous dissipation and can be neglected, the third term
plays an important role only, in the atmosphere, in the case of change of
phase and it is customary to write it then in the simple form $L\dot{\rho}^c$ where
L is the latent heat of evaporation and $\dot{\rho}^c$ the condensation rate.

Molecular fluxes (proportional to the molecular diffusivities ν, λ,
λ^α) are negligible and have only be included for the sake of completeness.

One particular feature of the GHER model is the transformation of the
vertical coordinate and vertical velocity by means of the density distribution of the geofluid in the reference state.

While this affects very little the equations in the case of the ocean,
it makes a signficant difference in large scale atmosphere models which
can be conveniently extended to infinity in x_3-space while keeping a
finite-height atmospheric layer in z-space.

Reducing the height of this layer to $[0,1]$ by an appropriate change
of variable, it is then possible to define suitable eigenfunctions (of,
say,the turbulent diffusion operator) and to resolve the vertical variations of the state variables by means of series expansions in these eigenfunctions.

The same technique is used for the ocean. The result is a "multi-
mode" three-dimensional model, rather different from the more classical
multi-layer models developed by other authors (e.g. 2, 9).

Although the variability of the atmosphere and the ocean is universally recognized (e.g. 3, 4, 5), nothing like an unanimous spectral
description of geohydrodynamic phenomena appears to be available in the
literature.

For the purpose of this study, however, - which is the definition
of a spectral window for climate models - the following simple spectral
scenario is probably sufficient.

One recognizes first that the basic equations contain three characteristic frequencies :
(i)   the Brunt-Väisälä frequency n is a measure of the stratification
       (n^2 is defined as the vertical gradient of buoyancy). The maximum
       Brunt-Väisälä frequency is of the order of 10^{-2} s^{-1} in the ocean
       and a similar value may presumably be used for the atmosphere.
(ii)   the Coriolis frequency f is a measure of the effect of the Earth's
       rotation (f is defined as twice the vertical component of the
       Earth's rotation vector). In mid-latitude, $f \sim 10^{-4}\,s^{-1}$.
(iii) the Kibel frequency j is a measure of the effect of the Earth's
       curvature. If β denotes the gradient of f and ε the energy
       dissipation rate, the Kibel frequency is given by (e.g. 5)

$$j \sim \varepsilon^{1/5}\,\beta^{2/5}$$

i.e., for $\beta \sim 10^{-11}$ m^{-1} s^{-1}, $\varepsilon \sim 10^{-9}$ $m^2 s^{-3}$ in the ocean at corresponding scales (5), $\varepsilon \sim 510^{-4}$ $m^2 s^{-3}$ in the atmosphere (3)

in the ocean $\qquad j \sim 0.6 \ 10^{-6} s^{-1}$ $\qquad (t_\beta \sim 18$ days$)$

in the atmosphere $j \sim \qquad 10^{-5} s^{-1}$ $\qquad (t_\beta \sim 1$ day$)$

The characteristic wave-number associated with the Kibel frequency
is (5) $\qquad \kappa_\beta \sim \varepsilon^{-1/5} \beta^{3/5}$, i.e.

in the ocean $\qquad \kappa_\beta \sim 1.6 \ 10^{-5}$ $\qquad (\ell_\beta \sim 60$ km$)$

in the atmosphere $\kappa_\beta \sim \qquad 10^{-6}$ $\qquad (\ell_\beta \sim 1000$ km$)$.

Schematically, one can distinguish between
(i) microscale motions, for frequencies larger than $n \sim 10^{-2} s^{-1}$,
unaffected by the earth's curvature, the earth's rotation and the
stratification and constiuted mainly by three-dimensional turbulence ;
(ii) mesoscale motions, in the range of frequencies, between the Coriolis
and Brunt-Väisälä frequencies, affected by the earth's rotation
and by the stratification and constituted by inertial oscillations,
tides, diurnal variations, internal waves and bliny-turbulence* ;
(iii) synoptic motions, for frequencies smaller than the Coriolis frequency
$(f \sim 10^{-4} s^{-1})$ and larger, say, that the frequency of annual variations $(\sim 10^{-7} s^{-1})$, constituted mainly of the so-called "synoptic
eddies" of pseudo-dimensional rossby-turbulence* ;
(iv) climatic motions, for frequencies in the range $10^{-7} s^{-1}$ $10^{-8} s^{-1}$,
corresponding to periods of 1 to 20 years, characteristics of
man's capabilities of exploiting a climate forecast ;
(v) "paleo-climatic" motions for still smaller frequencies.

To study geohydrodynamic phenomena in a particular range of scales,
one must smooth out smaller scale fluctuations by appropriate time averaging and find a suitable representation of larger scales'forcing.
From the basic equations, one thus derives
(i) primitive equations by filtering microscale "eddy" turbulence" ;
(ii) synoptic (weather) equations by filtering mesoscale waves,
bliny-turbulence and microscale turbulence ;
(iii) climate equations by filtering out rossby turbulence, bliny-turbulence and eddy turbulence.
The time of averaging T must be chosen sufficiently large for the
fluctuations to roughly cancel over a time of that order and, in the same
time, sufficiently small for the phenomenon under study to pass through
the averaging almost unaffected. T must correspond to a valley in the
energy spectrum.

---

* A "bliny" (from the Russian blini) is a pancake-shaped eddy, contributing to an energy cascade to smaller scales via epidermic instabilities
and internal waves. A "rossby" (from the scientist Rossby) is a
column-shaped "synoptic" eddy of diameter of the order of the Rossby
Radius of Deformation (5,6).

For the atmosphere, a time of averaging of the order of a month
(2.5 10^6 s) seems to be appropriate. It is more than ten times larger
than the characteristic time of the rossbies ("synoptic eddies").

In the ocean, however, the rossbies have a much longer life time and
a much lower mobility than their atmospheric analogues (travelling cyclones and anticyclones in mid-latitudes). A meaningful statistical procedure which involves a succession of these oceanic features locally would
require an averaging period of at least two or three years. It therefore
would not resolve the annual cycle of various air-sea interaction processes which are essential for our understanding of the climate system
as a whole.

As far as length scales are concerned, the situation is just the
opposite : the characteristic horizontal area of the oceanic rossbies
tends to be more than a hundred times smaller than that of their atmospheric counterparts. One therefore can get a reduced representation of
the large scale distribution of climate variables in the ocean by appropriate horizontal averaging.

Such an average is always included in the numerical modelling process
as the selection of a numerical grid implies the filtering out of sub-grid
scale processes. A spatial resolution of a few degrees of the great
circle* ($\sim$ 300-600 km) smoothes out most of the perturbations introduced
by travelling ocean rossbies and appears to be adequate.

Thus, the combination of time and space averaging, eliminating most
of the variance within time intervals of about one month and within
distances of about 300-600 km, provides the appropriate low pass filter
for the definition of the climatic state variables and the formulation
of ocean-atmosphere models.

3. Parameterization of "Reynolds stresses"

The geohydrodynamic equations are fundamentally non-linear. As a
result, time averages cannot eliminate smaller scale fluctuations completely. Averaging the non-linear terms, one gets indeed two contributions,
the first of which is related to the product of the means while the
second contains the mean product of the fluctuations.

The fluctuations associated with the primitive equations are the
three-dimensional fluctuations of microscale "eddy" turbulence and their
non-linear contributions are known as the "Reynolds stresses".

The fluctuations associated with weather or climate equations encompass a broader range of motions which, although simular to three-dimensional turbulence to a large extent, contain more "organized" components
to which typical turbulent closures are not always applicable.

If the concept of turbulent diffusivity can reasonably be extended to
mesoscale Reynolds stresses, in weather models and to the transport of
buoyancy and scalar constituents in climate models, the same approximation
is more disputable for the non-linear synoptic diffusion of momentum in
climate models. Synoptic turbulence with eddies involving the whole water
column (the so-called "rossbies") has some of the characteristics of two-
dimensional turbulence and may in unforeseeable places transfer energy
from small to large scales (a negative eddy viscosity effect).

---

* For instance, in the models of Washington et al (9) and Marchuk et al
(2), the horizontal resolution is 5°   ; the grid size of the GHER model
is 381 km in the 60° N stereographic plane (7).

The problem of parameterizing the synoptic Reynolds stresses in climate models has not yet received a completely satisfactory solution.

The trend in recent studies has been to resolve - in some approximate but ultimately statistically significant way - the dynamics of synoptic processes, taking advantage of their predominant kinetic energy, to "compute", - rather than parameterize -, the synoptic Reynolds stresses.

For instance, in the GHER ocean-atmosphere climate model, the quasi-geostrophicity of macroscale atmospheric and oceanic motions is exploited to compute explicitly satisfactory approximations of the synoptic Reynolds stresses from the statistics of the successive weather pressure fields.

The success of the approach is subordinate to finding by this method synoptic Reynolds stresses presenting a sufficient amount of statistical reproducibility, at least to the extent to which they affect the solution of the climate equations.

Encouraging results, in this respect, have been obtained by comparing horizontal synoptic Reynolds stress vector fields calculated for the same winter and summer months of several consecutive years : although there are differences in the monthly stress pattern from one year to another, the marked - and hopefully determinant - features of the field seem fairly recurrent and characteristic of each particular month (7).

Fig. 2 shows for instance the synoptic Reynolds stress vector field calculated in the atmosphere over the Atlantic and Europe for January 1982 by the GHER model. The same calculation made for January 1980 and 1981 confirms the existence, in the same regions, and with comparable intensity, of similar bundles of large stress vectors and gyres while a completely different - altogether less intense - pattern prevails in the summer with again strong similarities between several consecutive years (7).

4. Sensitivity studies of European climate forecasts with an ocean-atmosphere "regional" climate model

Fig. 2 shows the area studied by the GHER ocean-atmoshere climate model. The state variables are monthly averages. The spatial resolution is 381 km in the 60°N stereographic plane. The model is "regional" (the grid covers the North Atlantic and Europe) and may be regarded as a coupled ocean-atmosphere climate submodel embedded in some appropriate global General Circulation Model providing, with complementary experimental data, the necessary boundary conditions.

The model is time dependent, three-dimensional. The vertical resolution is sought by means of series expansions in the eigenfunctions of the vertical eddy diffusion operator.

As emphasized before, the synoptic Reynolds stresses produced by the non-linear interactions of sub-grid scales processes are explicitly calculated taking advantage of the quasi-geostrophicity.of large scale motions and exploiting the statistics of observed successive weather pressure fields for any given month with the assumption that the stress pattern so obtained is sufficiently characteristic of the chosen month - in its cogent features at least - to be used in ulterior forecast.

The regional (Atlantic + Europe) model was conceived to test the validity of this approximation by assessing the sensitivity of climate forecasts for Europe to year-to-year variations of the synoptic Reynolds stress field versus other factors such as the conditions on the lateral boundaries which picture the effects on the European climate of planetary changes and the exchanges at the Atlantic-Atmosphere interface and the related control on the atmosphere by the general circulation in the Atlantic Ocean.

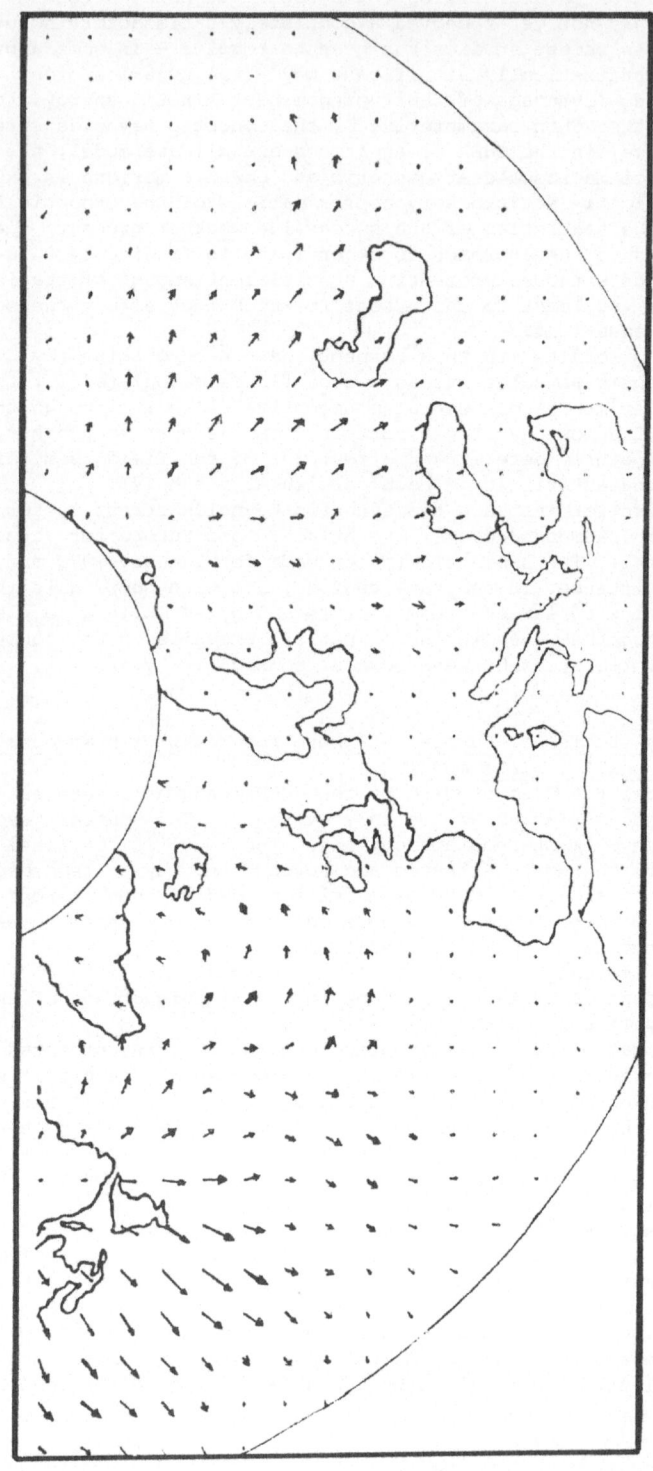

Figure 2 : Horizontal synoptic Reynolds stress.
Vectors for January 82.

The sensitivity studies, conducted as calibration exercices, also serve as a try-out of the parameterization schemes used to represent essential physical mechanisms.

Preliminary results for the winter atmospheric circulation show significant ocean-atmosphere interactions effects on the allocation of evaporation and precipitation in the Western and Eastern Atlantic and the maintenance of the storms' thermomachinery, in agreement with recent studies (e.g. 2).

REFERENCES

1. GUO, Yu-Fu, WANG, Xiao-Xi, CHEN, Ying-Yi and CHAO, Jih-Ping (1985).
    An atmosphere-ocean coupled model for long-range numerical forecasts.
    Proc. WMO-CCCO - 16th International Liège Colloquium on Ocean Hydro-
    dynamics (Liège, May 7-11 1984), edited by J.C.J. Nihoul, Elsevier
    Publ., Amsterdam (to be published).
2. MARCHUK, G.I., DYMNIKOV,V.P., LYKOSOV, V.N. and ZALESNY, V.B. (1985).
    Climate simulation by a general global atmosphere-ocean circulation
    model. Proc. WMO-CCCO - 16th International Liège Colloquium on
    Ocean Hydrodynamics (Liège, May 7-11 1984), edited by J.C.J. Nihoul,
    Elsevier Publ., Amsterdam (to be published).
3. MONIN, A.S. (1972). Weather forecasint as a problem in Physics.
    MIT Press, Cambridge, Mass., USA, 199 pp.
4. MONIN, A.S., KAMENKOVICH, V.M. and KORT, V.G. (1974). Variability of
    the oceans. John Wiley & Sons, New York, 241 pp.
5. NIHOUL, J.C.J. (1980). Marine Turbulence. Elsevier Publ., Amsterdam,
    378 pp.
6. NIHOUL, J.C.J. (1982). Basic equations of geophysical fluid dynamics
    and their applications to ocean-atmosphere weather and climate models.
    University of London, Special University Lectures, Belgian Visiting
    Professorship 1980-1981, Department of Mathematics, Imperial College.
7. NIHOUL, J.C.J. (1984). Geohydrodynamic prediction at decennial
    scale. Société Belge de Physique, Réunion Scientifique Générale,
    Louvain-La-Neuve, June 7-8, 1984.
8. SMAGORINSKY, J. (1984). The problem of climate and climate variations.
    WMO, WCP-72.
9. WASHINGTON, W.M., SEMTNER, A.J. Jr., MEEHL, G.A., KNIGHT, D.J. and
    MAYER, T.A. (1980). J. Phys. Oceanogr., 10, 1887.
10. WOODS, J.D. (1983). Satellite monitoring of the ocean for global
    climate research. Phil. Trans. R. Soc. Lond., A309, 337-359.

CLIMATE MODELLING ACTIVITIES AT THE MAX-PLANCK-INSTITUTE OF METEOROLOGY, HAMBURG

K. HASSELMANN
Max-Planck-Institut für Meteorologie,
Hamburg

Summary

The hierarchy of climate models in development at the Max-Planck-Institute of Meteorology extends from low order empirical linear prediction and stochastic models to high resolution models of the coupled ocean-ice-atmosphere-biosphere system. Simple stochastic models are able to explain many of the qualitative features of natural climate variability and can be applied in the inverse modelling mode to quantify important interaction processes within the climatic system. The statistical analysis techniques developed for inverse modelling and empirical model construction are useful also for analysing numerical climate response experiments with atmospheric general circulation models. The main emphasis in the development of high resolution models has been placed on the global ocean circulation. The tropical oceanic response to observed atmospheric wind forcing could be well simulated over a 25 year period with a high resolution primitive equation equatorial ocean model. A large scale, quasi-geostrophic global ocean circulation model has been used for the computation of the heat transport and storage in the oceans and the development of a global carbon cycle model incorporating a more realistic treatment of the storage, transport and biochemical conversions of carbon in the ocean. Models of sea ice and ice sheet dynamics are also in development and will be coupled in future experiments with high resolution atmospheric and ocean circulation models.

1. Introduction

The wide range of space and time scales and many different processes involved in the dynamics of climate calls for a multi-tiered approach to climate modelling. A model optimised for one class of climate investigations will normally differ significantly from a model developed for another climate problem. In this paper we give a brief overview of the climate model hierarchy in development at the Max-Planck-Institute of Meteorology in Hamburg. We present several examples of numerical experiments made with these models and discuss briefly various statistical aspects involved in the construction of models and the analysis of model experiments.

The models reviewed in the following can be divided into three main classes: semi-empirical statistical models, used to forecast natural climate variability (Section 2); stochastic models, used both for forecasting and, more importantly, in the inverse modelling mode, to identify and characterize important physical processes from the analysis of data (Section 4); and high

resolution deterministic models (Sections 5 - 7). In the latter class
emphasis has been placed on the development of general circulation models
of the global ocean, as this is generally regarded as the most urgent task
for a number of climate problems - for example, for the development of high
resolution coupled ocean-atmosphere models, or for the construction of
realistic global carbon cycle models (cf. Section 6). In Section 3 we
describe the application of statistical multivariate techniques to determine
the significance of atmospheric response patterns inferred from numerical
experiments with high resolution GCMs, or from the analysis of observations,
using essentially the same methods as applied to construct empirical climate
prediction models in Section 2.

A summary of the models in development at the Max-Planck-Institute is
given in the following table:

Technique	Application
Statistical models	
linear forecast methods (multivariate, cyclo-stationary)	El Niño North America Eurasia sea ice (Arctic, Antarctic)
atmospheric response pattern analysis	Pacific mid-latitude SST anomaly, GISS, GCM experiment
	El Niño tropical and mid-latitude response, ECMWF GCM experiment, observations
inverse modelling of climatic variability	SST anomalies
	sea ice variability
inverse modelling of atmospheric weather variability	barotropic model, time domain
	baroclinic model, frequency domain
principal oscillation patterns (POPs) of climate variability	El Niño - southern oscillation
	mid-latitude blocking and oscillation
Ocean models	
primitive equations	equatorial dynamics, El Niño
large scale quasi-geostrophic dynamics	global ocean circulation
	sensitivity studies (World Ocean Circulation Experiment)
carbon cycle	ocean CO_2 storage model, inorganic interactions only
	ocean CO_2 storage with biochemical cycle
	coupled ocean-atmosphere-terrestrial biosphere model

2. Statistical climate models

The construction of empirical prediction models by multivariate linear regression is a time honoured method in climate prediction. A frequent criticism of this approach in the past has been that insufficient attention has often been given to the question of statistical significance. The main emphasis in Hamburg has therefore been directed toward the development of techniques for establishing well defined trade-off criteria between model skill and significance. These qualities are generally in conflict. Whereas the model skill always increases monotonically with the number of predictors, the statistical significance of the model generally decreases. The optimal model, which maximizes skill while still remaining above a prescribed statistical significance level, cannot be defined *a posteriori,* for example by selecting or screening the most effective predictors from a larger set of potential predictors (a common practice in traditional regression model construction). It must be defined through an *a priori* selection criterion. This can take the form, for example, of an appropriate significance cut-off criterion applied to a previously defined hierarchy of models, in which predictors are successive introduced into the model in a prescribed sequence (3, 4).

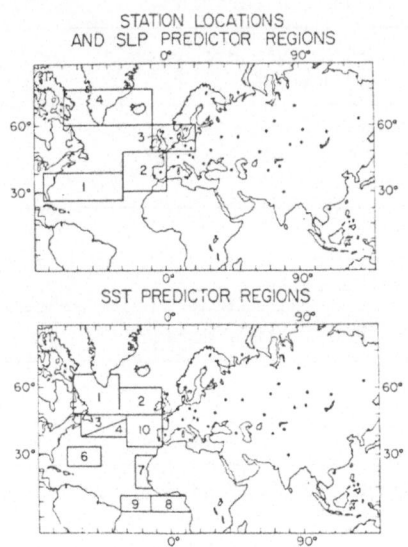

FIG. 1

Surface air temperature stations (solid dots), predictor averaging areas for sea level pressure (SLP, upper panel) and sea surface temperature (SST, lower panel) used for Eurasian forecast study (from (5)).

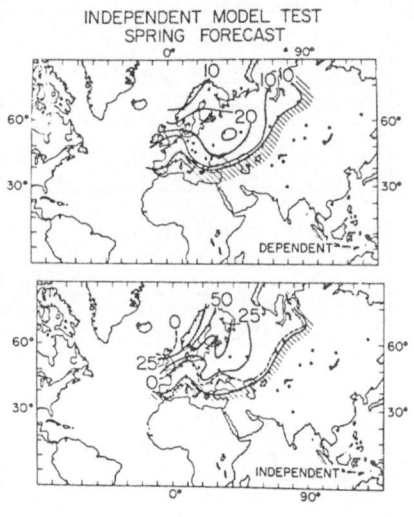

FIG. 2

Forecast experiment on independent data. Upper panel: hindcast skill (percent of variance) associated with the dependent model construction for the period 1902 – 1971. Lower panel: forecast skill over the 1972 – 1978 period. The heavy dashed line represents the 'zero skill' line associated with the independent test. Over most of the region it follows the zero line derived from the dependent test (from (5)).

The classical regression methods for stationary processes were also generalized to the cyclo-stationary case, following the methods developed in the analysis of economic time series (18), to include the important effects of seasonal modulation.

The methods have been applied to the prediction of climate variations in the tropics (3), North America (2, 18), the Eurasian continent (5), and to ice anomalies in the Arctic and Antarctic (21). Figs 1 and 2 (from (5)) show as example the predictors used for forecasting surface temperatures over the Eurasian continent and the skill achieved for a three month forecast for spring (the best forecast season).

In general, the skill values achieved with these models are fairly good for the tropics (typically of order 0.5 for a 6 month forecast) but are not very high in mid-latitudes. In mid-latitudes the natural internal variability of the atmosphere (which unavoidably folds energy into low frequencies through the nonlinearity of the system) largely masks the predictable low frequency component of the signal associated with the mean response of the atmosphere to the slowly changing boundary conditions (primarily the sea surface temperature and snow and ice cover).

Another contributing factor to the relatively low skill evident in Fig. 2, however, was probably the *a priori* choice of predictors. This reflected various published hypotheses on the cause of Eurasian interannual climate variability, which for the most part were not supported by the data. In the following section we discuss recent evidence suggesting that mid-latitude climate variations are more strongly influenced by teleconnection interactions with the tropical Pacific (not included in the regression model) than mid-latitude SST variations.

3. Statistical significance of GCM response experiments

The techniques for constructing statistically significant multivariate regression models for climate forecasting can be carried over with only minor modifications to the problem of assessing the statistical significance of inferred changes in the atmospheric circulation induced by changes in the boundary conditions or external forcing (SST, ice and snow cover, vegetation, CO_2 concentration, etc.). The methods can be applied both to the analysis of time series of observed data and to the interpretation of response experiments with atmospheric general circulation models (GCMs). The latter problem is a little more tractable, as the changes in the boundary conditions or external forcing introduced into models are generally time independent, and, in contrast to the real climate, the numerical experiment can be designed to yield a satisfactory signal-to-noise ratio.

The basic approach is to apply multivariate statistical tests to determine which part of the complete response pattern derived from the numerical experiment or the analysis of real data can be regarded as statistically significant. In contrast to the traditional uni-variate tests applied to individual grid points or observation stations (7), the technique takes full consideration of the spatial correlation structure of the natural low-frequency noise of the (real or model) atmosphere (16). The technique has been applied to the analysis of a numerical GCM experiment, made with the GISS model, on the response of the atmosphere to a typical ('Namias') mid-latitude Pacific SST anomaly pattern (13).

Von Storch and Kruse (40) have recently also applied the method, together with some alternative multivariate test techniques proposed by von Storch and Roeckner (38, 39), to study the mid-latitude atmospheric response to an El Niño SST anomaly pattern (Fig. 3). The statistical pattern analysis was applied both to a GCM response experiment, carried out by the European

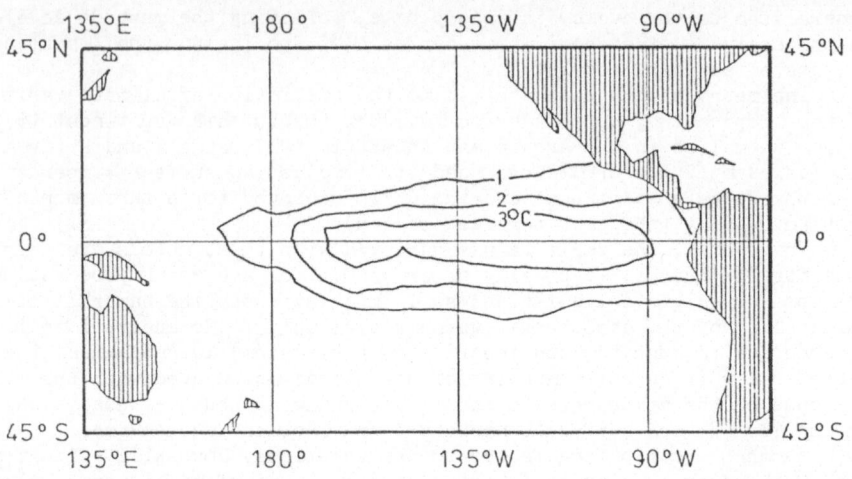

FIG. 3

Intensity and shape of the El Niño SST anomaly used in the ECMWF
GCM experiments (from (8)).

Centre for Medium Range Weather Forecasts (8), and to a 15-year ensemble of
real observations, which included the extreme 1982/83 El Niño event as the
anomaly case. Both the GCM experiment and the real observations yielded a
statistically significant atmospheric response pattern at mid-latitudes
(Figs 4 and 5). Moreover, the model and observed response fields yielded a
statistically significant pattern correlation coefficient. The analysis
presents perhaps the first statistically substantiated evidence of a mid-
latitude atmospheric response to tropical SST anomalies and opens the pro-
spect, in conjunction with the development of predictive El Niño models
(cf. Sections 1 and 7), of achieving useful climate predictions, at least
for extreme climatic events, in the time scale range of months to a year.

4. Inverse stochastic modelling

    Many of the basic properties of natural climate variability can be
explained by the simple concepts of Brownian motion. If an inert system,
such as the oceans, cryosphere or biosphere, is driven by short time-scale
noise from a rapidly varying system, such as the atmosphere, the spectral
response of the inert system to the effectively white noise input is red:
the energy of the response spectrum increases as the frequency decreases,
until a maximal spectral level is reached at very low frequencies of the
order of the inverse of the natural time constant of the inert system (14).
If the slow system contains a number of well separated time constants, the
response consists of a sequence of red-noise spectral segments separated
by white noise plateaus.
    The simplest model for such a system is the linear Langevin equation,
or its multi-dimensional generalisation

$$\frac{dy_i}{dt} = \sum_j L_{ij}y_j + n_i(t) \tag{1}$$

- 176 -

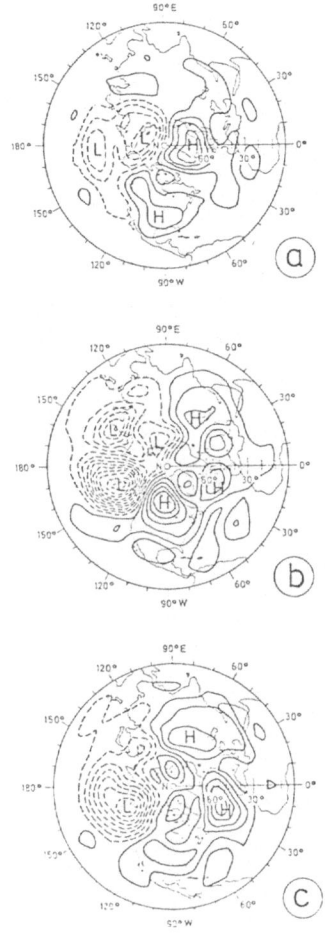

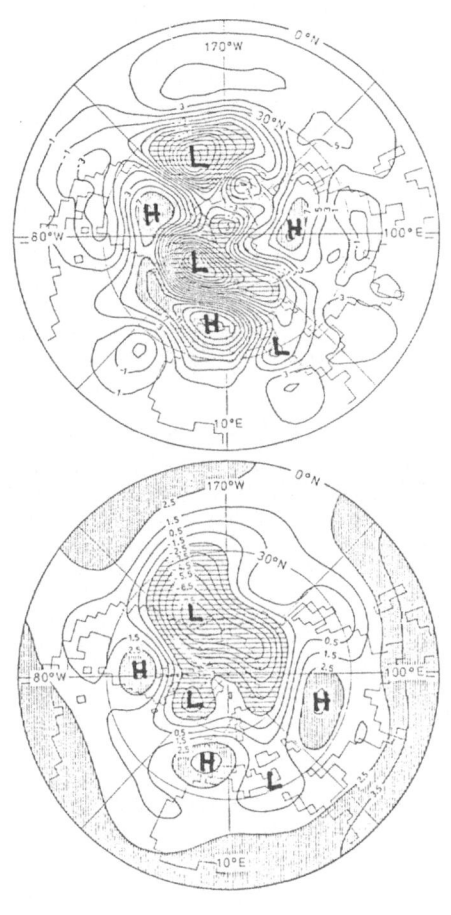

FIG. 4

Significant north-hemispheric
response of the ECMWF GCM to the
warm El Niño SST anomaly for three
different initial conditions
(from (8), (40)).

FIG. 5

Upper panel: The mean 500 mb geopoten-
tial height anomaly field observed in
January 1983. Contour interval 40 m.
Lower panel: Three-winter mean stationary north-hemispheric response pattern
simulated by the ECMWF GCM in full
spectral resolution (from (40)).

where y_i denotes the climate state vector (more precisely: the vector
characterizing the state of the slow components of the climate system), L_{ij}
is a constant matrix, and $n_i(t)$ is the (white) noise input.

In inverse stochastic modelling, a climate model of the form (1), or
some generalization thereof, including, for example, nonlinear terms, is
tuned to observe data. The fitting procedure yields the coefficients L_{ij}
which, for a specific model, can be identified with certain physical parameters, such as positive or negative feedback coefficients, or advection
velocities and diffusion coefficients. The main advantage of the technique

is that it provides quantitative estimates of processes which are often not readily amenable to direct measurements.

The tuning of a stochastic model of the form (1) to observed time series clearly bears some similarity to the construction of a linear multivariate regression model, which also has the general structure (1). However, the optimal fitting procedures are different in the two cases. In linear regression models constructed for climate prediction, the differential form (1) is normally transformed into an integral Green function form, and the model is optimized by minimizing the mean square residual noise $n_i(t)$. No requirement is made that the noise should be white (although the whiteness of the residual is often used as an indicator to decide whether it is meaningful to extend the model to a higher number of predictors). In inverse stochastic modelling, however, the assumption is made *a priori* that the forcing n_i is white, and the model is tuned by maximizing the agreement between the observed and theoretical auto- and cross-spectra, rather than by minimizing the forcing (15). The statistical acceptance criteria are normally also different for the two modelling approaches. For a linear regression model, the statistical significance is generally based simply on the statistical significance of the set of coefficients L_{ij} (or their Green function equivalents), while the acceptance of a stochastic model requires not only the statistical significance of the derived model coefficients, but also the consistency, within given statistical error bounds, of the theoretical spectra with the observed spectra.

Inverse stochastic modelling methods have been applied, using advective-diffusive models of the general linear form (2), to SST anomalies in the Pacific, Atlantic and Indian oceans, and to sea ice anomalies in the Arctic and Antarctic. The goal of the investigations was to derive fields of feed-back coefficients, advection velocities and diffusion coefficients, together with the energy levels and spatial correlation scales of the white noise forcing.

The prototype model was of the form

$$\frac{\partial T}{\partial t} + \lambda T + \sum_i u_i \frac{\partial T}{\partial x_i} - \sum_{i,j} \frac{\partial}{\partial x_i} (D_{ij} \frac{\partial T}{\partial x_i}) = n \qquad (2)$$

where T represents either the two-dimensional SST anomaly field or the sea ice area in a given longitudinal sector (in this case the spatial dependence of T, and the advection velocity and diffusion tensor, involve only the single dimension of longitude). Spatial discretization of (2) yields a linear system of the form (1).

As examples, Figs 6 – 9 show the relaxation time $\tau = $ (feedback coefficient $\lambda)^{-1}$, advection velocity u_i and white noise forcing-level n inferred from North Pacific SST anomalies (20), while Figs 10 and 11 show the corresponding functions for Antarctic sea ice anomalies (27). Where data are available, the distributions shown are reasonably consistent with traditional pictures (cf. Figs 8 and 11). In general, the parameter fields are believed to be more reliable for use in climate models than previous estimates based on the rather sparse direct measurements. It should be noted, however, that the derived fields represent the net effect of the projection of different processes on to a model of the form (2). Thus the advection velocity, for example, represents the superposition of the effects of the displacement of the anomaly patterns by ocean currents and the possible shift induced through interactions with prevailing wind systems.

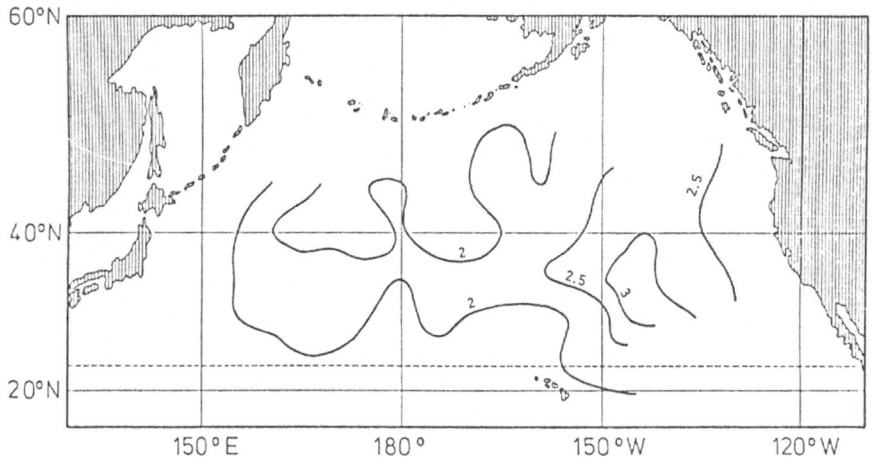

FIG. 6

Relaxation time τ (months) for North Pacific SST anomalies (from (20))

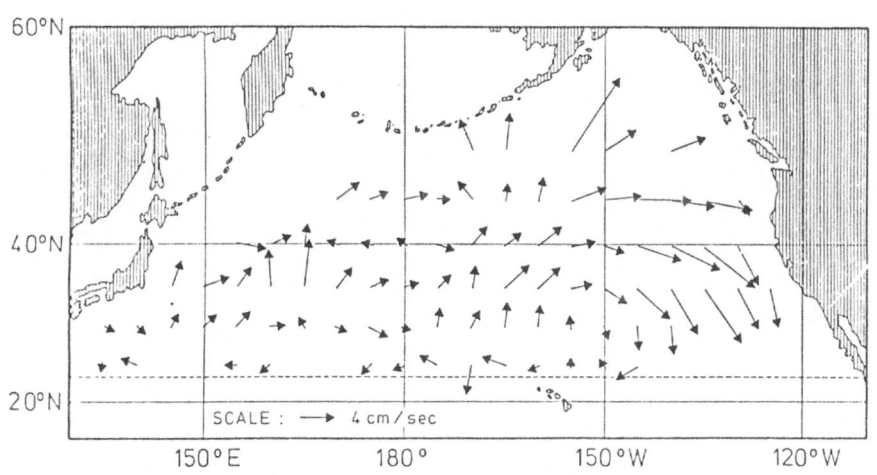

FIG. 7

Advection velocity field inferred from analysis of North Pacific SST
anomalies (from (40)).

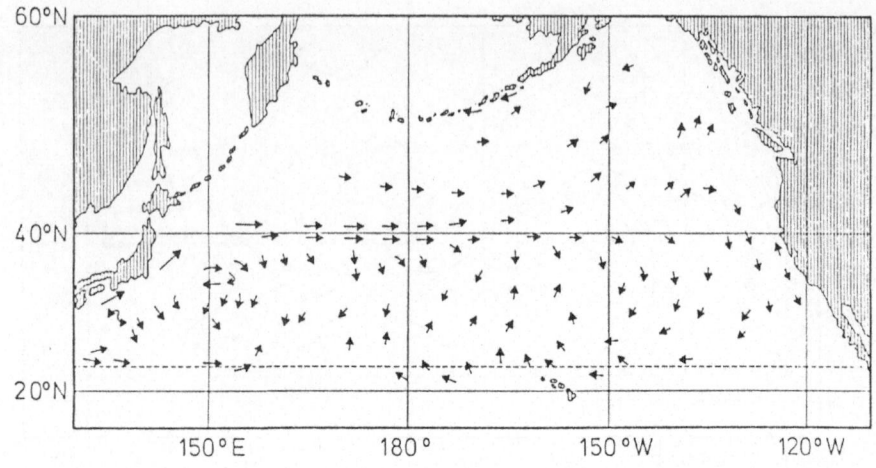

FIG. 8

August surface velocity field (directional information only) for
North Pacific inferred from ship drift data (from (12)).

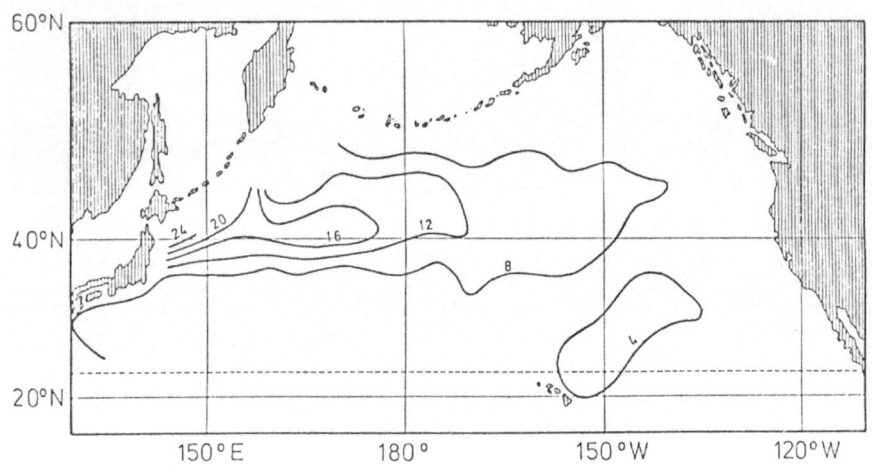

FIG. 9

White noise forcing levels (arbitrary units) driving North Pacific
SST anomaly fields (from (20)).

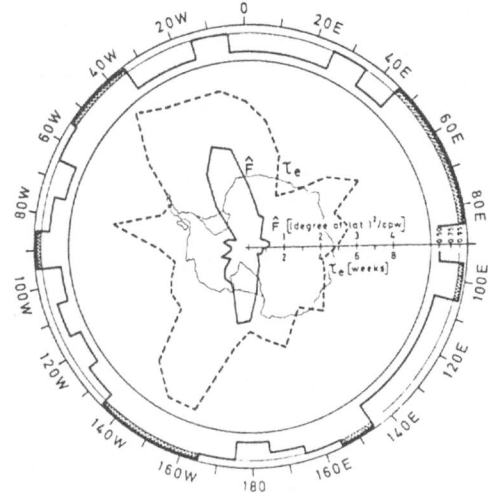

FIG. 10

Longitudinal distribution of
white-noise excitation level
$\hat{F}$ and relaxation time τ_e in the
Antarctic (from (27)).

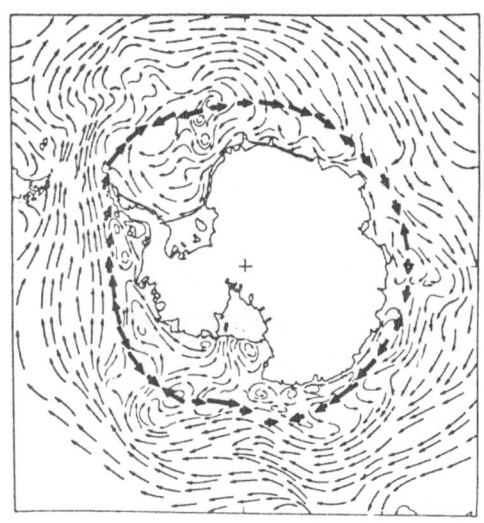

FIG. 11

Southern Ocean surface circulation (after Treshnikov, 1967)
and Antarctic sea ice model
advection pattern inferred from
a stochastic model (thick
arrows) (from (27)).

    Kruse (23), Kruse and Hasselmann (24), and Bruns (6) have applied
inverse stochastic modelling techniques also to more complex atmospheric
systems. As models they considered the nonlinear barotropic and two-mode
baroclinic quasi-geostrophic vorticity equations of the atmosphere in
spectral representation. Although the models were spectrally strongly truncated, typically at a total wavenumber of 8, the models nevertheless still
contained of the order of 50 - 100 nonlinearly coupled degrees of freedom.
The purpose of the analysis was to determine to which extent the evolution
of the lower order spectral harmonic components of the atmosphere could be
explained by interactions between the components themselves and with the
lower boundary of the atmosphere, and which of the processes included in
the model were the most important.

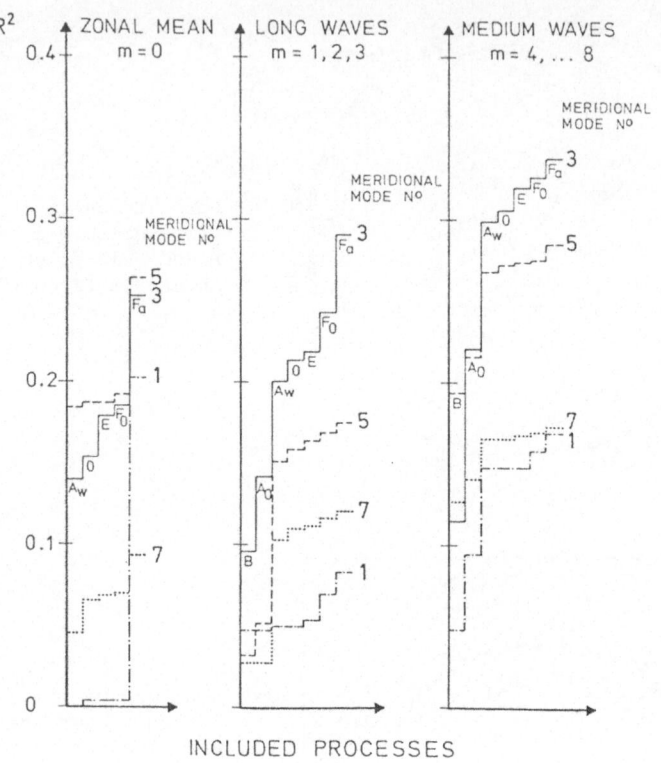

FIG. 12

Fraction of explained variance for a hierarchy of models constructed
by sequentially introducing the seven source functions: B = linear
β-term, A_0 = nonlinear wave-mean flow interaction, A_w = nonlinear wave-
wave interaction, O = orography interaction, E = Ekman friction, F_0 =
mean heat flux forcing, F_a = seasonal modulation of heat flux (from
(24)).

Fig. 12 (from (24)) shows an example of the relative contributions of
different processes to the variance of the zonal flow and the long and
medium scale zonal components. An example of the spectral decomposition of
the explained variance in the frequency domain is shown for selected spherical harmonics in Fig. 13 (from (6)). On average, it is found that less than
half of the total variance of the low order spectral components can be explained by the strongly truncated model, although higher ratios are attained
for individual spectral components in particular frequency bands (for
example, in the neighbourhood of 5 day periods, cf. Fig. 13). Thus an important element in the realistic simulation of atmospheric variability with
strongly truncated atmospheric models is the residual stochastic forcing
representing the interactions with the nonresolved components of the system
(which are unpredictable within the framework of the truncated model). It
appears from these investigations that, in contrast to recent hypotheses
on the dominant role of low-order spectral interactions in the dynamics of

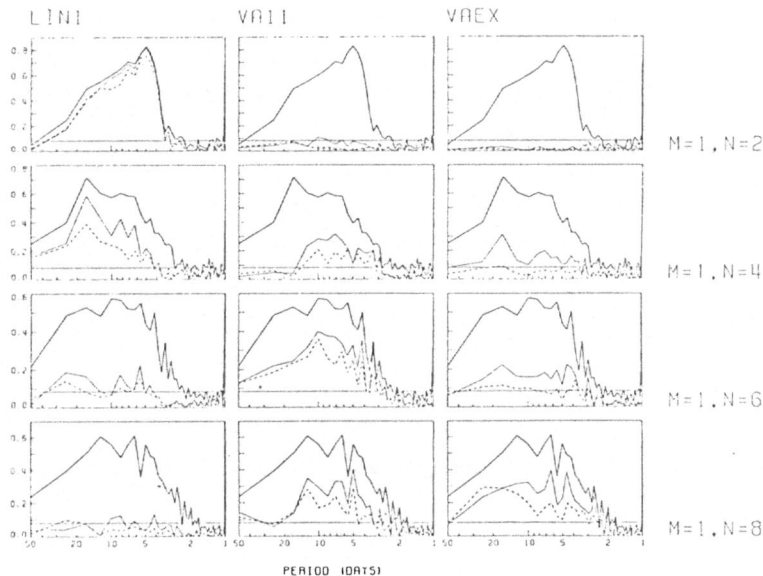

FIG. 13

Examples of coherence spectra between the barotropic stream function
mode M (zonal wavenumber), N (total wavenumber) and the three predictors
LIN1 (all linear terms), VA11 (all resolved nonlinear interactions) and
VAEX (nonlinear interactions involving nonresolved modes up to total
wavenumber N = 15). The thick solid line in each frame is the multiple
squared coherence of the full model including all three predictors. The
broken and solid thin lines represent partial coherence and individual
coherence, respectively (from (6)).

low frequency atmospheric vacillations, the stochastic interactions with
short waves play a significant role in the low-order spectral dynamics of
the atmosphere at all frequencies (cf. also (9)).

5. Oceanic general circulation models

     A global oceanic general circulation model has been developed by Maier-
Reimer et al. (32) under the simplifying approximations appropriate for
large scale, quasi-geostrophic flow, following the general concepts outlined
in (17). The numerics have since been slightly modified (31) to remove a
weak numerical instability which in the first model version produced a too
rapid overturning of the deep ocean, with a resultant slow warming in the
lowest model layers. Figs 14 and 15 show examples of the current field and
density distribution predicted by the present model, with 10 vertical layers
and $5^{\circ} \times 5^{\circ}$ horizontal resolution (from (31)), for observed wind stress
forcing (from (19)) and prescribed air temperature at the sea surface (from
(29)).

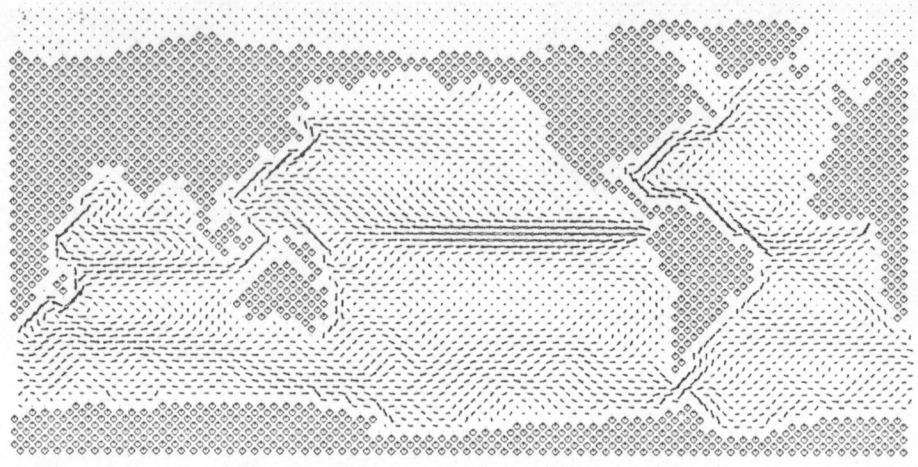

FIG. 14a   ABSOLUTE VELOCITIES IN   75 M DEPTH

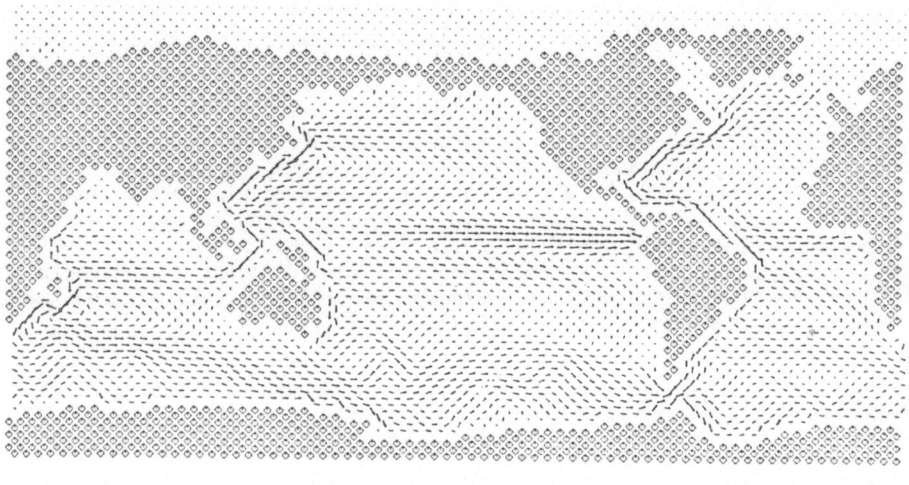

FIG. 14b   ABSOLUTE VELOCITIES IN   150 M DEPTH

Current velocities predicted by global ocean circulation model in 75 m
depth (upper panel) and 150 m depth (lower panel). The main ocean current
systems (including the equatorial undercurrent) are well reproduced (from
(31)).   Arrow areas are proportional to current speed u, arrow lengths
proportional to $\sqrt{u}$. The transports of the Gulf Stream and Antarctic
Circumpolar Current through Drake Passage are 55 and 145 Sverdrup, respectively.

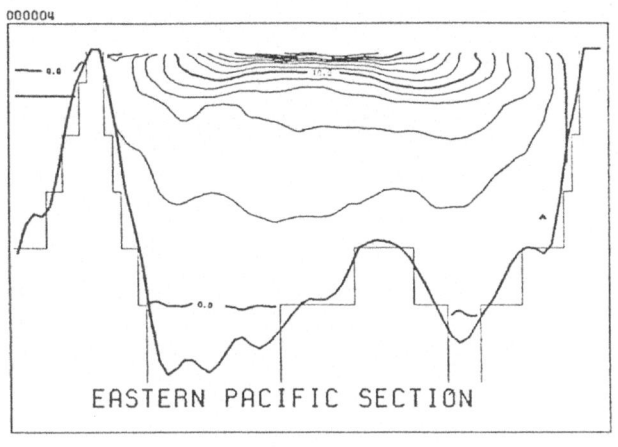

EASTERN PACIFIC SECTION

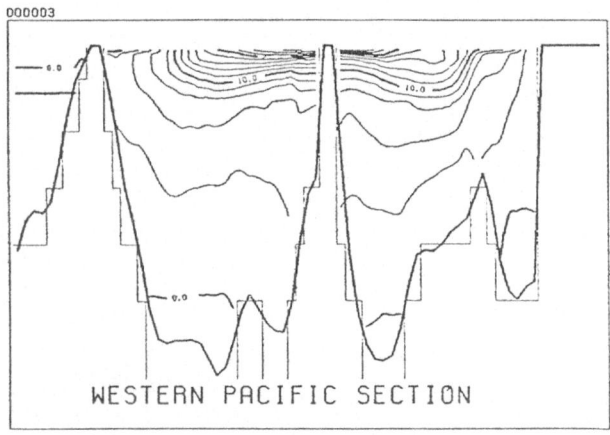

WESTERN PACIFIC SECTION

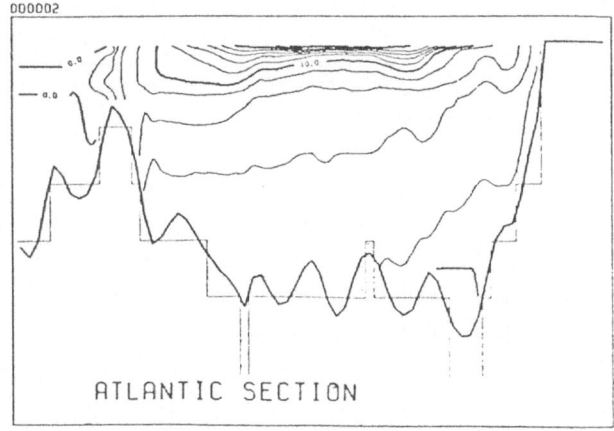

ATLANTIC SECTION

FIG. 15

Temperature sections running N – S after 200 years of integration.
Contour interval is 2K (from (31)).

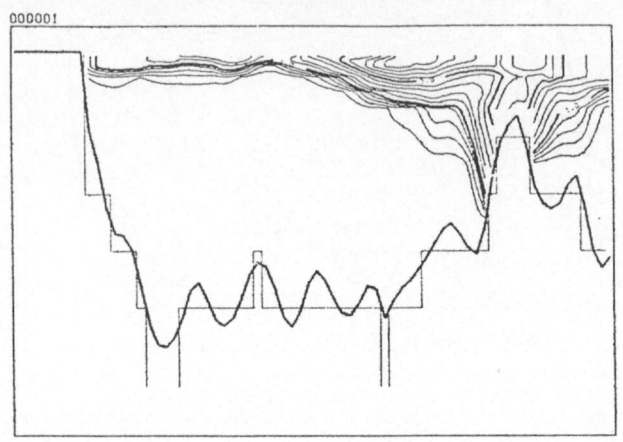

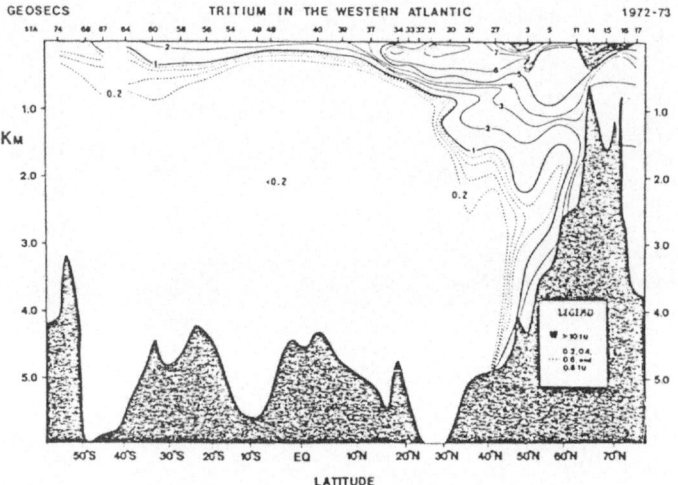

FIG. 16

Tritium sections. Upper panel: model (from (32))
Lower panel: observed (from (36).

The model has also been applied to compute tracer distributions. The tritium (Fig. 16) predicted by the model is in good general agreement with observations. Together with the realistic simulations of the current fields and the temperature and salinity distributions shown in Figs 14, 15, this suggests that the overall transports and residence times of the model are sufficiently realistic for application in climate studies.

Work is also currently in progress on an alternative global ocean model based on an isopycnal rather than a fixed layer description of the vertical density structure (35) and on improved descriptions of the mixed layer, including the mixed layer under ice (28), which should improve the coupling characteristics of the ocean models when used in coupled ocean-atmosphere-ice models.

6. Carbon cycle models

The global ocean circulation model has been extended to investigate the storage of carbon in the ocean. The transport and storage of the various carbon constituents within the ocean was treated in the same manner as in the tracer studies. The transformation of carbon constituents occurs in the uppermost mixed layer, where CO_2 is exchanged with the atmosphere, and was computed by solving the chemical interaction equations. In a first version of the model, only inorganic interactions were considered (34). The model has since been extended to include a simplified biochemical cycle in terms of one species of biota and nutrients (1). A number of experiments were also carried out with a full ocean-atmosphere-terrestrial carbon cycle model, using the terrestrial biosphere model developed by Lieth and Esser (30) and Esser (10).

Some examples of these studies are given in Figs 17 - 20. The ocean carbon cycle model is able to reproduce the observed pCO_2 concentrations in the upper layer of the oceans (Fig. 17) and predicts an airborne CO_2 fraction for the present CO_2 increase, assuming a terrestrial source of 1 GT carbon/yr, of the order of 65 % (Fig. 18). Inclusion of the terrestrial biosphere yields an airborne fraction of the order of 57 %, which is closer to estimates from observations (Fig. 19). Also shown for interest is the effect on the CO_2 concentration of a deforestation of the Amazonian basin, computed with the full carbon cycle model. It should be pointed out, however, that the computations with the full model including the terrestrial biosphere are rather sensitive to the response of the terrestrial biota to an increase in atmospheric CO_2 concentration, which is not very well known.

PARTIAL PRESSURE CO2 1860   FIG. 17

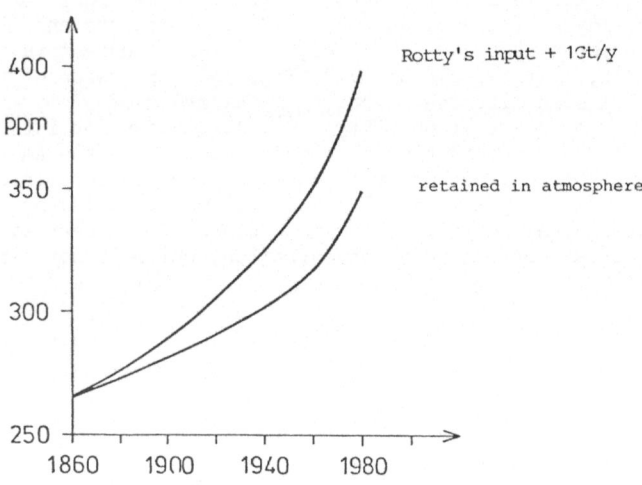

Rotty's input + 1Gt/y

retained in atmosphere

FIG. 18

Total CO_2 emission into atmosphere (Rotty's 1981 input + 1 GT/y input from biosphere) and model airborne fraction, computed from ocean carbon storage model with inorganic chemistry only (from (34)).

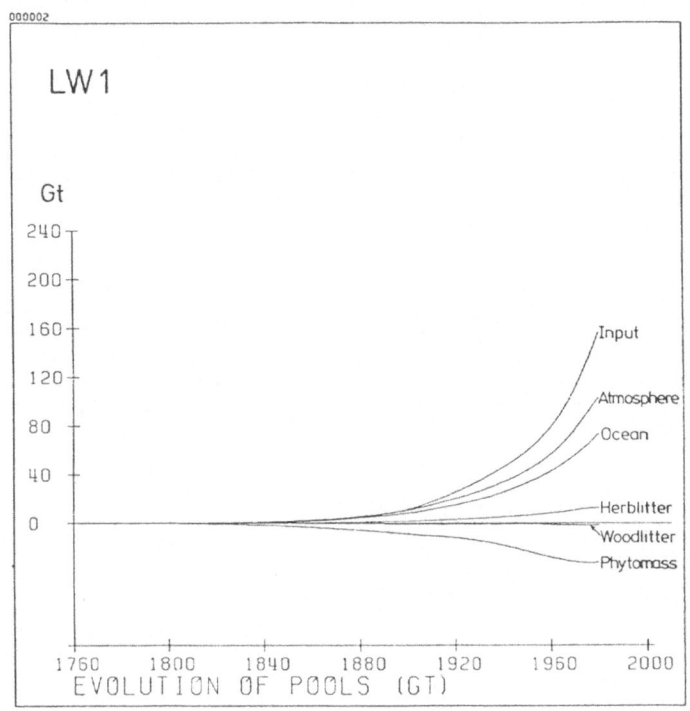

FIG. 19

Change in carbon storage of different pools since 1760 for given CO_2 industrial input (37) computed with full ocean-atmosphere-terrestrial biosphere model (from (33)).

ATMOSPHERIC RETENTION

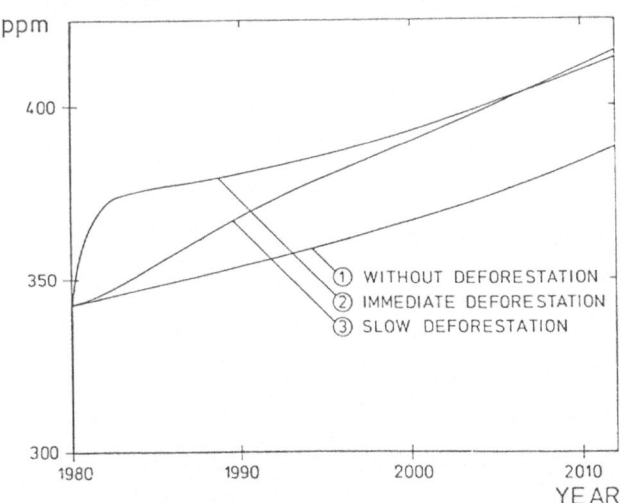

FIG. 20

Computed impact on atmospheric CO_2 concentration of deforestation of Amazonian basin (from (33)).

7. Equatorial ocean model

In addition to the global, quasi-geostrophic ocean model, a primitive
equation model has been developed for the equatorial regions (25, 26),
where the large-scale quasi-geostrophic approximation is no longer valid.
In the global model, the dynamics in the equatorial region are distorted by
effectively filtering out or retarding the rapidly travelling Kelvin and
Rossby wave components. This approach is justified for studies of global
climate variations on long time scales of the order of $1 - 10^3$ years, but is
no longer applicable to investigations of short term climate processes such
as El Niño involving characteristic time scales of a month to a few years.
The equatorial ocean model designed for these time scales yielded a realistic simulation of the mean distributions and annual variations of the density and current field (Fig. 21). It also reproduced remarkably well the
observed time series of SST anomaly patterns in the equatorial Pacific,
computed as the response to the observed wind field anomalies (Fig. 22).

As next step it is plannd to couple the equatorial model with an
atmospheric model in order to investigate the natural variability and
potential predictability of the coupled system.

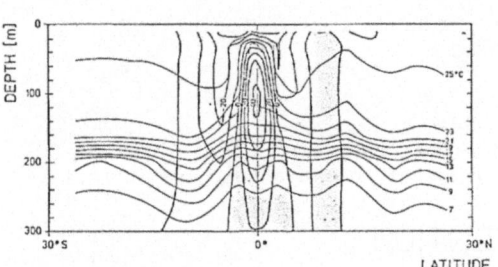

a) computed

FIG. 21

(a) Meridional cross section of
    zonal velocity and tempera-
    ture along the central
    meridian as computed by the
    model. Shaded area repre-
    sents eastward velocities
    (from (25)).

(b) Observed meridional cross
    section of zonal velocity
    (dashed lines, contour
    interval 25 cm/s, maximum
    speed: 125 cm/s, and tem-
    perature at 140° W (from
    (22) and (11)).

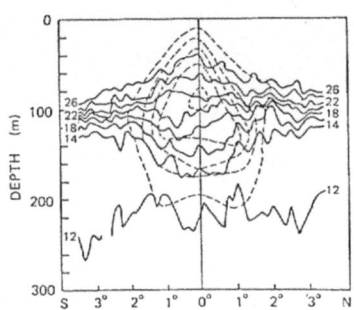

b) observed

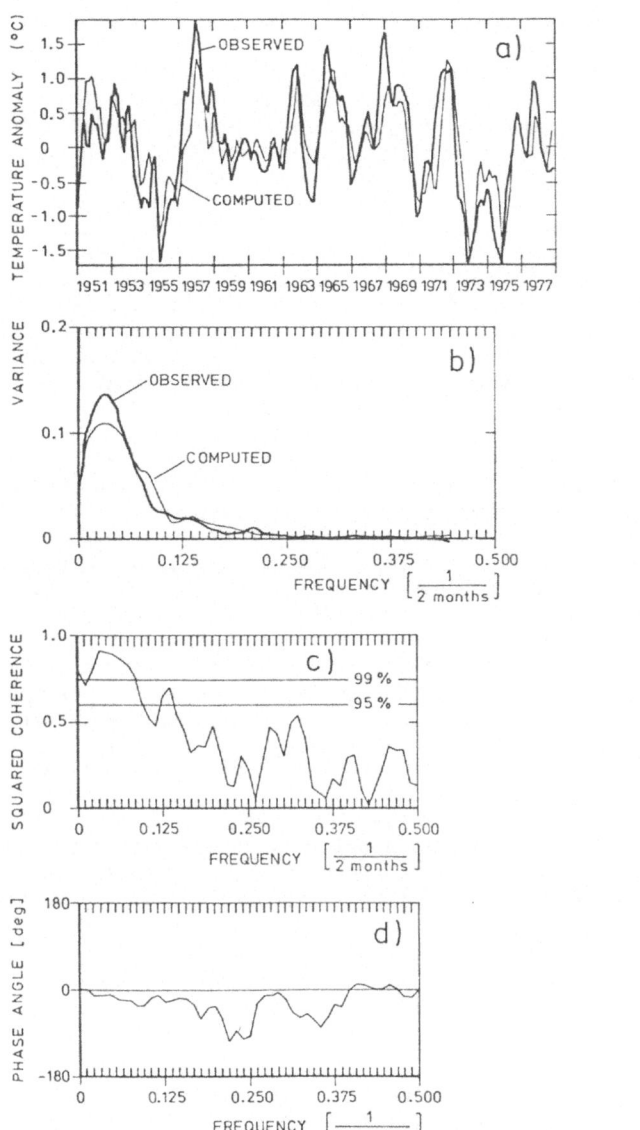

FIG. 22

(a) Time series of observed (heavy line) and computed (thin line)
    SST anomaly near the dateline on the equator.
(b) Variance spectra of the two time series.
(c) Coherence spectrum.
(d) Phase spectrum (from (26)).

8. Conclusions and outlook

The modelling work currently in progress at the Max-Planck-Institute of Meteorology ranges from simple statistical models and inverse modelling methods to the development of complex high resolution GCMs. Most of the effort in high resolution modelling has been devoted in the past to oceanic models. The next goal is now to couple high resolution models of the ocean with high resolution atmospheric and cryospheric models (which have already been implemented or are being developed by other groups in Hamburg).

Such high resolution coupled models should provide the necessary tools for investigating quantitatively the natural variability and predictability of the coupled climate system, using the statistical techniques which have been developed and applied so far only to relatively simple stochastic models. In particular, it will become meaningful to address the question of how much of the natural variability of the coupled climate system arises from unstable interactions within the system itself, through feedback loops between the slow and fast components of the system, and how much is due to the stochastic forcing generated by the natural short term variability of the fast component, the atmosphere. The ratio of these two contributions is the key quantity determining the degree of predictability of natural climate variability.

Coupled models are equally important, however, for a realistic assessment of the time dependent impact of man's activities on global climate. In particular, the incorporation of realistic models of the ocean, cryosphere and biosphere, together with a high resolution atmospheric model, within a comprehensive coupled climate model is an essential prerequisite for evaluating the important time delay effects associated with the gradual adaption of the slow components of the climate system to changing external forcing, such as an increase in CO_2 concentration.

Detailed models of the complete coupled climate system are also necessary to exploit the information which is expected to become available from planned new earth observing satellites in the next years. Together with existing meteorological and land satellites, these will provide a comprehensive view not only of the atmosphere and land surfaces, but also of the oceans and ice covered regions of the earth, which exert a controlling influence on climate. Coupled climate models will be needed both to assimilate the satellite data and to provide a conceptual framework to interpret the data.

REFERENCES

1. BACASTOW, R. and MAIER-REIMER, E. (1985). Circulation model of the ocean carbon cycle (in preparation).
2. BARNETT, T.P. (1981). Statistical prediction of North American air temperatures from Pacific predictors. Mon.Wea.Rev. 109, 1021-1041.
3. BARNETT, T.P. and HASSELMANN, K. (1979). Techniques of linear prediction, with application to oceanic and atmospheric fields in the tropical Pacific. Rev.Geophys.and Space Phys. 17, 949-968.
4. BARNETT, T.P., PREISENDORFER, R.W., GOLDSTEIN, L.M. and HASSELMANN, K. (1981). Significance tests for regression model hierarchies. J.Phys. Oceanogr. 11, 1150-1154.
5. BARNETT, T.P., HEINZ, H.-D. and HASSELMANN, K. (1984). Statistical prediction of seasonal air temperature over Eurasia. Tellus 36A, 132-146.

6. BRUNS, T. (1985). On the contribution of linear and nonlinear processes to the long term variability of large scale atmospheric flows (submitted for publication).
7. CHERVIN, R.M. and SCHNEIDER, S.H. (1976). On determining the statistical significance of climate experiments with general circulation models. J.Atmos.Sci. $\underline{33}$, 405-412.
8. CUBASCH, U. (1983). The response of the ECMWF global model to the El Niño anomaly in extended range prediction experiments. Tech.Rep. 38, European Centre for Medium Range Weather Forecast, Reading, Berkshire, U.K.
9. EGGER, J. and SCHILLING, H.-D. (1983). On the theory of the long-term variability of the atmosphere. J.Atmos.Sci. $\underline{40}$, 1073-1085.
10. ESSER, G. (1984). The significance of biospheric carbon pools and fluxes for atmospheric CO_2: A proposed model structure. Progress in Biometeorology $\underline{3}$, 253-294.
11. GILL, A.E. (1975). Models of equatorial currents. In: Numerical Models of Ocean Circulation. National Academy of Sciences, Washington, D.C.
12. GORSCHKOV, C.G. (1976). World Ocean Atlas, Vol. 1: Pacific Ocean (in Russian with English introduction). Pergamon.
13. HANNOSCHÖCK, G. and FRANKIGNOUL, C. (1985). Multivariate statistical analysis of a sea surface temperature anomaly experiment with the GISS general circulation model (J.Atmos.Sci., in press).
14. HASSELMANN, K. (1976). Stochastic climate models, Part I, Theory. Tellus $\underline{28}$, 473-485.
15. HASSELMANN, K. (1979). Linear statistical models. Dyn.Atmos.Oceans $\underline{3}$, 501-521.
16. HASSELMANN, K. (1979). On the signal-to-noise problem in atmospheric response studies. In: Meteorology of the Tropical Oceans, D.B. Shaw, editor. Roy.Met.Soc. 251-259.
17. HASSELMANN, K. (1982). An ocean model for climate variability studies. Prog.Oceanogr. $\underline{11}$, 69-92.
18. HASSELMANN, K. and BARNETT, T.P. (1981). Techniques of linear prediction for systems with periodic statistics. J.Atmos.Sci. $\underline{38}$, 2275-2283.
19. HELLERMAN, S. (1967). An updated estimate of the wind stress on the world ocean. Mon.Wea.Rev. $\underline{95}$, 606-626.
20. HERTERICH, K. and HASSELMANN, K. (1985). Extraction of SST advection, relaxation and atmospheric forcing patterns from the statistical analysis of North Pacific SST anomaly time series (to be submitted to JPO).
21. JOHNSON, C.M., LEMKE, P. and BARNETT, T.P. (1985). Linear prediction of sea ice anomalies. J.Geophys.R. (in press).
22. KNAUSS, J.A. (1960). Measurements of the Cromwell Current. Deep-Sea Res. $\underline{6}$, 265-286.
23. KRUSE, H. (1983). A statistical-dynamical low-order spectral model for tropospheric flows. Hamburger Geophysikalische Einzelschriften No. 59.
24. KRUSE, H. and HASSELMANN, K. (1985). Investigation of processes governing the large-scale variability of the atmosphere using inverse modelling techniques (submitted for publication).
25. LATIF, M., MAIER-REIMER, E. and OLBERS, D.J. (1985). Climate variability studies with a primitive equation model of the equatorial Pacific (acepted for publication).
26. LATIF, M. (1985). Western and eastern variability in a model of the tropical Pacific (in preparation).
27. LEMKE, P., TRINKL, E.W. and HASSELMANN, K. (1980). Stochastic dynamic analysis of polar sea ice variability. J.Phys.Oceanogr. $\underline{10}$, 2100-2120.

28. LEMKE, P. and MANLEY, T.O. (1984). The seasonal variation of the mixed layer and the pycnocline under polar sea ice. J.Geophys.Res. 89, 6494-6504.
29. LEVITUS, S. and OORT, H.A. (1977). Global analysis of oceanographic data. Bull.Am.Met.Soc. 58, 1270-1284.
30. LIETH, H. and ESSER, G. (1985). The attempt to simulate the global carbon flux from 1860 to 1981 using the Osnabrück Biosphere Model. In: Transport of carbon and minerals in major world rivers, Part 3, Mitt. Geol.-Paläontolog. Inst. Univ. Hamburg, SCOPE/UNEP Sonderband (in press)
31. MAIER-REIMER, E. (1985). A large scale global ocean circulation model (in preparation).
32. MAIER-REIMER, E., MÜLLER, D., OLBERS, D., WILLEBRAND, J. and HASSEL-MANN, K. (1982). An ocean circulation model for climate variability studies. Report Max-Planck-Institut für Meteorologie.
33. MAIER-REIMER, E., HASSELMANN, K., LIETH, H. and ESSER, G. (1984). Entwicklung eines globalen Kohlenstoff-Kreislauf-Modells. Report Max-Planck-Institut für Meteorologie and Fachbereich Biologie/Chemie, Universität Osnabrück.
34. MAIER-REIMER, E. and HASSELMANN, K. (1985). Transport and storage of CO_2 in the ocean (in preparation).
35. OBERHUBER, J.M. (1985). An isopycnal ocean circulation model (in preparation).
36. ÖSTLUND, H.G., DORSEY, H.G. and BRESCHER, R. (1976). GEOSECS Atlantic radiocarbon and tritium results (Miami), Rosenstiel Sch.Mar.Atmos.Sci., Univ.Miami, Data Rep. No. 5.
37. ROTTY, R.M. (1981). Data for global CO_2 production from fossil fuels and cement. In: Bolin, B. (ed.), Carbon Cycle Modelling, 121-127, SCOPE 16, John Wiley and Sons.
38. STORCH, H.v. and ROECKNER, E. (1983a). Methods for the verification of general circulation models applied to the Hamburg University GCM. Part I: Test of individual climate states. Mon.Wea.Rev. 111, 1965-1976.
39. STORCH, H.v. and ROECKNER, E. (1983b). On the verification of January GCM simulations. Proc. II, International Meeting on Statistical Climatology, Sept. 26-30, 1983, Lisboa, Portugal, Instituto de Meteorologica e Geofisica.
40. STORCH, H.v. and KRUSE, H.A. (1985). The extra-tropical atmospheric response to El Niño events - a multivariate significance analysis. Tellus (in press).

OCEAN MODELLING IN CAMBRIDGE

PETER D. KILLWORTH
Department of Applied Mathematics and Theoretical Physics
Silver Street
University of Cambridge
Cambridge CB3 9EW
England

Summary

The work of the Cambridge Ocean Modelling Group during the past few years
is described. Research has fallen into seven main categories: midlatitude
circulation, topographic effects, mesoscale features, convection, storms,
tropical ocean-atmosphere problems, and modelling techniques. A bibliography of work produced by the group, plus related papers, is provided.

1. Introduction
      The Ocean Modelling Group in Cambridge usually comprises 4 postdoc-
toral staff, 1-2 programmers and some research students. The group maintains a keen interest in most areas of geophysical interest, since DAMTP
contains staff concerned with laboratory modelling as well as with practical geophysical problems. Its main function, however, lies with understanding questions relating to oceanic and atmospheric dynamics, mainly
on climatic time scales (weeks to decades). Studies in the last few years
have been devoted to a variety of topics to be discussed here.   These
fall into seven categories:
general midlatitude ocean circulation;
the effects of topography upon circulation;
the dynamics and mixing effects of small-scale features like eddies;
the role of oceanic deep convection;
the effects of storms upon the ocean beneath them;
the response of the tropical ocean and atmosphere to each other;
general numerical and analytical modelling techniques.

2. Midlatitude ocean circulation
      It is well known that northwestern Europe enjoys a generally warmer

- 195 -

climate than any other land or sea area at its latitude, either north or south of the equator. It has been proposed (62) that this warm climate owes its existence to a meridional circulatory 'pump' in the Greenland Sea. Deep convective events in this sea, it is suggested, force a surface replacement inflow of water to the Greenland Sea. This inflow forces the northward movement of warm water past the British Isles and so warms the climate. Another view has it that the Gulf Stream extension fans out near the British Isles and thus provides the surface warming. Yet another view would suggest the northward movement of warm water by eastern boundary currents. Given the evidence that in comparatively recent geological times there have been sudden (100-year time scales) shifts in surface temperature of the Atlantic to and from an ice-covered region, it is important to find out what are the control mechanisms in this climatic structure.

Modelling provides useful and varied methods to study this problem. We have used models ranging from the very simple to the very complex for North Atlantic simulations, and all seem capable of reproducing the northeastern warming without the need for deep convective driving.

The simplest model (4) uses a two-layer ocean driven solely by prescribed surface heating and cooling (i.e. no wind). Such a simple model allows various features to be added or subtracted and their effects studied, whereas more complex, time consuming models cannot exploit this feature. The model shows the effects of Kelvin and Rossby waves on the steady solution, which possesses a wide warm layer on the eastern boundary, with a balance between surface heating and wave propagation effects. Time dependence, nonlinearity, and diffusion have all been studied within the framework of the model. There is a tendency for flattening of isopycnals along the coast, as in many models.

The recent return to consideration of the large-scale thermocline problem (e.g. 58, 60) prompted several slightly more complex models, although the thermocline system does have certain undesirable, and not understood, features (40). A linear thermocline model (25), in which geostrophic dynamics are combined with the linearised buoyancy equation (vertical advection balancing vertical diffusion), permits similarity solutions in which the effects of wind and surface buoyancy driving are coupled in a manner specified by the solution. Thus totally free specification of boundary conditions is not permitted. Nonetheless, the wind-driven case is quite realistic. In the subtropical gyre, isotherms slope up towards the east, so that surface temperatures increase westwards. In the subpolar gyre, however, isotherms slope up towards the west and the maximum surface temperature is at the eastern boundary.

A nonlinear thermocline model (47), which retains its simplicity by limiting vertical resolution to two levels, permits the specification of both Ekman wind pumping and surface buoyancy conditions. This model, too, shows a warming in the northeast, with warmest water in the southwest. Convective overturn must occur in the north of the basin. The density structure is fairly insensitive to the specified surface density, responding to a Rossby-wave equation driven by Ekman pumping and vertical diffusion from the surface (assuming this diffusion to be included). The magnitude of the baroclinic flow is far larger than the imposed barotropic, although reflecting its shape. Apparently the steady solution is not unique. It is thought that this may be an artefact of the two-level assumption, and work is currently proceeding to examine this. Particle trajectories show that much of the deep subtropical gyre is never in direct contact with the atmosphere, which is important for tracer studies, particularly so for potential vorticity which is only created near-surface.

A three-level version, allowing a dynamically active, rather than passive, Ekman layer, is currently being produced (51). This will yield a totally free specification of surface boundary conditions. Tests show a similar structure to the two-level case. A diffusive boundary condition which removes the necessity for flat boundary isopycnals is being tested.

All the models discussed so far have used subsets of the total dynamics to examine certain features more closely. Additionally, one may employ the large-scale ocean circulation model due to Bryan, Cox and Semtner at the Geophysical Fluid Dynamics Laboratory (GFDL) at Princeton, U.S.A. This model is implemented in various forms at the U.K. Met. Office, the Cray in London, and the IBM 3081 in Cambridge.

The model has been used in a variety of ways to study the North Atlantic climate problem. For example, the relaxation of an ocean initially at rest but possessing a North-South temperature gradient was studied (55). The immediate onset of an East-West geostrophic flow produces up- and downwelling at the vertical walls. This in turn produces damped Kelvin and Rossby waves, evident throughout the relaxation process. For the first year of the relaxation, a single-vertical-mode model along the lines of the two-layer model discussed above seems to describe the results well, but later more complicated dynamics come into play.

The GFDL model has also been used to study the response of an ocean to the onset of heating and cooling, but without wind-driving (50). The model has fine horizontal resolution and low (artificial) viscosities; one of the intentions was to see whether eddy effects would occur and modify the north-south heat flux of the mean flow. Computer time limits the length of the run to around two years. At this time, currents have reached 11 cm/s in the western boundary layer, but there is no sign of eddying. Again, the northeastern corner is the warmest (at intermediate depths), apparently due to a combination of downwelling and arrested Kelvin waves along the eastern boundary. The surface density is tied strongly to the imposed atmospheric values. A single anticyclonic gyre is produced (unlike the double wind-driven gyre seen previously). Curiously, the barotropic (depth-independent) flow is cyclonic, but very weak.

Another use for the GFDL model was to study the 'dam-break' problem (referred to in 33) with realistic geometry and topography for the northern North Atlantic. Warmer, stratified water was placed south of an artificial barrier stretching from Greenland to Iceland to the Shetlands to Norway, with colder, stratified water to the north of the barrier. With the flow initially at rest, the barrier was withdrawn, yielding an initial flow of small cyclonic and anticyclonic gyres anticlockwise around the Norwegian and Greenland coasts, but little interchange of water masses across the site of the barrier otherwise.

3. Topographic effects

Oceans are frequently assumed by modellers to possess flat bottoms. Partly assumed for convenience, this obscures the fact that the important interaction between bottom topography and stratification is not well understood. A numerical study of adjustment of an initial density front in a channel with and without a sloping sidewall has been made (33). Without topography, most of the initial potential energy is converted to baroclinic kinetic energy. With topography, nearly barotropic shelf waves are produced. Since these are faster than the baroclinic waves, they are preferentially selected by the adjustment process.

Another model (7) examined the effects of a step in topography which impinges normally on a vertical boundary. Coastal waves (e.g. Kelvin waves) will propagate until they meet the step from the south (for an eastern boundary, say). What then occurs depends on whether the step is up or down to the north. This is because there exist trapped waves along the topography which may only propagate in one direction. Hence a step up will allow waves to carry information away from the boundary, whereas a step down can only transport information into the boundary from the open ocean, thus blocking the flow from propagating further along the boundary. The response in the two cases is thus very different; the latter case needing a boundary layer near the coast in which either nonlinear or frictional processes come into play. Laboratory experiments with two-layer systems confirm the extension of the numerical and analytical results, which were for barotropic fluids.

4. Mesoscale features and fronts

A ubiquitous feature of the ocean circulation is the presence of structures whose length scales resemble the internal deformation radius (i.e. with length scales of order 30 to 150 km). Despite a great deal of observational and theoretical work in the '70s, little is still known of the mechanisms which create, maintain and dissipate these features, or of what governs if or how they propagate. It is not known whether these features are efficient at transporting heat, momentum, potential vorticity, or any other active or passive tracer; in particular, their role in maintaining the global heat budget of the ocean-atmosphere system is not understood. Work in Cambridge has examined two types of mesoscale features: isolated eddies, such as those shed by intense currents like the Gulf Stream; and fronts, where a great deal of surface exchange and dissipation is known to occur.

A simple one-layer model of an isolated eddy was constructed (38). It was shown how eddies on the scale of the deformation radius are slightly distorted from radial symmetry by the beta-effect (the northward gradient of the vertical component of Coriolis force). This distortion yields a westward movement of the eddy; bounds may be placed on this velocity, typically a few cm/s. A lower bound may also be placed on the radius of such an anticyclonic eddy (2.8 times the deformation radius).

The isolated eddy above has no interaction with the fluid around it; a more realistic model would include the surrounding fluid, and seek a time-dependent problem rather than a steadily propagating solution. This would enable the eventual dissipation of the eddy to be studied. A one-layer model (6) shows that such eddies are easily produced (in this case by a forcing term in the continuity equation) and propagate westwards and (usually more slowly) north or south according as the eddy is cyclonic/anticyclonic respectively. Cyclones move west slower than the long Rossby wave speed, anticyclones faster. The more nonlinear the eddy, the longer it appears to be able to maintain its structure against the competing effects of mass shedding and vorticity conservation. Using potential vorticity and marked water particles as tracers, together with simple theory, estimates for mixing and dispersion produced by these eddies have been made. East-west diffusions of up to 10^4 m^{2}s^{-1} have been deduced, with much smaller northward values. Strong eddies have evidence of irreversible deformation (wave breaking) at their fringes, rather like observations in the stratosphere (59) or in the northeast Atlantic 'Tourbillon' eddy (57).

A later study (48) has examined the relevant parameters of such eddies with an eye to producing an even simpler model. Quasigeostrophy, a dispersal-advective equation set, and a model involving nested vortex rings, have all been shown to reproduce some but not all of the behaviour of the one-layer primitive equation model.

The other main topic in this area has been the instability of oceanic fronts (29, 35, 39, 44, 46). A front is here defined as a region where an isopycnal either surfaces or impinges on the ocean floor. A variety of frontal configurations has been shown to be unstable no matter what the velocity or vorticity distribution, in contrast with the standard - but inapplicable - quasigeostrophic result. Typical bottom-trapped propagation of dense water, which can form a double-front situation as in the Denmark Strait or a single front against a boundary as in the northwest Atlantic, is shown to be actively unstable. Even a single front in isolation is normally unstable.

All these calculations were for a one-layer system, which intrinsically omits the possibility of baroclinic instability. Another calculation (44) includes an active lower layer. These extra dynamics yield a much more active instability (again, no matter what the structure of the front). The calculations give excellent agreement both with laboratory studies (56) as well as oceanic observations, and may help to explain features of the seasonal variation of the Loop Current in the Gulf of Mexico.

Another application of the theory is to instabilities of ice-frontal regions in the marginal ice zone (46). Observations show whorls of ice of 20 to 50 km scale shed from the ice edge into warmer water, with corresponding disturbances to the oceanic near-surface structure beneath. Since ice melt is an important part of the heat budget of the marginal ice zone, we need to understand what mechanisms produce these instabilities. Both barotropic and baroclinic instability of the underlying water have been suggested, and the truth probably involves a combination of the two. However, the theory for fronts carries over to the dynamics of just the ice, without consideration of the water beneath. It is shown that typical ice fronts are unstable to disturbances of about the right length scale, but that the growth rates are probably too small compared with those involving the ocean.

5. Convection in the ocean

The manner in which the effects of wind and buoyancy combine to force the ocean is still somewhat mysterious. The effects of wind driving are probably better understood than those of buoyancy driving; mixed layer models are almost always one-dimensional, with only a few papers including advective effects. The retreating springtime mixed layer leaves behind it a stratified fluid which also possesses large scale potential vorticity. Since both heat and potential vorticity are conserved quantities, and both are functions of the temperature field (or more properly of the density field), understanding precisely how their near-surface values are produced is of importance for the understanding both of the large-scale oceanic circulation and for climate. The planned World Ocean Circulation Experiment has as one of its main aims the elucidation of the physics of water mass transformation. The way in which information about surface events propagates through the mixed layer into the deep ocean is also not well understood. For example, the seasonal cycle of deep eddy activity is well correlated with surface wind activity but with a lag which increases with depth (9). Attempts with a layered model suggest that even with convection, there is no apparent way that energy on these time scales can propagate to the observed depths. It is suggested that deeper convective events occur elsewhere in the ocean and propagate along topography to the observed sites. A review of deep convective processes has appeared (36).

The mixed layer in the Arctic has been a source of enquiry for some time among observationalists. Below about a 30m thick mixed layer, the temperature field remains roughly uniform (and near freezing) for another

70–100m, while an intense halocline is evident in the salinity field, which increases by several parts per thousand in a hundred meters. The phenomenon occurs over virtually the entire Arctic basin, and is horizontally very uniform. Attempts to explain this by vertical mixing between incoming (deep) Atlantic water and overlying fresh surface water fail, as no mixing ratio can yield the observed temperature and salinity structure simultaneously. It has thus been suggested that the source of the cold, salty water at 100m is brine produced on the wide Russian continental shelves by intense winter cooling. This brine is believed to sink down the continental slope and spread laterally below the mixed layer. A model of this process was produced (42) which considered the Arctic as a filling-box, with inflow and outflow in geostrophic balance through Fram Strait (between Svalbard and Greenland, the main inflow to the Arctic basin from the Atlantic). Under no circumstances could the observed temperature-salinity structure be produced. It was found necessary to include a hitherto-ignored portion of the water budget in the calculations: the inflow from the Pacific through the Bering Straits. Although providing negligible impact on the Arctic heat and salinity budget, this water nonetheless can feed (perhaps indirectly) the strong outflow through Fram Strait, and thus relieve the main Arctic Ocean of the necessity of strong upwelling to feed that outflow. However, the model remained incapable of simulating the temperature-salinity structure at around 200–300m depth.

Current work includes midlatitude simulations of heating and cooling, to see under what circumstances deep convection will occur and when advection will remove the necessity for convection to great depths. An allied study is examining a simple advective mixed layer model, with a view to seeing what determines the magnitude of the large-scale potential vorticity created at or near the surface.

6. Storms

Most calculations of energy transfer from atmosphere to ocean, especially those on climatic time scales, use average winds for the estimation. However, a great deal of energy can be transferred by a single storm, especially at low latitudes. Several studies have been aimed at understanding the effects of this transfer.

The response of an ocean to a moving storm has been studied (26, 27, 28). It has been shown that a first-order definition of the response is given in a wide range of cases by neglecting the horizontal pressure gradient terms in the momentum equations. This idea has allowed the elucidation of both nonlinear effects and the scale and parameter dependence, together with a feel for the effects of turbulent mixing. It is possible to determine such features as the sea-surface temperature, and to decide whether Ekman divergence is important in a given circumstance, or whether local mixed layer calculations will give correct answers.

The decay of near-inertial waves in the mixed layer after a storm has passed has also been studied (21). These waves contain half the energy in the internal wave spectrum and their is much interest in their generation and behaviour. The study explains many observed features, namely: intermittency; the strongest signal is immediately below the storm track, but lasting long after the storm has passed; a decrease in both horizontal and vertical coherency with time; slow vertical group propagation; and a tendency for bottom intensification. Methods of parameterizing the loss of energy from the mixed layer following a storm were developed.

7. Tropical oceans

The planned Tropical Ocean – Global Atmosphere (TOGA) draws attention to the importance of the tropics on (at least short time scale) global climate. The El Nino phenomenon is now seen as part of a wider Southern Oscillation, whose features can be detected far into midlatitudes.
Although the rapid equatorial timescales make numerical simulations more
tractable than their midlatitude equivalents, the ocean-atmosphere
interaction must be included in the modelling if we are to explain fully
the observed events. This added complexity means that observations must
be examined carefully for clues as to the physics.

Some work has been done to analyse El Nino events (13, 15, 18, 19,
20). Anomalous atmospheric heating induced by sea-surface temperature
(SST) anomalies, followed by convection, precipitation, and latent heat
release, appears to be a fundamental mechanism (and the SST is itself
affected by atmospheric movement). El Ninos, which involve westward pro-
pagatioin of warm surface anomalies, strongly alter the preferred site of
atmospheric convection; in 1982-3, this led to a major drought in Australia and Indonesia, and disastrous floods in Ecuador.

A great deal of progress has been made using the long-wave approximation, which maintains the east-west oceanic and atmospheric velocities
in geostrophic equilibrium (15, 18, 19, 20, 24). This permits the linear
ocean or atmosphere response to be split into an infinite sum involving
an orthogonal set of horizontal modes (31) beginning with the Kelvin,
Yanai and Rossby waves. The orthogonality means that each mode may be
treated in isolation, thus grossly simplifying the dynamics. Using this
approach, results for sea level and zonal current along the equator agree
well with the available observations. Arguments linking SST anomalies
with anomalous zonal and vertical advection can be used to infer the SST,
again with good agreement. The long-wave approximation also permits more
rapid numerical integration of models, and this has been used to study
how the transmission and reflection of equatorial waves are affected by a
pre-existing east-west sloping thermocline (53).

The long-wave model has also been employed to study the initialisation of equatorial models, using sparse observations (2). Since large-
scale observations in TOGA are imminent, initialisation studies are vital
if we wish to produce models which can accept data in real time and make
predictions like atmospheric forecast models.

The interaction of the ocean back on the atmosphere has been studied
(10, 23, 8). A moist equatorial atmospheric model has been run in
shallow-water form to study the baroclinic response of the tropical atmosphere. Because moisture is included explicitly, latent heat effects and
precipitation can be taken into account. The forcing is usually taken as
a given sensible heating of the same pattern as the underlying SST. Realistic wind and precipitation patterns were produced. Results turn out to
be sensitive to the saturation moisture level. If this is increased, more
intense precipitation occurs on smaller scales, together with irregular
oscillations, similar to nonlinear phase changes in coupled dynamical
systems.

8. Modelling and numerical techniques

A number of effects have been found to depend on features of the
model or even of the numerical scheme used to solve the model, so that it
is important to understand how these effects occur. An important feature
of the transient response found in large-scale ocean simulations has been
the coastal Kelvin wave. In large-scale simulations, this tends to flat-
ten the isopycnals along a coast, but it has not been clear how much of
this is due to poor numerical resolution. The structure of these in
modelling terms has been studied (5), in particular how the wave speed,

decay time and offshore structure depend on the (artificial) frictions
and diffusions used in the models. Such frictions are often rather high
in order to maintain numerical stability. Wave speeds can be reduced
markedly, with barotropic waves principally affected by bottom friction
and baroclinic waves by lateral viscosity. Finite-difference effects were
also investigated (32). It was found, for example, that the speed of
inviscid waves is increased by poor resolution on one type of finite
difference grid but unchanged on another.

While running the GFDL model, it was found that the barotropic part
of the solution, which formally should remain small in the absence of
wind driving, was doubling every ten model days. The form of this errone-
ous solution resembles a 'basin mode', i.e. a natural oscillation of the
system on a rotating earth. A detailed study of this was made (43), as
the effect has plagued other researchers. This shows how the combination
of extrapolation and relaxation for the barotropic stream function yield
a gradually growing numerical mode which equilibrates eventually at a
magnitude which is proportional to the truncation criterion in the relaxation scheme, and also depends on the amount of bottom and lateral friction present in the scheme, being especially damped by bottom friction.
Finer resolution, which permits lower viscosities, makes the problem more
pronounced. Some methods for overcoming the problem have been produced.

Recent work mentioned above has shown that a similar type of instability may affect the barotropic mode if islands are present, owing to a
different relaxation scheme being used for such regions. This possibility
is being investigated currently.

The use of large-scale circulation models for the study of climate
involves a great deal of computer time. On climatic scales one is seldom
worried about the short-time scale oceanic response, which is conveniently usually combined with short length scales. Methods for speeding
up the existing models have also been studied (52). This study began by
examining the possibility of rescaling the time derivative in the momentum equations so as to allow a larger time step than is normally allowed
by numerical stability requirements. This alters the operant physics, but
not in a damaging way. However, some new numerical restrictions come into
play, and the study ended up by listing all known numerical constraints
and the new ones discovered by the study. Models with poor horizontal
resolution, as currently used to study world climate, can benefit from
the timestep rescaling; finer resolution models cannot. In the latter
case only a redefinition of the physics can lead to decreased computer
time.

Other studies have sought to compute the oceanic circulation given
its observed density structure (the 'inverse' problem). One method (37)
assumes geostrophy and the hydrostatic and mass balances, and seeks solutions with least noise in the buoyancy equation. It turns out that this
method, given only density on a north-south section, can derive the
best-fit three-dimensional velocity field on that section, as well as an
estimate of the density gradient normal to the section. Calculations for
the North and South Atlantic yielded quite realistic solutions on sections ranging from 10 to 50° latitude.

Similar dynamics allow the use of the 'beta spiral' method introduced some years ago (61). A detailed analysis of the accuracy of this
method for deducing velocity structure has been lacking. By using computer generated velocity and density fields from the GFDL program, it has
been possible (1) to investigate how dependent the solutions to the beta
spiral method are on the diffusivities assumed for the problem. Other,
more crude, methods are also being studied: a box model of the North and
South Atlantic has been devised for studies of nuclear waste disposal at

sea (54). In this, best-fit transports between neighbouring boxes were computed to match the observed temperature and salinity structure as closely as possible. Once the transports have been determined, one may then inject tracers into the system and analyse their behaviour extremely cheaply and rapidly compared with more usual computer models.

REFERENCES TO CAMBRIDGE GROUP WORK

1. BIGG G.R. (1984) On the beta spiral method. (submitted to Deep-Sea Research).

2. BIGG G.R. and A.E. GILL (1984) Preliminary results from a search for equatorial long waves (in preparation).

3. BLUNDELL J.R. and A.E. GILL (1983) Equatorial ocean response to wind forcing. Ocean Modelling No. 53.

4. DAVEY M. K. (1983) A two-level model of a thermally-forced ocean basin. J. Phys. Oceanogr. 13, 169-190.

5. DAVEY M.K., W.W. HSIEH, and R.C. WAJSOWICZ (1983) The free Kelvin wave with lateral and vertical viscosity. J. Phys. Oceanogr. 13, 2182-2191.

6. DAVEY M.K. and P.D. KILLWORTH (1984) Isolated waves and eddies in a shallow-water model. J. Phys. Oceanog., to appear.

7. DAVEY M.K., A.E. GILL, E.R. JOHNSON and P.F. LINDEN (1984) Steady flow over a step perpendicular to a vertical boundary. Ocean Modelling No. 58.

8. DAVEY M.K. (1984) Results from a moist equatorial atmosphere model. To appear in Proceedings of 16th International Liege Colloquium on Ocean Hydrodynamics, J.C.J. Nihoul (ed.), Springer-Verlag (to be published).

9. DICKSON R.R., W.J. GOULD, P.A. GURBUTT and P.D. KILLWORTH (1982) A seasonal signal in ocean currents to abyssal depths. Nature 295, 193-198.

10. GILL A.E. (1982) Studies of moisture effects in simple atmospheric models: the stable case. Geophys. Astrophys. Fluid Dyn. 19, 119-152.

11. GILL A.E. (1982) Spontaneously growing hurricane-like disturbances in a simple baroclinic model with latent heat release. In: "Intense Atmospheric Vortices: Proceedings of the Joint Symposium held at Reading (UK) July 1981". L.Bengtsson & M.J.Lighthill (eds). pp.111-129. Springer-Verlag, Berlin.

12. GILL A.E. (1982) Atmosphere-Ocean Dynamics. Academic Press, New York. 662 pp.

13. GILL A.E. (1982) Changes in thermal structure of the equatorial Pacific during the 1972 El Nino as revealed by bathythermograph observations. J. Phys. Oceanogr. 12, 1373-1387.

14. GILL A.E. (1983) Eddies in relation to climate. Chapter 19 in "Eddies in Marine Sciences", A.R.Robinson (ed.). Springer-Verlag, Berlin. 609pp.

15. GILL A.E. (1983) An estimate of sea-level and surface-current anomalies during the 1972 El Nino and consequent thermal effects. J. Phys. Oceanogr. 13, 586-606.

16. GILL A.E. (1983) Patterns of interannual variability associated with the ocean and the atmosphere. in: "Large-scale Oceanographic Experiments in the WCRP. CCCO/JSC Study Conference, Tokyo, May 1982". Vol.II, 57-75, WCRP Publ. Ser. No.1. WMO/ICSU, Geneva.

17. GILL A.E. (1983) Report of the JSC/CCCO Study Group on Interannual Variability of the Tropical Oceans and the Global Atmosphere (TOGA). (Ed.) WCRP/ICSU/IOC/WMO, Geneva, WCP-49, July 1983.

18. GILL A.E. and E.M. RASMUSSON (1983) The 1982/3 climate anomaly in the equatorial Pacific. Nature 306, 229-234.

19. GILL A.E. (1983) Thermal anomalies in the central Pacific during the 1972 El Nino. Ocean Modelling No. 48

20. GILL A.E. (1983) The 1982/83 warm event in the Pacific. Ocean Modelling No. 49.

21. GILL A.E. (1984) On the behaviour of internal waves in the wakes of storms. J. Phys. Oceanogr. 14, (July issue).

22. GILL A.E. (1984) Walter, Aristotle and the tides of Euripus. In "A Celebration of Geophysics and Oceanography - 1982. In honour of Walter Munk", SIO Ref. Series 84-5, pp.96-99.

23. GILL A.E. (1984) Elements of coupled ocean-atmosphere models for the tropics. To appear in Proceedings of the 16th International Liege Colloquium on Ocean Hydrodynamics. J.C.J. Nihoul (ed.) Springer-Verlag (to be published).

24. GILL A.E. (1984) Coastal and equatorial waveguides and the 1982/83 El Nino event. Springer-Verlag (to be published).

25. GILL A.E. (1984) An explicit solution of the linear thermocline equations (submitted to Tellus).

26. GREATBATCH R.J. (1983) On the response of the ocean to a moving storm: the nonlinear dynamics. J. Phys. Oceanogr., 13, 357-367.

27. GREATBATCH R.J. (1984) On the response of the ocean to a moving storm: parameters and scales. J. Phys. Oceanogr., 14,59-78.

28. GREATBATCH R.J. (1984) On the response of the ocean to a moving storm: temperature response. J. Phys. Oceanogr. (submitted).

29. GRIFFITHS, R.W., P.D. KILLWORTH and M.E. STERN (1982) Ageostrophic instability of ocean currents. Journal of Fluid Mechanics 117, 343-377.

30. GRIFFITHS R.W. and A.E. GILL (1982) Why should two anticyclonic eddies merge? Ocean Modelling No. 41.

31. HECKLEY W.A. and A.E. GILL (1984) Some simple analytical solutions to the problem of forced equatorial long waves. Quart. J. Roy. Met. Soc. 110, 203-217.

32. HSIEH W. W., M.K. DAVEY, and R.C. WAJSOWICZ (1983) The free Kelvin wave in finite-difference numerical models. J. Phys. Oceanogr. 13, 1383-1397.

33. HSIEH W.W. and A. E. GILL (1984) The Rossby adjustment problem in a rotating stratified channel with and without topography. J. Phys. Oceanogr. 14, 424-437.

34. KILLWORTH P.D. and E.C. CARMACK (1982) Arctic Ocean modelling. Report by Comite Arctique and SCOR WG58 meeting, Cambridge, June 1982; University of Bergen Technical Report.

35. KILLWORTH P.D. and M.E. STERN (1982) Instabilities on density-driven boundary currents and fronts. Geophysical and Astrophysical Fluid Dynamics 22, 1-28.

36. KILLWORTH P.D. (1983) Deep convection in the world ocean. Reviews of Geophysics and Space Physics 21, 1-26.

37. KILLWORTH P.D. (1983) Absolute velocity calculations from single hydrographic sections. Deep-Sea Research 30, 513-542.

38. KILLWORTH P.D. (1983) On the motion of isolated lenses on a beta-plane. Journal of Physical Oceanography 13, 368-376.

39. KILLWORTH P.D. (1983) Long-wave instability of an isolated front. Geophysical and Astrophysical Fluid Dynamics 25, 235-258.

40. KILLWORTH P.D. (1983) Some thoughts on the thermocline problem. Ocean Modelling No. 48.

41. KILLWORTH P.D. (1983) Lecture notes, GFD Summer School, Woods Hole Oceanographic Institution, Woods Hole, Mass., U.S.A.

42. KILLWORTH P.D. and J.M. SMITH (1984) A one-and-a-half dimensional model for the Arctic halocline. Deep-Sea Research 31, 271-294.

43. KILLWORTH P.D. and J.M. SMITH (1984) Gradual instability of relaxation-extrapolation schemes. Dynamics of Atmospheres and Oceans, 8, 185-213.

44. KILLWORTH P.D., N. PALDOR and M.E. STERN (1984) Wave propagation and growth on a surface front in a two-layer geostrophic current. Journal of Marine Research (in press).

45. KILLWORTH P.D. (1984) Comment on 'The level-of-no-motion in an ideal fluid' Journal of Physical Oceanography 14, 213.

46. KILLWORTH P.D. and N. PALDOR (1984) A model of sea-ice front instabilities. J. Geophys. Res. (Oceans) (in press).

47. KILLWORTH P.D. (1984) A two level wind- and buoyancy-driven thermo-cline model. Journal of Physical Oceanography (submitted).

48. KILLWORTH P.D. and M.K. DAVEY (1984) Parameter regimes for studying isolated eddies. Journal of Physical Oceanography (submitted).

49. KILLWORTH P.D. and M.E. MCINTYRE (1984) Does a nonlinear Rossby wave critical layer absorb or reflect? (to be submitted)

50. KILLWORTH P.D. (1984) Spin-up of a thermally-driven ocean (in preparation).

51. KILLWORTH P.D. (1984) A three-level thermocline model driven by wind and buoyancy.

52. KILLWORTH P.D., J.M. SMITH and A.E. GILL (1984) Speeding up ocean circulation models. Ocean Modelling No. 56.

53. KING B.A. (1984) Ph.D. Thesis, University of Cambridge (in prepara-tion).

54. MOBBS S.F., P.A. GURBUTT, M.D. HILL, J.G. SHEPHERD, and P.D. KILL-WORTH (1984) Development of an ocean model for use in radiological assessments of sea disposal of solid radioactive wastes. (to be submit-ted)

55. WAJSOWICZ R.C. (1984) Adjustment of the ocean under buoyancy forces. Ph. D. Thesis, University of Cambridge.

OTHER REFERENCES

56. GRIFFITHS R.W and P.F. LINDEN (1982) Laboratory experiments on fronts. Part I. Density-driven boundary currents. Geophysical and Astro-phys. Fluid Dyn., 19, 159-187.

57. LE GROUPE TOURBILLON (1983) The Tourbillon experiment: a study of a mesoscale eddy in the eastern North Atlantic. Deep-Sea Research, 30A, 475-512.

58. LUYTEN J.R., J. PEDLOSKY and H. STOMMEL (1983) The ventilated thermo-cline. J. Phys. Oceanogr., 13, 292-309.

59. MCINTYRE M.E. and T.N. PALMER (1983) Breaking planetary waves in the stratosphere. Nature, 305, 593-600.

60. RHINES P.B. and W.R. YOUNG (1981) A theory of the wind-driven ocean circulation I. Mid-ocean gyres. J. Marine Res., 40 (suppl.), 559-596.

61. SCHOTT F. and H.STOMMEL (1978) Beta spirals and absolute velocities in different oceans. Deep-Sea Res., 25, 961-1010.

62. WORTHINGTON L.V. (1970) The Norwegian Sea as a mediterranean basin. Deep-Sea Research 17, 77-84.

SUBPOLAR CIRCULATION, DEEP WATER FORMATION AND AIR – SEA – ICE
INTERACTIONS IN LABRADOR AND GREENLAND SEAS

J. C. GASCARD
Laboratoire d'Océanographie Physique
Muséum National d'Histoire Naturelle PARIS

Summary

The growth, drift and decay of the sea ice margins are closely related to the
dynamics of the subpolar oceanic regions. This is especially true in the
Greenland and Labrador Seas where warm northward currents (Norwegian
and Irminger currents) encounter rapidly cooling atmospheric conditions
and southward advancing sea ice and cold currents (East Greenland and
Labrador currents). Location of the ice margin is partly due to the large
oceanic heat flux from the deeper ocean which occurs primarily in winter
when deep convection brings up warm water from below and prevents ice
formation in early winter. After an historical data set analysis of Greenland
and Labrador Seas hydrological characteristics, we are presenting some
recent results about oceanic deep convection events and mesoscale
dynamics close by an ice edge which emphasized the role of the ocean in
air – sea – ice interactions.

1. Introduction

The heat flux from the ocean to the atmosphere and the meridional oceanic
heat and salt transport are key problems for the understanding of earth climates.
From this point of view, the North Atlantic subpolar circulation is quite important
simply because in the northern hemisphere, this circulation represents the most part
of the total exchange between polar and subtropical oceanic areas.
This circulation is characterized by:
– The North Atlantic current and the North East Atlantic drift, branching into
Norwegian and Irminger currents, and carrying on subtropical central water masses
warm and salty towards high latitudes.
– The Greenland and Labrador currents carrying on fresh and cold polar waters
towards low latitudes.
The horizontal circulation of these water masses is largely interacting with
deep convection occurring in some part of the gyre under winter meteorological
conditions and resulting in deep water formation. Our research program aimed to
estimate meridional heat and salt fluxes between the North Atlantic and the Arctic
Ocean and the contribution of oceanic deep convection in Labrador and Greenland
seas to the heat flux towards the atmosphere. Since both convection and circulation
occur there in close connection with sea ice formation (in particular in the marginal
zone) we have to deal strongly with full interactive processes between ocean – ice and
atmosphere in order to find right order estimation of these fluxes and, more
important, how they vary and what are reasons for their variability.
In the following, we are presenting our studies and progresses during the

last three years according to three distinct types of research on the subject:
- Statistical analysis of historical data: preliminary results on θ–S characteristics of water masses within Labrador and Greenland Seas.
- Study of deep convection in Labrador Sea.
- Study of Air – Sea – Ice Interactions : the MIZEX Experiment.
For each topic, we will present separately materials and methods used, main results obtained, conclusions and suggestions for future research.

2. Results:

2.1 Statistical Analysis of water masses historical data:

A large set of stations carried out from the beginning of the century up to now have been analyzed for two domains belonging to the North Atlantic Subpolar circulation: The Norwegian Sea (NS) (about 6700 "historical" stations) and the Labrador Sea (LS) (about 7500 "historical" stations). The spatial coverage of these two domains is : 70°N–80°N and 20°E–20°W for NS, 50°N–70°N and 40°W–70°W for LS. To look at the space-time distribution of the stations, they have been gathered into a limited number of geographical zones (48 for NS and 108 for LS), each of them covering an area of 18 to 40. 10^3 Km², and also distributed depending on the season (01-01 to 31-03, 01-04 to 30-06, 01-07 to 30-09, 01-10 to 31-12), whatever the year they were made.

2.1.1. Space-time distribution of stations

a/ The Norwegian Sea

Highest concentrations of stations are encountered along the Eastern boundary, near the coast and also in the Lofoten Basin. Lowest concentrations are present in the central part of the Greenland Basin, one of the most interesting area of the domain where the cyclonic circulation of an important volume of quasi-homogeneous water takes place during the whole year and makes this region very sensitive to the surface cooling in winter. The boundary currents of the Norwegian Sea (the East Greenland current (EGC), the West Spitzbergen current (WSC) and the Norwegian current (NC)) are much better sampled, both over the continental shelf and beyond it. Finally, the seasonal distributions seem to follow the ice distribution in the region: concentrations are minimum in winter when ice covers the major part of the domain and maximum in summer when ice retreats towards the Northeast.

b/ The Labrador Sea

The same remarks can be made for this region where concentrations are higher on the eastern and western continental shelves and where the influence of the seasonal cycle and of the fluctuating ice cover are prominent in the time evolution of the concentrations. Besides, there is a zone, centered on the OWS BRAVO (56°30'N–51°W), which contains as much as 2064 stations for the whole period, due to multi-year routine sampling at this point.

2.1.2. Some mean hydrological features. (Fig. 1a and 1b)

At each station, the observed hydrological properties T, S and O² were considered, at any given standard depth, as random variables whose different realizations are values obtained at each of the stations belonging to the same zone season. After a statistical validation of the samples thus constructed (Ferguson test), a mean station, for each zone season, was obtained by averaging all the values of a same sample.

As an example, we discuss some results mainly related to summer conditions. Two different θ–S distributions characterize the eastern and western boundaries of both the NS and LS, and this difference appears in a similar way in the two domains:

– On the western boundary of the NS, in the EGC, a strong halocline characterizes the polar surface layer (0–200m) with very low temperatures increasing with depth up to 0°C when reaching the Atlantic water layer. Then temperatures decrease rapidly in the deep layer. In the LS, the Labrador current (LC) exhibits a rather similar θ–S relationship except for a global shift towards higher temperatures.

– On the eastern side of the NS, the Atlantic water in the WSC is in direct contact with the atmosphere. Consequently, the surface temperature is relatively high (often > 2oC) and the vertical salinity gradient in the first 200 m layer is not as strong as in the EGC. Nevertheless, the weakening of the stratification during winter in the WSC is mainly induced by the disappearing of this halocline and not by a decrease of the surface water temperature which does not seem very sensitive to the atmospheric cooling. In the eastern LS, the structure of the West Greenland current beneath 200 m is very similar to the one described in the WSC but, in the surface layer, there is still a strong halocline due to the presence of polar water which recirculates around the southern tip of Greenland passing by the Cape Farewell and then flowing towards the North along the Western coast of Greenland.

– Finally, the θ–S relationship in the central part of the two basins reduces to very tight values due to the high homogeneity of the water involved in the cyclonic gyre of the basins.

In conclusion, we plan to build an atlas of vertical sections and horizontal distributions of the observed characteristics of the water masses (T, S, O^2) and some derived quantities (potential temperature and density, dynamic height) by an objective analysis of the "reconstructed mean" fields. These maps will be used as reference real fields for comparison with the results of the subpolar circulation modeling which is one of our current work, and will help to the understanding of the new data set collected in this area (mainly Greenland Sea) during the MIZEX 84 Experiment.

2.2 Deep Convection Studies

The estimation of the contribution of oceanic deep convection to the heat flux towards the atmosphere is seriously affected by the small extent (both in time and space) of the areas where and when the convective processes do occur. This is one of the most striking features of the phenomenon. From what we know, after having observed it in the Labrador Sea in 1976 and 1978 (Clarke and Gascard, 1983; Gascard and Clarke, 1983) the strong areal limitation comes from a very constraining preconditioning phase which requires three conditions have to be satisfied simultaneously, namely: a warm and salty subsurface water masses have to get trapped in a tight cyclonic gyre (200 Kms – 1 month) together with surface waters. This gyre is clearly identifiable because it produces a characteristic doming structure of the isopycnals. Within this gyre, intense cooling and evaporation at surface, controlled by the atmosphere, modify quite deeply the vertical density structure, ultimately leading to the formation of deep waters. Inside the gyre, which appears in general on the offshore side of a boundary current, various features are observed such as:

– unstable mesoscale features (50 kms – 1 week)

– eddies (20 kms – days) resulting from a baroclinic instability of the mesoscale structure.

– convective cells with vertical and horizontal scales quite comparable (1 km – hours).

All these features have been identified in the Labrador Sea in March 1976 in a region centered around 57oN–54oW leading to the formation of homogeneous waters from surface down to 2 kms depth (Fig. 2, 3 and 4).

The baroclinic instability is a powerful mechanism for preconditioning an area to a deep mixing because this mechanism forces the warm and salty subsurface layer to come up to the surface. In doing so, this layer is first dragged mechanically in a mixing process with the surface layers and second and more important, it is then exposed thermodynamically to the atmosphere where it releases heat and water vapour (increasing its salinity). The mechanical effect (due to the instability) converts the potential energy: this is the spreading phase. The thermodynamical effect creates potential energy since the surface water masses increase in density.

Most results concerning the Labrador deep water formation are strikingly comparable with the western Mediterranean deep water formation in the so-called "Medoc" area (Gascard, 1978). The horizontal scales involved in the Labrador Sea are twice the "Medoc" ones. So, in one month of the winter season, the Labrador Sea can form 10^5 km³, that is 4 times more deep water than the Mediterranean Sea one, since the mixing depth is comparable in both cases (2 kms). In both regions, we have been estimating evaporation rate of about 2 cm/day during deep water formation. This corresponds to a surface heat loss of approximately 400 W. m^{-2}.

One of the most important questions to be addressed now is how the deep convection is coupled to the general circulation of the ocean? We know that the "normal" conditions which lead to deep water formation can be strongly disturbed by persistent large scale atmospheric anomaly. It looks like the mesoscale features are related to topographic Rossby waves propagating offshore from the continental slope region and interacting positively with the upper layer through the main pycnocline which separates the Labrador Sea water from the bottom layer. Other important questions remain to be explored concerning the way the cyclonic gyre might be set up, the effect instabilities generated at the edge of the Labrador current might have on cooling the water to the states found in March 1976, the effects, if any, that the Labrador ice pack might have on the processes. Besides Labrador Sea, two regions are also quite relevant for such a study: the Greenland and Weddell gyres.

2.3 Air- Sea - Ice Interaction Studies

The exchanges which take place near the open ocean boundaries of polar ice fields and which determine the advance and retreat of the sea ice edge affect climates on the global scale. The overall problem of understanding the annual and interannual variability of the polar ice margins, and of relating these to the large-scale behavior of the atmopheric and ocean circulations is to be addressed by a long term monitoring and modeling program. Complementary to this program, there was a need for a mesoscale experimental program to study physical processes occurring within the Marginal Ice Zone (MIZ), and to develop models of these processes. This is known as the Marginal Ice Zone EXperiment (MIZEX).

For logistic reasons, it was decided that this program should take place first in the Fram Straits, north and west of Svalbard, which handles most of the heat, salt and water exchanges between the Arctic Ocean and the rest of the world. Obviously, this experiment, scheduled in June-July 1984, is not going to give us, at once, all the answers to a very complex problem and there are already other complementary programs being studied in other places (Bering Sea), and other seasons (winter time), since we know already there is a great variability both in space and time of physical processes controlling the MIZ.

In the context of the MIZEX experiments, we have centered our efforts on two aspects. One concerned the oceanographic conditions met in the MIZ, the other concerned the developement and use of remote sensing techniques which are essential for monitoring globally sea - ice on a long run. The Fig. 5 is showing conditions we met in the Fram Straits during the pre-MIZEX experiment in July 1983. This region is characterized by permanent and transient frontal systems, eddies, and upwelling events along the ice edge and mesoscale structures formed by interleaving of Polar and Atlantic water intrusions. Salinity fronts off the ice edge develop strongly during Summer due to melt water input; eddies along the ice edge shed ice off into warmer water, thereby providing an ice export mechanism. So there is a strong coupling between the ice pack and the oceanic mixed layer.

On Fig. 5 we see clearly thermal fronts such as the East Greenland Polar front which separates the cold low salinity southward-flowing East Greenland current from the more saline warm water in the Greenland Sea. A surface drifter we deployed on the 10th of July in this region, was tracked by satellite via the Argos system until the 11th of August. The one month drift showed the float following polar Atlantic frontal features. It indicated eddy-like structures of about 30 kms and 5 days period and meanders of 80 kms with a 10 days period. The sea-ice is embedded further away into polar waters and can be observed by different kinds of remote-sensing techniques. The nature of the MIZ is quite different from the interior pack, mainly because it is broken up by waves and swell into floes which are about 30 m diameter, and also because the interaction with the nearby turbulent oceanic mixed layer. Many problems are known to exist in summer, mainly due to snow melt, when one attempts to use remote sensing techniques for observing sea-ice.

During the MIZEX 84 experiment, we operated three types of radar mounted separately on ship, helicopter and airplane to improve our ability in using remote sensing for summer observations of sea-ice in the MIZ and to get synoptic pictures of the ice field at the mesoscale (100 x 100 kms). At the same time, we deployed 10 surface drifters and 10 underwater floats equipped with thermistors, to map objectively the oceanic currents (mainly the Atlantic current) at surface and at depth and the sea ice motions, to get informations about variability and mean components of motions and their dependance on atmospheric forcing, bottom topography, frontal structures (3D)...

Three types of floats and buoys were launched depending on their use at surface or at depth:

at depth: 10 acoustic floats like SOFAR floats. They are freely drifting Swallow floats ballasted in order to be neutrally buoyant at preselected depths : 4 floats stabilized around 110 to 130 m depth and 6 around 200 to 260 m. The 10 floats were launched from the POLARQUEEN (ship or helo) between 80°14,4N - 80°43,5N and 3°30E - 6°05,5E from June 10 until June 20. Each float, equipped with a 1562 Hz transducer, sent a pulse every hour for tracking and was telemetering temperature and pressure in situ once every second day. Five listening stations received floats signals, and floats locations were computed by considering time of arrival and position of listening stations. Three stations were operating in real time from three different ships: POLARQUEEN, POLARSTERN and KVITBJORN, and two autonomous listening stations, F1 and F2, were deployed, at 700 m depth, from ice floes tagged with Argos buoys and tracked by satellite. Later on (post-MIZEX cruises), two listening stations were mounted first on POLARSTERN (late July-early August) and later on LANCE (late August) extending the float tracking over a one month period in addition to the 50 days MIZEX period.

As a preliminary result, we can tell that we have been able to map some features of Atlantic currents in the Fram Straits, below the ice, during this summer 84 season: four features are already appearing related to the branching of the northward Atlantic current towards the southward East Greenland current:

   - northward Atlantic current flowing along the continental slope West of Yermak plateau and entering in the Arctic basin.
   - strong topographic features above Yermak plateau: one float has been trapped during the whole MIZEX Experiment above the 500 m isobaths Northwest of Svalbard.
   - recirculation of Atlantic waters west of the Yermak plateau along the continental slope and in the deepest part of the Straits.
   - southward flowing Atlantic current trapped in the East Greenland current.

In addition to this experiment, we launched three VCMs (Vertical Current Meter) from the H. MOSBY close to the ice edge. Like the other floats, properly ballasted, they become neutrally buoyant at some depth. They are equipped with vanes around the main axis of the cylinder, 1.5 m high. Then they rotate when drifting according to the vertical component of motion (if any). We had a lot of trouble during this deployment for two reasons: float depth equilibrium and listening equipment too sensitive according to ship noise due to variable pitch. Nevertheless, after quite a bit of work, we succeeded in deploying two floats, at 200 and 700 m depth, in the Molloy deep area and one further Southwest along the ice edge, at about 78°40N and 1°W. Later on, we were not able to recover the 700 m float because of a failure in the release mechanism. The float at 200 m, equipped with a SOFAR 1562 Hz acoustic (in addition to its proper 5 kHz short range acoustic tranducer) was tracked during 11 days above the Molloy deep. Launched on July 4, 30 miles away from the ice edge, at approximately 79°15N – 3°E, it was recovered on July 15 by the KVITBJORN at about 79°12N – 3°30E, among thick ice floes. During these 11 days, this float described an anticyclonic eddy trajectory with a 10 miles radius of curvature, more or less similar to the one described by the 700 m float. The third VCM, deployed along the ice edge, followed a track parallel to the ice edge during 2 days (July 10 to 12) at a depth around 125 m. It has been drifting Southwest at a mean speed of 20 cm. s^{-1} and was recovered at about 78°35N – 2°W in loose ice.

at surface: Other drifters were deployed, both on the ice (7 drifters) and in the nearby ocean (7 drifters). They used Argos transmitter for location via satellite. Ocean drifters used temperature sensor for measuring sea surface temperature within 0.1°C accuracy.

Initially, three of them were deployed in the Ocean, at the end of May from the LYNCH, two above the Molloy deep, at about 79°20N – 3°E, close to the ice edge, and one West of Svalbard, at 79°N – 8°E, in the West Spitzbergen current. This last buoy remained more or less in the same area for almost the whole MIZEX period, going back and forth, North and South, and describing a lot of loops. At the end, it started drifting rapidly northwards and disappeared after passing by the Northwest corner of Svalbard. The two other floats remained in the Molloy deep area but did not last more than two weeks. Early June, 4 buoys were deployed in the Ocean, two from KVITBJORN and two by helos from POLARQUEEN, few miles away from the ice edge. One of these failed rapidly for unknown reasons, two others, deployed from KVITBJORN at 80°20N–7°E and 80°30N–8°E, lasted only few weeks, while the last one is still working after three months. On the average, it has been drifting Southwest along the ice edge from about 79°45N–7°E (July 6) to 74°45N –8°30W (September 18). During the first month, this buoy remained at approximately 79°30N – 5°E, trapped in eddy motions east of the Molloy Deep.

Seven Argos drifters were deployed on ice floes, in order to follow the drift of ice stations: C2, C3, F1 and F2, and later on, at a currentmeter station with Bergen toroid, which was recovered by POLARSTERN during post–Mizex. All these Argos buoys deployed on the ice floes worked perfectly during the whole MIZEX period. The recovery of these stations was not that simple and we are quite grateful to the pilots and chief scientists on POLARQUEEN for their patience and great help.

In conclusion, we are generally quite satisfied with our results considering our initial planning. The SOFAR floats program has been accomplished to nearly 100% and we are quite pleased with the offer of our German colleagues on board POLARSTERN and our Norwegian colleagues on board LANCE for extending successfully the float tracking from late July to late August. Due to the range limitation (100 miles) in sound propagation, related mainly to ice coverage and bottom topography, we have not been able to receive all signals from SOFAR floats at each listening station, but since our 5 stations were redundant, we hope to recover most of the data for float tracking computations.

PUBLICATIONS

GASCARD J.C. (1981).- North Atlantic Subpolar circulation and Deep Water Formation in Labrador and Greenland seas. C.E.C. Contact Group "Climate Models". Extended Abstracts of Presentations at the first meeting, Brussels, 26-27th May 1982.

HOUSSAIS M.N. and J.C. GASCARD (1982).- Quelques éléments sur la validation des données appliquée à la mer du Groenland et à la mer du Labrador. C.E.C. Contact Group "Climate Models". Abstracts of Presentations at the second meeting. Brussels, 11th May 1982.

GASCARD J.C. (1982).- Deep Convection in the Ocean and the general circulation. Summer School of Space Physics, Space Oceanography, Grasse (France), 1-28th July 1982.

CLARKE R.A. and J.C. GASCARD (1983).- The formation of Labrador sea Water. Part I: Large Scale Processes. J. Phys. Oceanogr., 13, 1764-1778.

GASCARD J.C. and R.A. CLARKE (1983).- The Formation of Labrador sea Water. Part II: Mesoscale and Smaller scale Processes. J. Phys. Oceanogr., 13, 1779-1797.

HOUSSAIS M.N. (1983).- Premiers résultats du traitement statistique des fichiers de données historiques de la mer de Norvège et de la mer du Labrador. (unpublished manuscript).

COMMUNICATIONS

GASCARD J.C. (1981).- North Atlantic subpolar circulation and Deep Water formation in Labrador and Greenland seas. C.E.C. Contact Group "Climate Models", Brussels 26-27th May 1981.

GASCARD J.C. (1981).- Open Ocean deep convection. NATO Advanced Research Institute on "Large Scale Transport of Heat and Matter in the Ocean". Bonas (France) 22-30th September 1981.

HOUSSAIS M.N. and J.C. GASCARD (1982). Quelques éléments sur la validation des données appliquée à la mer du Groenland et à la mer du Labrador. C.E.C./Contact Group "Climate Models", Brussels, 11th May 1982.

GASCARD J.C. (1982).- Deep Convection in the Labrador Sea. SCOR Working Group 58 on "Arctic Ocean Modelling Meeting". Cambridge (G.B.) 21-25th June 1982.

GASCARD J.C. (1982).- Deep Convection in the Ocean and the general circulation. Summer School of Space Physics: Space Oceanography, Grasse (France), 1-28th July 1982.

GASCARD J.C. (1983).- Labrador sea Water Formation. I.U.G.G. XVIII General Assembly, Hamburg, 15-27th August 1983.

GASCARD J.C. (1983).- Preliminary Results on the French MIZEX 83 Test Experiment: Lagrangian description of currents in the MIZ. MIZEX Science Group Meeting, Bremerhaven (RFA), 19-21th November 1983.

Fig. 1 (a et b) . Seasonal mean temperatures and salinities at 10m depth
in the Norwegian Sea extracted from a validated historical data set.

MER DE NORVEGE

TEMPERATURE EN °C
10 mètres

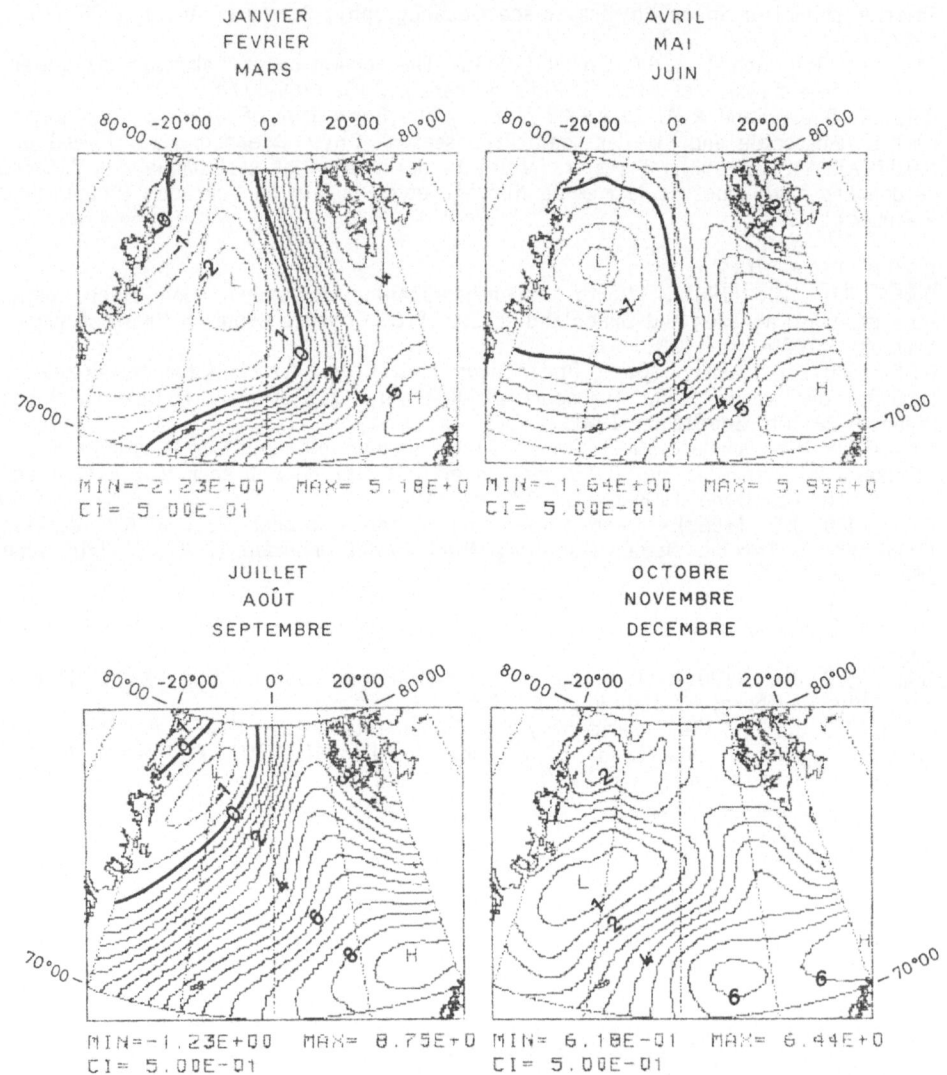

Fig. 1a

MER DE NORVEGE

SALINITE ‰
10 mètres

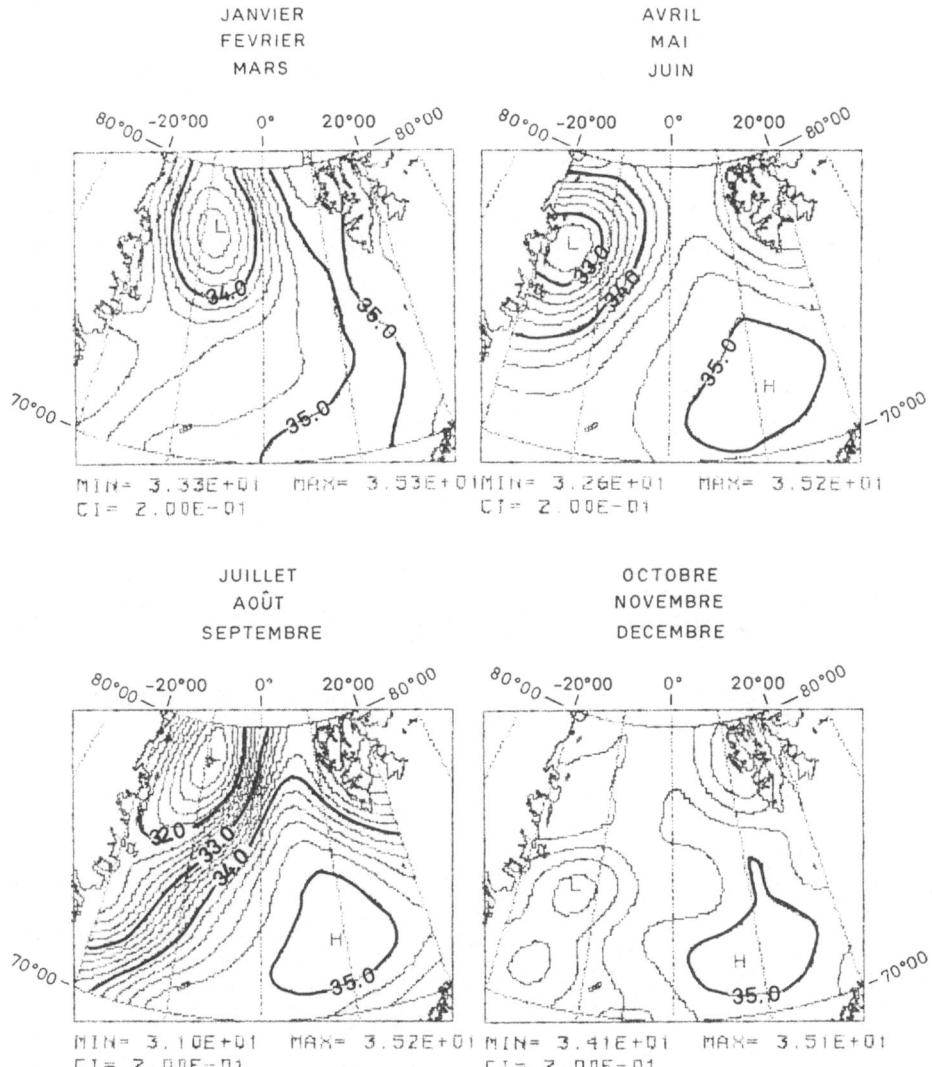

Fig. 1b

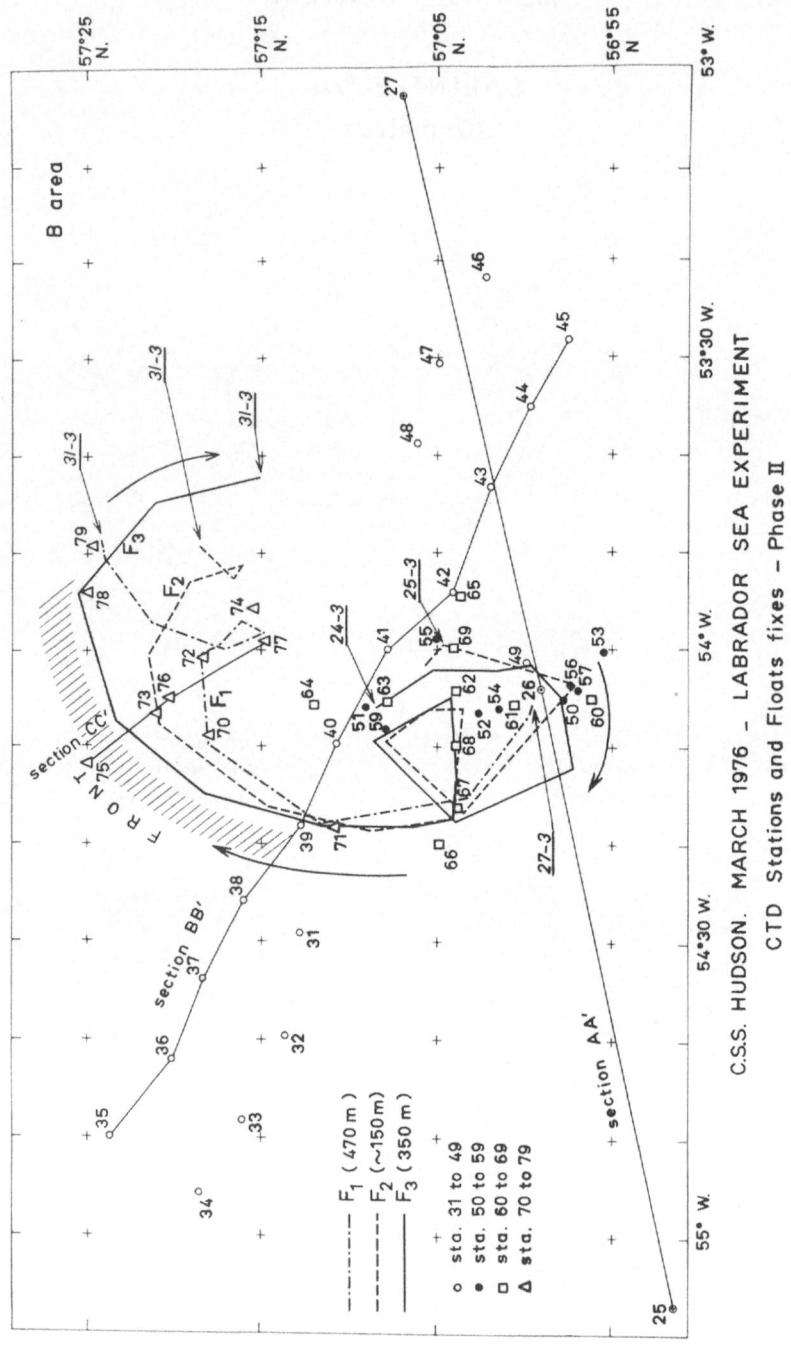

Fig. 2 Trajectories of three vertical current meters "VCM" tracked in the western Labrador Sea, March 1976, along with positions of CTD Stations occupied at the same time.

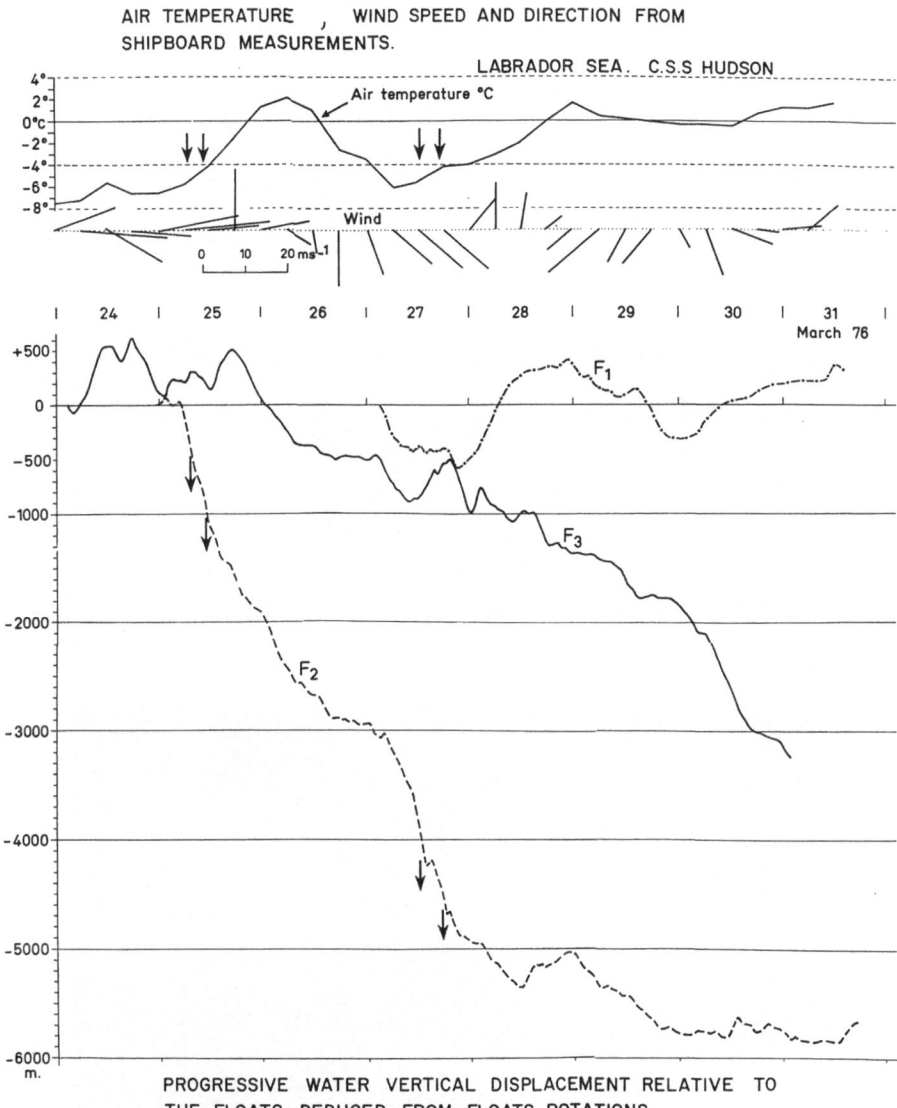

AIR TEMPERATURE , WIND SPEED AND DIRECTION FROM
SHIPBOARD MEASUREMENTS.

LABRADOR SEA. C.S.S HUDSON

Air temperature °C

Wind

0   10   20 ms^{-1}

PROGRESSIVE WATER VERTICAL DISPLACEMENT RELATIVE TO
THE FLOATS DEDUCED FROM FLOATS ROTATIONS.

Fig. 3   Integrated vertical flow past each of the VCMs. Wind and air
temperature as measured every six hours on CSS Hudson.

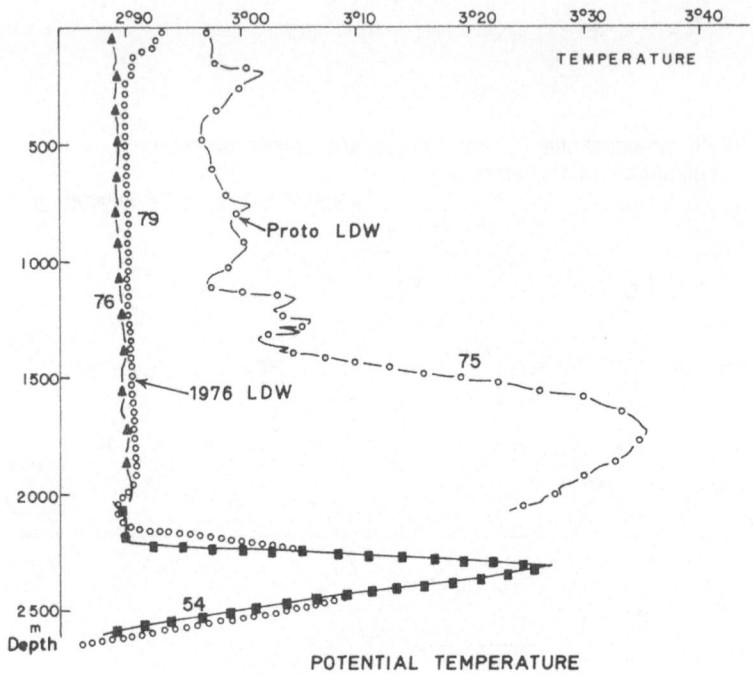

Fig. 4 Potential temperature versus depth for stations across the thermal
front

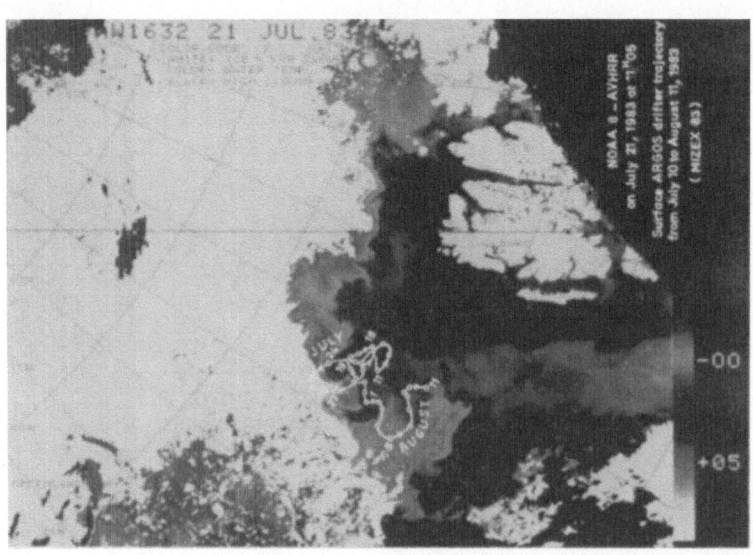

Fig. 5 IR image taken from AVHRR on NOAA 8 on July 21, 1983 at 11h05. On this
picture is superposed the trajectory, in the open ocean, of a surface drifter
localized with Argos, from July 10 to August 8, 1983 (MIZEX 83)

DEVELOPMENT OF AN ECONOMICAL SOIL MODEL FOR CLIMATE SIMULATION

H. BAUER, E. HEISE, J. PFAENDTNER[1] and V. RENNER[2]
Deutscher Wetterdienst, Offenbach, FRG

SUMMARY

The prediction of surface temperature and moisture over continents is an important problem in climate simulation. The
paper describes the development of an economic two-layer soil
model. Special emphasis has been given to a proper simulation
of the diurnal and the annual period. Moreover, the model permits a gross representation of the effects of different soil
types and vegetation coverings.

1 Introduction

About fifty percent of the solar energy reaching the earth-
atmosphere system is taken up by the ground and returned to
the atmosphere in the form of long wave radiation, sensible or
latent heat fluxes. For the simulation of these essential energy
fluxes, general circulation models (GCMs) need a proper treatment of surface conditions. Whereas the relatively slowly
varying sea surface temperature is often prescribed from observations, temperature and moisture over land have to be
modelled as time dependent variables, if one wants to account
for their strong interaction with atmospheric processes e.g. in
the synoptic time scale or with regard to the diurnal variations.
Moreover, there are also important long-lived effects,
which call for an interactive modelling of land surface conditions. Experiments conducted with different GCMs produced
large and significant anomalies of the atmospheric circulation,
when initial anomalies of soil moisture content (1,2,3,4) or
snow cover (5) were prescribed. The effects lasted for weeks
or months of simulated time. This may indicate a basis for some
success in dynamical long range prediction beyond the limits
of pure atmospheric predictability.

[1] Present affiliation: Goddard Space Flight Centre, Greenbelt,
Maryland, USA

[2] Speaker

The state of the parameterization of land surface processes in GCMs at about the time, when the EC climatology program
began, has been described in an overview paper by CARSON (6).
Most models predicted surface temperature by simple 1-parameter
models (force- or force-restore method, see DEARDORFF (7)) or
diagnosed it from the surface energy balance by neglecting the
soil heat flux, as has first been done in the GFDL model (8).
Simulation of soil moisture largely followed MANABE (9), who
applied simple balance equations for the water content of a
relatively thick (and therefore slowly reacting) soil layer and
for the mass of snow cover. Generally, no allowance was made
for different soil types or the effects of different vegetation
coverings. The most prominent exception at that time being the
NCAR model, which applied a 2-parameter formalism for the prediction of surface temperature and moisture with explicit consideration of vegetation, as derived by DEARDORFF (7).

Today, two-layer soil models are also applied e.g. at the
ECMWF (personal communication) and in the GISS model (2), where
moreover the influence of regional vegetation distribution (as
given by MATTHEWS (10)) has been accounted for in a rather
simplified manner. Much work has been devoted to a better specification of evaporation E, which in MANABE's formalism is a
fraction of the potential evaporation E_p in the form $E= \beta E_p$, β
being a simple, piecewise linear function of soil moisture
content. RANDALL (11) for the GLAS model and LAVAL et al. (12)
for the LMD model report on improved simulations especially in
low latitudes, when more elaborate formula for β are used (in
case of the GLAS model together with a redefinition of E_p).
DICKINSON (13) developed a new parameterization of evapotranspiration for the NCAR model, where evaporation from bare soil
is defined by either the potential evaporation rate or by the
maximum rate at which water can diffuse to the surface, depending on which rate is smaller. Thus, in the latter case, evaporation is effectively independent of its potential value. On
the other land, transpiration from plants is modelled thus that
it may markedly exceed the potential evaporation from a wet
soil.

Our own work aimed at the formulation of a two-layer soil
model, which is relatively economic, but on the other hand permits a gross representation of the effects of different soil
types and vegetation coverings and a proper simulation of
short- and long-term variability (with special emphasis put on
the diurnal and the annual variations). The following sections
give a short description of the model and some experiments done
so far.

2 Model description

The thermal part of the soil model is based on the extended
force-restore method (JACOBSEN & HEISE (14)). Under the assumption of a homogeneous soil and a prescribed harmonic surface
energy flux, this method correctly reproduces amplitude and
phase of the surface temperature for two arbitrary frequencies,
which are chosen for our model to correspond to the diurnal and
the annual cycle.

The structure of the newly developed hydrological part of the soil model is shown in the adjacent diagram. The water content W_O of a surface storage and of two soil layers (W_1, W_2) is simulated. Precipitation P partly fills the surface storage and partly infiltrates into ground (15). The infiltration rate I is bounded by a maximum permissable value, determined by soil properties and moisture content (16). Moisture fluxes between the two layers ($F_{1,2}$) and between the lower layer and the underground ($F_{2,u}$) are simulated by the well known Darcy equation. Evaporation from the surface storage (E_O) takes place with the potential rate, as determined by atmospheric conditions and surface temperature.

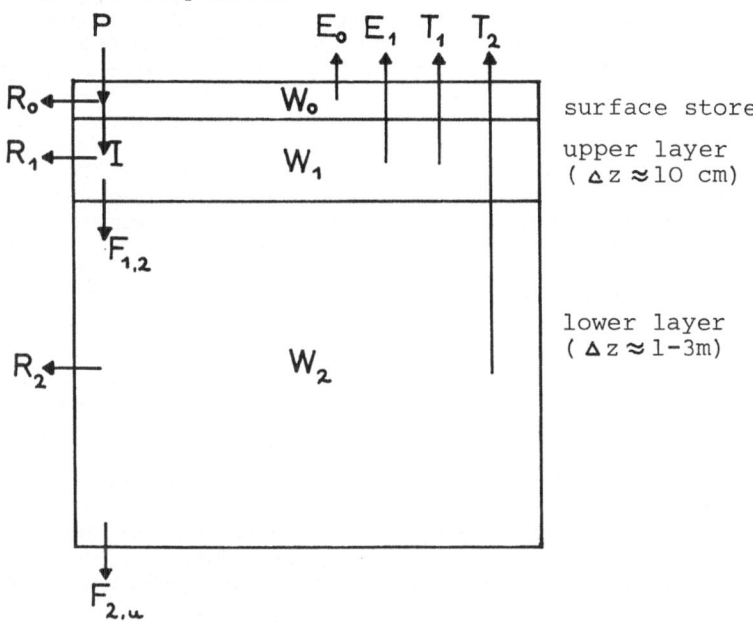

Evaporation (E_1) and transpiration (T_1, T_2) out of the soil depend on vegetation cover and soil moisture within certain soil dependent threshold values. Runoff (R_O, R_1, R_2) occurs, when the moisture holding capacity is exceeded.

Interdependencies between the two parts of the model are considered during the process of freezing/melting, in the surface energy balance and in the dependance of soil thermal properties on the moisture content. In addition, a simple budget equation for the water content of a variable snow cover is modelled.

Data concerning the regional distribution of soil properties were taken from the "Soil map of the world", FAO/ UNESCO (17). Since no global information on the actual land use was available during the course of the project, data for the potential natural vegetation after SCHMITHÜSEN (18) were prepared in cooperation with members of the University of Osnabrück (LIETH/ESSER).

3 Results

In order to test the overall performance of the hydrological part of the surface model, at first an experiment was
undertaken to simulate the soil moisture values measured at
Braunschweig (FRG). Starting on January 1st, 1970, 6 years
were simulated continuously. Figure 1 shows the results for
the last year of this run. Despite poor input data for this
experiment (surface ambient temperature and relative humidity
once per day and daily sums of precipitation and global radiation) the similarity of measured and simulated soil water
content is encouraging. An overall drying of the soil during
the six years of simulation is in accordance with the measurements and reflects the influence of unusual small precipitation (432 mm/year) in this period. Two remaining problems,
which deserve attention during the further development of the
model, are a rather poor simulation of cases with supersaturation and a somewhat rapid decrease of soil moisture in spring.
A January simulation has been performed with the hemispheric version of the German Weather Service general circulation model. One objective was to investigate the sensitivity
of the simulated climate with respect to regional differences
of prescribed soil parameters. Figure 2 shows diurnal cycles
of predicted temperatures and soil water contents together
with energy balance components for two different sites in the
central Sahara on the last day of the simulation run. Soil
types were sand and loam, respectively, differing in hydrological and thermal parameters as, e.g., field capacity and heat
capacity. Although the shortwave radiation balance at the surface (not shown separately) was nearly the same in both cases,
with loam the diurnal temperature wave was damped considerably. Reasons for this behaviour are the different thermal properties as well as the different moisture content of the soil,
which results in larger latent heat fluxes with loam. Even
with sand the diurnal temperature wave seems to be damped, as
compared to measurements, which is probably due to unrealistic
large latent heat fluxes. These may be caused by our simple
evaporation formula (compare (11) and (12)). Additionally,
our lower boundary conditions for soil water content may be
inadequate in the Sahara region, leading to rather large values
of the moisture content of the two soil layers. Unfortunately,
the diurnal cycle of the soil water content can't be compared
with real data due to the lack of measurements. At least it
does not look unreasonable.
Additional GCM experiments with (i) prescribed values of
surface variables and (ii) application of a one-layer soil model
were performed in order to allow the comparison of circulation
statistics obtained with soil treatment of varying complexity.

4 Concluding remarks

The two-layer soil model developed within the frame of the
project has produced reasonable results in experiments with
prescribed atmospheric conditionsand in a hemispheric January
simulation experiment with a GCM. Continuation of our work will

deal with (i) improvements of the model with respect to some
shortcomings, which did show up in these experiments, and (ii)
more general applications, especially in global GCM experiments
with explicit consideration of the vegetation cover.

Besides the development of more comprehensive parameterization schemes for the soil processes, progress in soil modelling requires activities in following fields:
- Sensitivity studies should be conducted in order to determine, which of the various surface parameters are most important for climate simulations and to which degree of detail the vastly complex surface processes have to be modelled (compare (13)).
- Horizontal variations of soil conditions within large areas
  as represented by the grid elements of typical GCMs must be
  adequatly accounted for (see e.g. DOOGE (19)). A pilot at-
  mospheric-hydrological experiment for the WCRP, designed
  particularly to this problem (20), will help to solve this
  difficult and largely unattacked task.
- Development, application and evaluation of soil models are
  strongly dependent on appropriate global data sets of surface conditions, which today are largely missing, but, hopefully, will become available as a result e.g. of the International Satellite Land Surface Climatology Project (ISLSCP).

REFERENCES

1. KURBATKIN, G.P., MANABE, S. and HAHN, D.G. (1979). The
    moisture content of the continents and the rate of summer
    monsoon circulation. Soviet Meteor. and Hydr., Nr.11, 1979;
    1-6
2. RIND, D. (1982). The influence of ground moisture condi-
    tions in North America on summer climate as modeled in the
    GISS GCM. Mon.Wea.Rev. 110; 1487-1494
3. ROWNTREE, P.R. and BOLTON, J.A. (1983). Effects of soil
    moisture anomalies over Europe in summer. In: Variations
    in the global water budget, A. Street-Perrott et al., Eds.,
    D. Reidel Publ. Comp.; 447-462
4. YEH, T.-C., WETHERALD, R.T. and MANABE, S. (1984). The
    effect of soil moisture on the short-term climate and hy-
    drology change - A numerical experiment. Mon.Wea.Rev. 112;
    474-490
5. YEH, T.-C., WETHERALD, R.T. and MANABE, S. (1983). A model
    study of the short-term climate and hydrologic effects of
    sudden snow-cover removal. Mon.Wea.Rev. 111; 1013-1024
6. CARSON, D.J. (1982). Current parameterization of land-sur-
    face processes in atmospheric general circulation models.
    Paper presented at the JSC Study Conference on Land Surface
    Processes in Atmospheric General Circulation Models, Green-
    belt, 5-10 January 1981. Cambridge University Press; 61-108

7. DEARDORFF, J.W. (1978). Efficient prediction of ground sur-
    face temperature and moisture, with inclusion of a layer
    of vegetation. J.Geoph.Res. Vol. 83, No. C4; 1889-1903

8. SMAGORINSKY, J., MANABE, S. and HOLLOWAY, J.L. (1965).
    Numerical results from a nine-level general circulation
    model of the Atmosphere. Mon.Wea.Rev. 93; 727-768
9. MANABE, S. (1969). Climate and the ozean circulation. I.
    The atmospheric circulation and the hydrology of the
    earth's surface. Mon.Wea.Rev. 97; 739-774
10. MATTHEWS, E. (1983). Global vegetation and land use: New
    high-resolution data bases for climate studies. J.Clim.
    and Appl.Met. 22; 474-487
11. RANDALL, D.A. (1983). Monthly and seasonal simulations with
    the GLAS climate model. Workshop on intercomparison of large
    -scale models used for extended range forecasts, 30.6.-
    2.7.1982, ECMWF; 107-166
12. LAVAL, K., OTTLE, C., PERRIER, A. and SERAFINI, Y. (1984).
    Effect of parameterization of evapotranspiration on climate
    simulated by a GCM. In: Berger, A.L. and C. Nicolis, Eds.:
    New perspectives in climate modelling. Elsevier; 223-247
13. DICKINSON, R.E. (1984). Modeling evapotranspiration for
    three-dimensional global climate models. Climate processes
    and Climate sensitivity. Geophysical Monograph 29, Maurice
    Ewing Volume 5; 58-72
14. JACOBSEN, I. and HEISE, E. (1982). A new economic method
    for the computation of surface temperature in numerical
    models. Beitr.Phys.Atm. 55; 128-141
15. BECKER, A. (1974). Applied principles of catchment simula-
    tion. IAHS-AISH Publ. 101, Vol. 2; 762-774.
16. HOLTAN, H.N. (1970). Representative and experimental basins
    as dispersed systems. Symposium IASH/UNESCO Publ. 96
17. FAO/UNESCO (1971-78). Soil map of the world. Paris
18. SCHMITHÜSEN, J. (1976). Atlas zur Biogeographie. Mannheim
19. DOOGE, J.C.I. (1982). Parameterizations of hydrologic
    processes. Paper presented at the JSC Study Conference on
    Land Surface Processes in Atmospheric General Circulation
    Models, Greenbelt, 5-10 January 1981. Cambridge University
    Press
20. ICSU/WMO (1984). Report of the meeting of experts on the
    design of a pilot atmospheric-hydrological experiment for
    the WCRP, Geneva, 28 November-2 December 1983. WCP-76

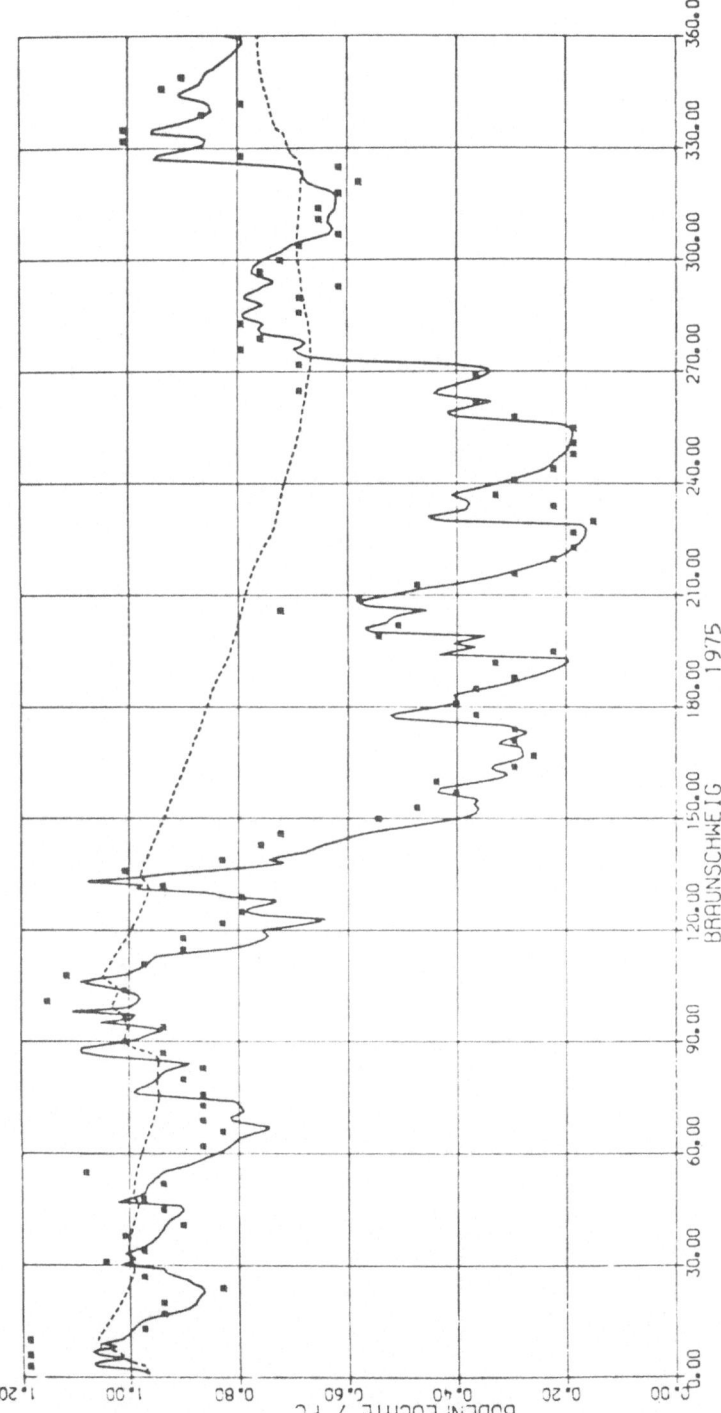

Figure 1: Simulated soil moisture content divided by the field capacity for the upper (———) and lower (-----) layer of the soil model as a function of time (days). Corresponding measurements for the upper layer at Braunschweig (FRG) are denoted by dots.

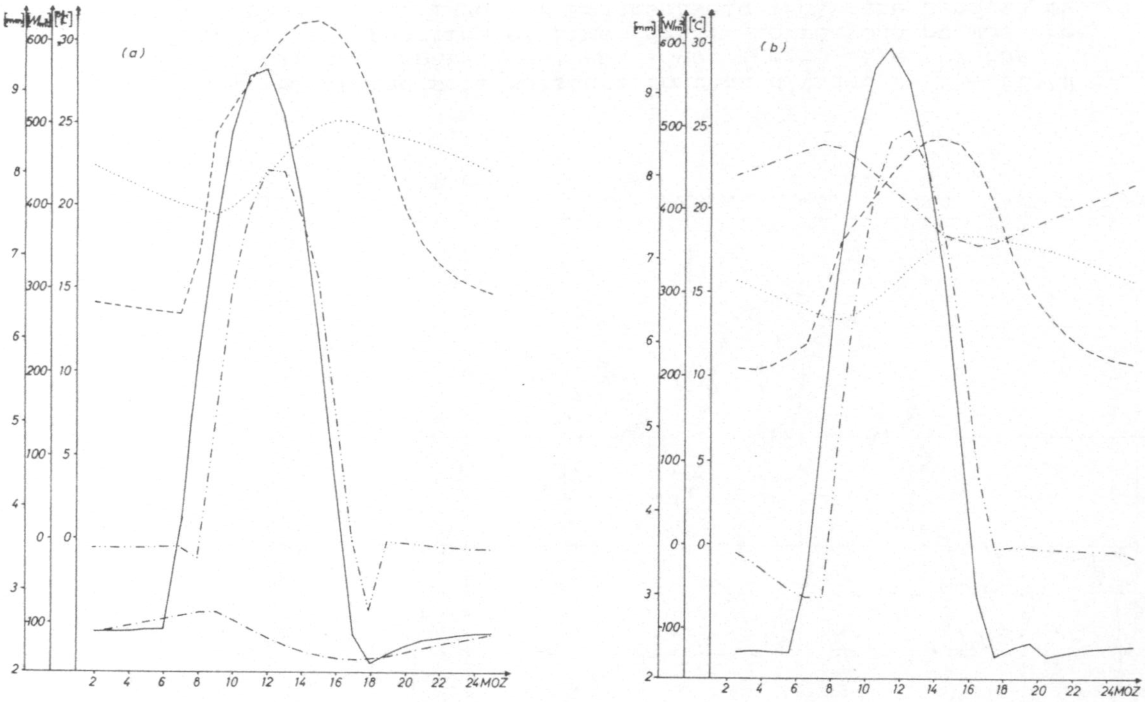

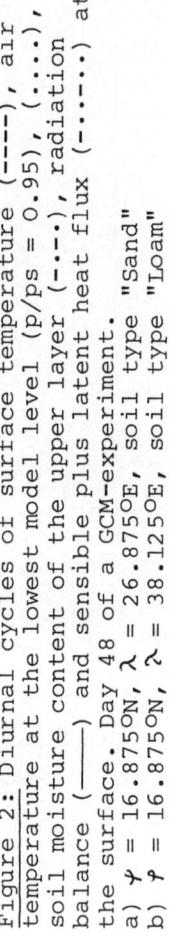

Figure 2: Diurnal cycles of surface temperature (————), air
temperature at the lowest model level (p/ps = 0.95), (·····),
soil moisture content of the upper layer (–·–·), radiation
balance (————) and sensible plus latent heat flux (–··–··–··) at
the surface. Day 48 of a GCM-experiment.
a) φ = 16.875°N, λ = 26.875°E, soil type "Sand"
b) φ = 16.875°N, λ = 38.125°E, soil type "Loam"

CHAPTER III : ANTHROPOGENIC CLIMATE
PERTURBATIONS

- On modelling the effects of CO_2 on climate

- CO_2 and climate : Information from Antarctic ice core studies

- Atmospheric CO_2 concentration in subantarctic countries

- The influence of man-made SO_2, and particle emission on the global aerosol concentration and the optical properties of the atmosphere

ON MODELLING THE EFFECTS OF CO_2 ON CLIMATE

J F B MITCHELL

Meteorological Office, Bracknell, England

Summary

The sensitivity of the Meteorological Office's 5-layer atmospheric general circulation model to enhanced atmospheric CO_2 concentrations with both a uniform and a latitudinally varying increase in sea surface temperatures is assessed. The changes in atmospheric temperatures depend on the sea surface temperature increases imposed, but the geographical distribution of other changes, for example in the hydrological cycle, are qualitatively similar whether a uniform or a latitudinally varying change in ocean temperature is applied. This relative insensitivity of certain aspects of similated climate to the details of the changes in ocean temperature is explained in terms of the physical processes involved. In contrast, when the first experiment is repeated with a second atmospheric model, although there are broad similarities, many aspects of the detailed regional response of the two models are found to be quite different. These regional differences can be attributed to differences in the unperturbed climates simulated by the two models.

Some of the problems of comparing simulated and observed data are illustrated using data from both sources over eastern England. The simulated climate over western Europe is assessed, and the changes in the more realistic integration with latitudinally varying changes in ocean temperature are presented. Finally, some of the outstanding problems in modelling the effects of CO_2 on climate are outlined.

1. Introduction

In recent years there has been a growing interest in the possibility that climate may be changed substantially due to increased concentrations of atmospheric carbon dioxide (CO_2) released by burning fossil fuels. Many of these conjectures are based on results from numerical models of climate. In this paper we consider various aspects of such results using examples from work carried out in the Meteorological Office supported by E.E.C. Contract CL1-030-81-UK(H).

In the first part we assess the response of an atmospheric model to increased CO_2 with prescribed increases in sea surface temperatures. The increases in sea surface temperatures were those expected to accompany the given increase in CO_2; in the first approximation a uniform increase was imposed, and in the second, the temperature increments varied with latitude. We then consider the response of a second model to increased CO_2 and a uniform rise in sea surface temperature. These two studies indicate the dependence of the simulated changes on the assumptions made in the experiment and the particular model used. The response of the hydrological cycle in summer is treated in detail as an example.

In the second part, methods of comparing model and observational data are described, and are used to evaluate the simulated climate over western Europe. Some of the changes in climate over Europe found in the experiment with a latitudinal increase in sea surface temperature are analysed. Finally areas for future research are identified.

2.   Numerical simulation to determine the effect of CO_2 on climate
2.1   Choice of sea surface temperature changes
      In assessing the effect of perturbations on climate, one must take
into account all the elements of climate which contribute on the relevant
time scale. As CO_2 is expected to increase over the next few decades, it
is necessary to simulate the response of the ocean, but the major ice
sheets may be regarded as 'fixed'. Although coupled dynamical models of
the ocean and atmosphere have been developed, they are expensive in terms
of computing resources. If the horizontal resolution of the model is
degraded in the interests of economy, it is found that the simulation,
particularly of the ocean, is degraded. Many recent studies (1,2,3) have
used an atmospheric model coupled to a static well mixed ocean 50 to 70
metres deep. Many important processes including advection of heat by the
oceans are neglected, with subsequent degradation of the model climate,
though one study has attempted to prescribe the horizontal heat convergence
in the ocean from climatological data (2).
      Here an alternative approach is pursued. The change in radiative
heating due to increasing CO_2 is small (a warming of order 5 Wm^{-2} of the
troposphere and surface due to doubling CO_2 compared to an ambient
tropospheric cooling of 100 Wm^{-2}), suggesting that the response of climate
to increased CO_2 may be regarded as a perturbation. In the control
integration, sea surface temperatures (and sea ice extents) are prescribed
from climatology. In the anomaly integrations, the sea surface
temperatures are derived from a series of perturbation experiments in which
the change in zonally averaged net surface heating accompanying the
increased CO_2 is progressively reduced towards zero (4,5). Since the
annual average net surface heating at each latitude represents an implied
divergence of heat in the ocean, no change in net surface heating implies
that the zonally averaged meridional advection of heat is unaltered.
2.2   The model
      Except where stated, all the results in this paper are from
integrations made with the Meteorological Office 5-layer model (5LM)(4).
It is a primitive equation model, using a quasi uniform 330 km horizontal
grid. The radiative fluxes are dependent on clouds, temperature, water
vapour, carbon dioxide and ozone; infra-red radiation is treated using the
emissivity approximation. Seasonally varying cloud amounts, sea surface
temperatures and sea ice extents are prescribed from climatology. Over
land, snow is accumulated and surface albedo is a function of snow depth,
where appropriate, and a soil moisture variable is used in the derivation
of evaporation and runoff.
      The model simulates all the main features of the global circulation,
the main shortcomings being excessive surface pressure in polar regions,
under and over estimation of the depth of mid-latitude depressions in
southern and northern winter mid-latitudes respectively, and heavier than
observed precipitation over the continents (see for example, Figure 1).
Further details of the model and its climatology are given in reference
(4).
2.3   Sensitivity of the simulated response to imposed changes in sea
surface temperatures
      In the initial experiment, CO_2 concentrations were doubled and sea
surface temperatures were enhanced everywhere by 2K, as a first
approximation to the oceanic response. Land surface temperatures increased
by 3K, and precipitation increased by 5% (Table I).

There was a marked increase in westerly flow in mid-latitudes in winter. Large increases in precipitation occurred in existing regions of low level convergence, notably the eastern coasts of continents in summer, with decreases in some regions of the sub-tropics (for example Figure 2a). There was a general decrease in soil moisture, although some regions became moister, including part of the tropics.

Evidence from this experiment and from studies carried out elsewhere indicate that the rise in sea surface temperature accompanying doubled CO_2 should be larger in high latitudes and smaller in the tropics (1,4). Hence an additional experiment in which the sea surface temperature change increased with latitude was carried out (5) (the sea temperature changes were chosen to reduce the zonally averaged surface heating towards zero, as explained in Section 2.1). CO_2 amounts were quadrupled, and the sea surface temperature increments, ranging from 2K in the tropics to over 5K in high latitudes, were chosen accordingly. In addition, the sea ice margins were moved 6° of latitude poleward.

The land surface temperature increased by 4K, and precipitation increased by 3.5% (Table 1). Not surprisingly, in the second experiment, the changes in land surface temperature in high latitudes were considerably enhanced relative to the first experiment (assuming a quadrupling of CO_2 amounts produces twice the temperature change expected from a doubling). On the other hand, the changes in zonally averaged westerly wind and precipitation were qualitatively very similar. For example, during June to August (Figure 1b) there were increases in precipitation over the southern and eastern coasts of North America and Asia, over the tropical and southern circumpolar oceans and poleward of 55°N. There was reduced precipitation in northern mid-latitudes and in parts of the subtropics. There were some areas where discrepancies occur, including the Sahara and the Middle East where the reductions in rainfall found in the first experiment (Figure 2a) are less extensive in the later integration (Figure 2b). However throughout the year the response in hydrological cycle was in general very similar.

The reason for this similarity is that in each study the atmosphere becomes warmer and moister (Table 1). In the absence of marked changes in circulation, this enhances the convergence and divergence of moisture in the control integration, so that regardless of the details of the warming, the patterns of change are similar. There is also a reduction in the transient eddy kinetic energy in mid-latitudes, implying weaker and/or less frequent disturbances in both studies which may contribute to the reduction in precipitation over the northern continents in summer. A more detailed discussion of these changes is given elsewhere (4,7).

In both experiments, much of the northern mid-latitude continents became drier in summer (7). The changes were more pronounced in the second experiment though this may be because the response to the larger perturbation shows more clearly above the year to year variations inherent in the model. Other studies made with atmospheric models coupled to a simple representation of the ocean produced a drying in mid-latitudes in summer (6). Reduced precipitation, increased radiative heating of the surface, and earlier snow melt in the warmer climate leading to a longer summer drying season all contribute to this phenomenon. Again, the responses are alike because the same physical processes operate, even though the changes in sea surface temperature or the models used are different.

2.4 Sensitivity of different models to the same perturbation

The experiment in which CO_2 was doubled and sea surface temperatures were increased by 2K was repeated using the Meteorological Office 11-layer model (11 LM, see Slingo and Wilson, this volume)). The main differences in formulation are listed in Table II. Note that the 11LM, by virtue of its regular latitude/longitude grid, has much finer resolution in high latitudes, and uses different boundary layer and convection schemes.

In the 11LM the southern circumpolar depression belt is deeper, and the westerly flow in northern middle to high latitudes is stronger. Furthermore the 11LM tends to produce its most intense precipitation over the tropical oceans, whereas the 5LM gives excessive precipitation over land (Figures 1, 3a). A comparison of the simulations of the 5LM and an older version of the 11LM on a 200 km quasi uniform grid is given in reference (8).

The increases in land surface temperature and precipitation were similar to those in the original experiment (Table I). Again there was an increase in westerly flow in middle latitudes in winter. However the increase is shifted poleward relative to the 5-layer model experiment, consistent with the maximum in westerly flow in the control integrations being further poleward in the 11-layer model than in the 5-layer model. The zonally averaged changes in precipitation were similar in both models, with increases in high latitudes and the tropics throughout the year and in middle latitudes in winter, with little change or slight decreases in the subtropics. There was less rainfall over the northern mid-latitude continents in the summer. However, there were considerable discrepancies in the detailed regional changes which again reflect the differences in the models' climatologies. Thus, for example, from June to August, increases in precipitation along the southern and eastern coasts of North America and Asia were predominantly over the ocean in the 11-layer model (Figure 3b) but over land in the 5-layer model (Figure 2a). It should be noted that this discrepancy is consistent from year to year, but some of the small scale changes in both Figures 2a and 3b arise from year to year variations which occur by chance. As in the 5-layer model experiments, the continental surface in northern mid latitudes in summer tends to dry out more in the anomaly integration.

In section 2.3 it was demonstrated that a climate model may respond in a similar manner to different perturbations as the same physical mechanisms may be involved. In this section we have seen that different models can produce a divergent response, particularly on smaller scales, if the unperturbed circulations are not the same. This is consistent with the argument advanced earlier that the projected increases in CO_2 may be regarded as a perturbation, so that the simulated response is strongly dependent on the unperturbed climate.

3. Interpretation of model data and simulations of climate over western Europe

3.1 Comparison of simulated and observed data

In the previous sections we have considered the large scale response of a particular model to increases in CO_2, and some of the accompanying uncertainties. Even if these uncertainties can be resolved, there remains the problem of comparing model gridpoint data, which can be regarded as a mean value over an area several hundred kilometres square, often averaged through the lowest 100 to 1000 metres of the atmosphere, with observational data at a given location, usually taken at a specific time a few metres above the ground. For climate impact studies, not only mean data are required, but also the frequency of extreme events and so on. We consider as an example the simulated time series of temperature and precipitation at a grid box over eastern England (9).

The annual mean simulated surface temperature, 3.8°C, is 5.6°C colder than the annual mean Central England Temperatures (averaged from 1959 to 1973, (10)). The Central England Temperature series is a mean of maximum and minimum temperatures, whereas the model temperatures are sampled at midnight. Observations of midnight temperature at Hemsby and Crawley indicate that 8°C is a more reasonable estimate of the annual mean midnight temperature for eastern England. The model temperature is for the surface, and more akin to a ground (grass) temperature rather than a screen temperature taken at a height of 1.5 metres, as in the Central England Temperature time series. A comparison of grass and screen temperatures at stations in eastern England indicate that the grass temperatures are typically 2°C lower than the corresponding midnight screen temperatures; hence we conclude that the model surface temperatures are about 2°C too cold. A similar comparison of temperatures in the bottom layer (at about 900 mbs) suggests that the model is only about 0.5°C too cold at that level. The frequency distribution of the simulated and observed daily temperatures (with the mean and annual and semi-annual cycles removed) are remarkably similar, except for the large number of very cold events in the modelled distribution (Figure 4). This 'cold tail' is a manifestation of the greater than observed frequency of easterly flow during late winter and spring, which leads to unrealistically low temperatures over much of western Europe.

It was noted earlier that the simulated precipitation is higher than observed over the continents, and this is true over western Europe, except for gridpoints bordering the Mediterranean. The seasonal variation is qualitatively realistic, except for a few isolated gridpoints in Central France and Germany where the observed summer maximum is not reproduced. As precipitation is poorly correlated in space, the model data for eastern England have been verified against daily precipitation averaged over 144 stations in England and Wales during 1971 to 1973. The model substantially overestimates the frequency of days with about 5 mm of precipitation (Figure 5). Note also that the heaviest precipitation produced in the model is considerably less than in the England and Wales rainfall. Although there are difficulties in measuring daily precipitation of less than a few tenths of a millimetre, it is clear that the model grossly underestimates the number of dry days (see also Table IV).

3.2 Simulated climate over western Europe and the response to increased CO_2

Having outlined some of the methods of comparing model and real data, we now consider some of the detailed changes in climate over western Europe found in the 4 x CO_2 integration described in section 2.3. It is evident from the paper so far that there is much to be done before predictions of regional climate from climate models can be used with any confidence. However, it is important that methods of translating changes in model gridpoint data to parameters from which the economic impacts can be assessed should be developed and tested so that they are available when more reliable forecasts can be made. Here, two examples are considered, namely changes in precipitation and in the frequency of occurrence of low temperatures. A more detailed account of changes over western Europe is given in reference (11). The stations used for comparison are Raunds (eastern England), Muenchen (West Germany) and Capo Palinuro (southern Italy).

Enhancing CO_2 concentrations and ocean temperatures produces a surface warming which throughout northern Europe is greatest in winter and spring, but around the Mediterranean is a maximum in summer or autumn. The overall rise in temperature leads to a decrease in frequency of low temperatures (Table III). However, this decrease is enhanced by stronger westerly flow in winter, which is associated with a reduction in frequency of cold easterly outbreaks. If, for example, the annual mean increase in surface temperature over eastern England were applied to the time series of temperatures from the control integration, the frequency of midnight temperatures below zero would fall from 76 to 10 per year. In the 4 x CO_2 integration, on average only 2 occurrences per year were found.

Although the frequency of days with measurable precipitation is grossly overestimated (Table IV), the model results can be used to indicate the sense of the changes in precipitation in the warmer climate. In general, northern and western Europe are wetter throughout the year except in summer, whereas southern and especially south eastern Europe receive less rainfall throughout the year. Eastern England and Central Germany have more precipitation in winter, although the frequency of daily precipitation changes little (Table IV). In contrast, rainfall at the gridpoint in southern Italy is reduced in the mean and occurs less frequently throughout the year. By applying a simple statistical model (12), one can estimate the statistical significance of the changes in both mean precipitation and the frequency of precipitation (Table IV). Thus for example, the changes in precipitation over southern Italy are significant at the 95 per cent level of confidence in both winter and summer, but only the change in mean summer precipitation is significant at the gridpoint in Central Germany.

4.   Concluding Remarks

We have examined some of the potential and shortcomings in the use of numerical models for the prediction of climate change. Recent experiments using climate models with a simple representation of the ocean have indicated an equilibrium global mean surface warming of about 4°C on doubling CO_2. These studies attribute a substantial fraction of the warming to either changes in cloud cover (2) or sea ice (3). Further research is needed to clarify the role of cloud cover and sea ice in climate change, and to use models which include full representation of the oceans.

Both studies to date have considered the equilibrium response to increased CO_2 concentrations and should not be regarded as predictions. The response of climate to the observed gradual increase in CO_2 amounts is expected to be slowed by the large thermal inertia of the oceans. This retardation depends critically on the rate at which heat is mixed down into the deep ocean, and many of the processes involved are poorly understood.

Undertaking research on climate and climate change is a long term commitment, and progress on a year-to-year basis may appear slow. It will be some years before we can predict regional changes in climate with any confidence. However, our understanding of the physical basis of climate has improved considerably over the last decade, and we can expect significant progress in the decade to come.

REFERENCES

1. MANABE, S AND   1980   Sensitivity of a global climate model to an
    STOUFFER, R J   increase in the CO_2 concentration in the
                                   atmosphere.

- 233 -

2. HANSEN, J; LACIS, A;   1984   Climate Sensitivity Analysis of feedback
    RIND, D; RUSSELL, G;   mechanisms. In 'Climate Processes and
    STONE, P; FUNG, I;   Climate Sensitivity'.
    RUEDY, R AND   (Maurice Ewing Series, 5, eds J E Hansen and
    LERNER, J   T Takahashi). A.G.U. Washington 368 pp.

3. WASHINGTON, W M AND   1984   Seasonal Cycle Experiment on the Climate
    MEEHL, G A   Sensitivity due to a Doubling of CO_2 with an
                                  Atmospheric General Circulation Model
                                  Coupled to a Simple Mixed Layer Ocean Model.
                                  To be published in J. Geophys. Res.

4. MITCHELL, J F B   1983   The seasonal response of a general
    AND LUPTON, G   circulation model to changes in CO_2 and sea
                                  surface temperature.
                                  Q.J.R. Met. Soc. 109, 133-152.

5. MITCHELL, J F B   1984   A 4 x CO_2 experiment with prescribed AND
    AND LUPTON, G   changes in sea temperatures. Progress in
                                  Biometeorology 3 353-374. Paper presented at
                                  CEC Symposium 'Interactions between climate
                                  and biosphere'. Leith, Fantechi and
                                  Schnitzler, eds., Osnabrueck March 21-23,
                                  1983.

6. MANABE, S;   1981   Summer dryness due to an increase of
    WETHERALD, R T;   atmospheric CO_2 concentration.
    AND STOUFFER, R J   Climatic change 3, 347-385.

7. MITCHELL, J F B;   1984   An evaluation of the impact of an increase
    WILSON, C A;   in CO_2 in the European climate. Final Report
    REED, D N; AND   of CEC Contract CL1-030-81-UK(H).
    LUPTON, G   Meteorological Office, Bracknell,
                                  Berkshire.

8. MITCHELL, J F B   1982   Some differences between the Met O 20 5 and
    AND BOLTON, J A   11-layer model annual cycle integrations.
                                  Met O 20 Tech Note II/176.
                                  Meteorological office, Bracknell, Berkshire,
                                  England.
                                  (A revised version appears in Proceedings of
                                  the ECMWF Workshop on the Intercomparison of
                                  large-scale models used for extended range
                                  forecasts (30 June-2 July 1982) 193-224).

9. REED, D N   1984   Model Simulation of Temperature and
                                  Precipitation in a multi-annual integration
                                  of a general circulation model.
                                  To be submitted to Journal of Climatology.

10. STOREY, A   1981   Homogenisation of the daily Central England
                                  Temperature series. Met O 13 Discussion
                                  Note No. 24, Meteorological Office,
                                  Bracknell, Berkshire.

11. WILSON, C A AND   1984   The 5-layer model climate over western
    MITCHELL, J F B   Europe and the frequency of occurrence of
                                  extreme values of temperature, precipitation
                                  and wind for selected grid boxes: the
                                  changes with 4 x CO_2 and prescribed sea
                                  surface temperatures. To be submitted for
                                  publication. (An early version is contained
                                  in reference 7).

12. KATZ, R W   1983   Statistical procedures for making inferences
                                  about precipitation changes simulated by an
                                  atmospheric general circulation model. J.
                                  Atmos. Sci. 46, 2193-2201.

Table I Changes due to increasing CO_2 and enhancing sea surface temperatures

Experiment	Tropospheric Temperature (°C)	Atmospheric Humidity (%)	Land Surface Temperature (°C)	Precipitation (%)
2 x CO_2 (5LM)	3.02	18	2.86	5.0
*4 x CO_2 (5LM)	2.05	16	4.00	3.5
2 x CO_2 (11LM)	3.08	20	3.05	5.6

* Differences halved to allow comparison.

Table II   Main differences in model formulations

5 LAYER MODEL	11 LAYER MODEL
GRID 5 layers, equally spaced	11 layers, concentrated near surface and tropopause
300 km in horizontal	2.5° x 3.75°

BOUNDARY LAYER

1 layer (up to σ = 0.8)	3 layers (up to σ = .79)
Explicit boundary layer height	Vertical diffusion, "Clarke" scheme
Bulk aerodynamic formula	Drag coefficient continuous
Stable/Unstable, land/sea drag coefficient	function of stability and roughness length
Full evap. when soil moisture = 10 cm	Full Evaporation — 5 cm
Run off when soil moisture = 20 cm	Run off — 15 cm

RADIATION

Temperature and humidity interpolated to 10 equally spaced layers for radiation.	Fluxes calculated on model layer boundaries
No absorption of reflected solar beam.	Reflected solar beam absorbed
Snowfree albedo a function of latitude	Constant (= .2)
Albedo over snow a function of snow depth	Constant (= .5)

Cloud amounts, albedos similar

PENETRATIVE CONVECTION

Detrains only at upper levels	May entrain and detrain at any level.
May affect a given layer more than once/timestep.	Only affects a given layer once/timestep.

DIFFUSION

Diffusion of potential temperature θ	Diffusion of aT + bθ
Non linear diffusion of humidity	Linear diffusion of humidity.

Table III   Number of days/year with OO GMT temperatures below O°C

GRID POINT	CONTROL	4 x CO_2
Eastern England	76	2
Central Germany	187	49
Southern Italy	55	6

Table IV   Average number of days per season on which precipitation exceeded a given threshold

GRID BOX	THRESHOLD (mm)	WINTER O	C	A	SUMMER O	C	A	
England	0.25	45	71	71 +	37	86	77	- *
Germany	0.1	47	86	88 *	49	89	88	-
Italy	1.0	34	52	35 - *	9	10	4	- *

O = Observations (see text), C = Control, A = 4 x CO_2

+/- increase/decrease in mean precipitation significant at 95% level of confidence
  * change in frequency of precipitation significant at the 95% level of confidence.

TOTAL PRECIPITATION (MM/DAY) JUNE, JULY, AUGUST

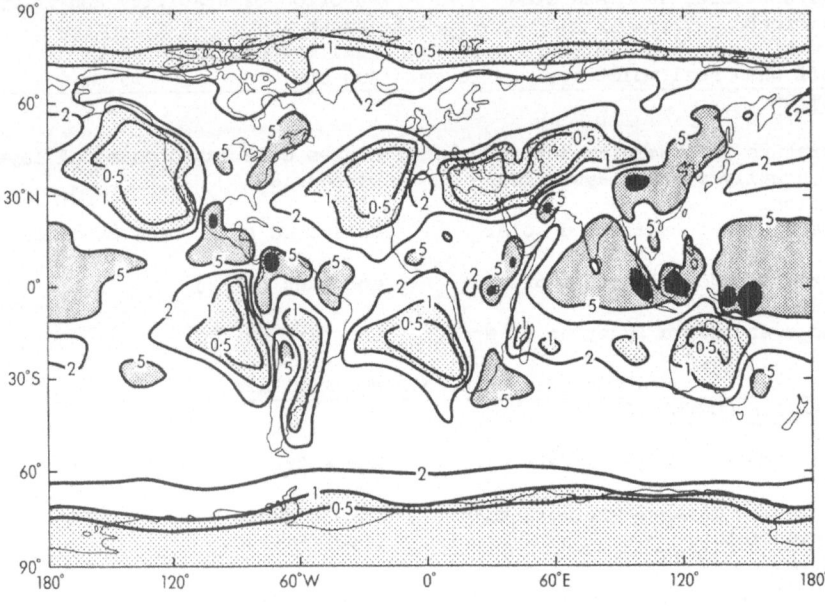

1.   Total precipitation simulated by 5LM during June, July and August (3 year mean). Contours at 0.5, 1, 2, 5 and 10 mm/day. Light stippling where less than 1 mm/day, heavy stippling where greater than 5 mm/day, solid shading where greater than 10 mm/day.

CHANGES IN PRECIPITATION DUE TO QUADRUPLING CO$_2$
AND ENHANCING SSTs (mm/day)
JUNE   JULY   AUGUST

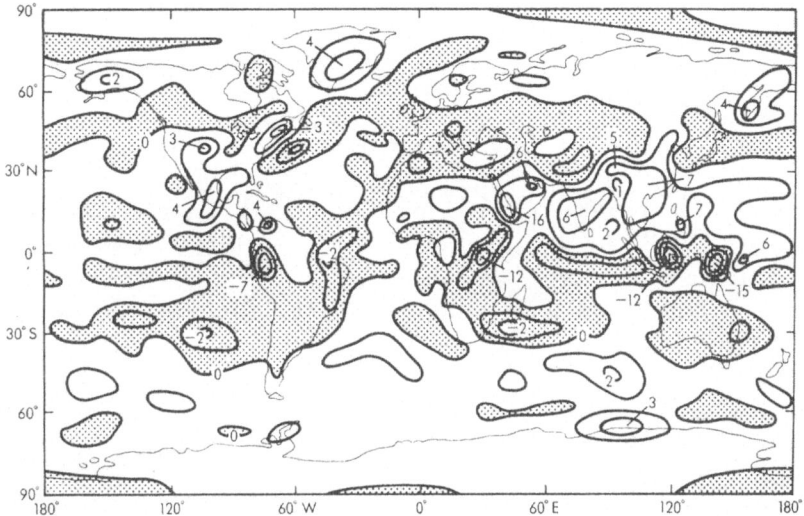

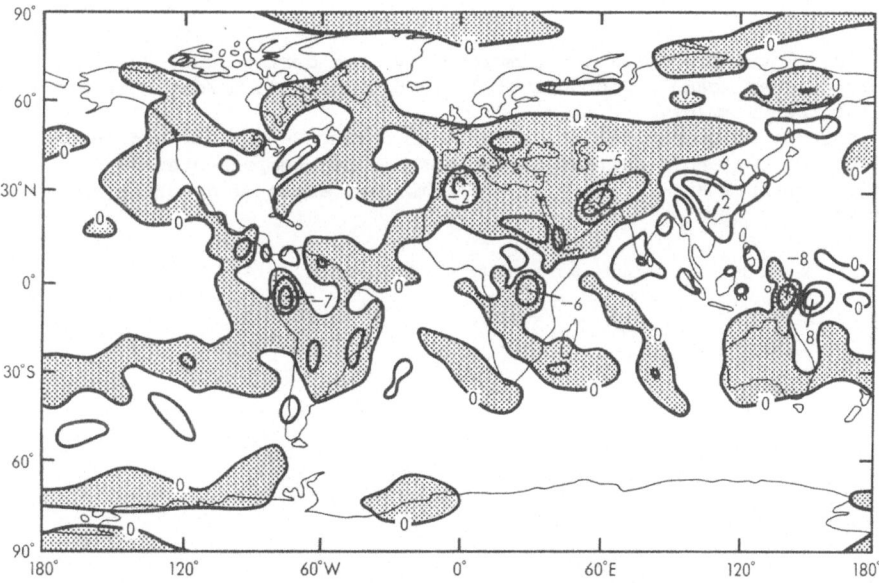

2.   Changes in precipitation simulated by the 5LM during June, July and
August (3 year mean).   Contours at −2, −1, 0, 1 and 2 mm/day.   Areas of
decrease are stippled.
(a) due to doubling atmospheric CO$_2$ and a uniform 2K increase in sea
temperatures.
(b) due to quadrupling atmospheric CO$_2$ and rises in sea temperatures which
increase with latitude.

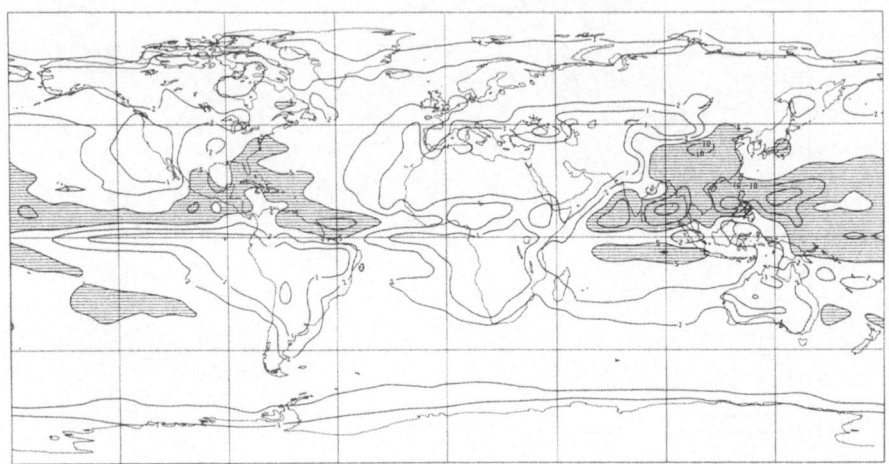

Figure 3a

3. (a) Total precipitation simulated by 11LM during June, July and
August (8 year mean). Contours at 1, 2, 5, 10 and 20 mm/day, hatched where
greater than 5 mm/day.
(b) Changes in precipitation in the 11LM during June, July and August due
to doubling CO_2 and a uniform 2K increase in sea temperatures (3 year
mean). Contours at -2, -1, 0, 1 and 2 mm/day, areas of decrease are
hatched.

CHANGES IN PRECIPITATION (MM/DAY) DUE TO 2XCO2 + 2K SEA TEMPS (11LM) JUN.JUL.AUG

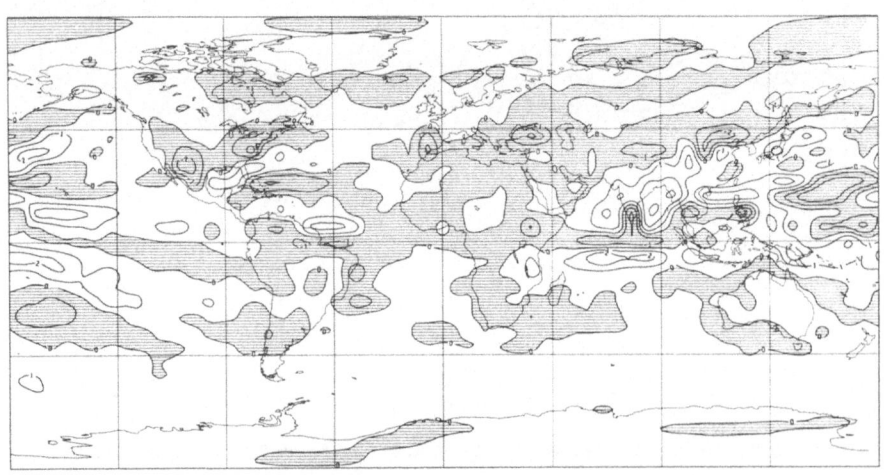

Figure 3b

FREQUENCY DISTRIBUTION OF SURFACE TEMPERATURE

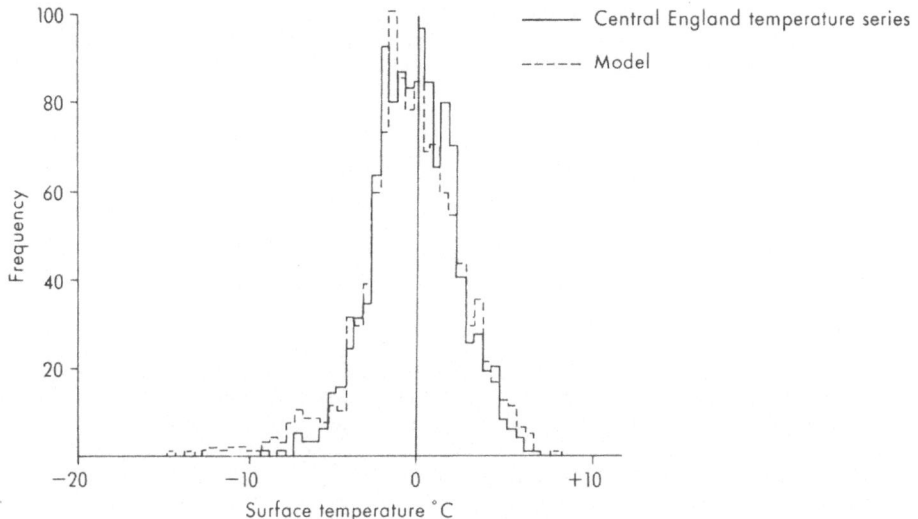

FREQUENCY DISTRIBUTION OF LOGARITHM OF DAILY PRECIPITATION

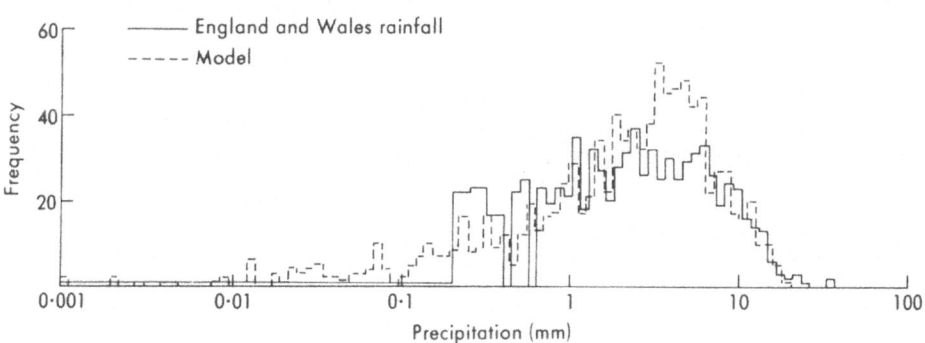

4. Frequency distribution of surface temperature (annual mean and annual and semi annual cycles removed). Solid line, daily Central England temperature series (1100 days of daily data commencing January 1st 1971). Dashed line, 1100 days of data simulated at a grid box in eastern England by the 5LM.

5. Frequency distribution of logarithm of daily precipitation. Solid line, 1100 days of daily data commencing January 1st 1971, averaged over 144 stations in England and Wales. Dashed line, 1100 days of data simulated at a grid box in eastern England by the 5LM.

CO$_2$ AND CLIMATE : INFORMATION FROM ANTARCTIC ICE CORE STUDIES

D. RAYNAUD and J.M. BARNOLA

Laboratoire de Glaciologie et Géophysique de l'Environnement
B.P. 96 38402 ST MARTIN D'HERES CEDEX, France

Summary

Ice cores provide the most direct tool for reconstructing the evolution of the atmospheric CO$_2$ during the last 40,000 years. The results obtained on antarctic cores indicate that the atmospheric CO$_2$ was increasing by a factor of about 1.3 at the end of the last Ice Age. They suggest a close CO$_2$-climate relation, with the CO$_2$ change starting almost simultaneously or even slightly before the temperature change at high latitudes. For the recent period (the last 500 years) the antarctic ice suggest that the "pre-industrial" CO$_2$ level was not constant and was in the 260-280 ppmv range. If so, it was significantly lower than the 290 ppmv adopted previously in modelling the evolution of the atmospheric CO$_2$ during the present period and the corresponding climatic response.

1.Introduction

When the ice is formed by settling and sintering of the snow, air bubbles become trapped. It was initially believed that, by extracting the gas thus trapped, it would be possible to get an unique information about the gaseous composition of the atmosphere during the past. The method would, of course, work only if the ice is formed and preserved without the occurrence of any melting. The large effort made, especially in Antarctica and Greenland, for ice drilling during the last 20 years provided promising cores for this purpose. Measurements performed until the last part of the 70's indicated that the content of the three major gaseous components (N$_2$ O$_2$ and Ar) in the air extracted from the ice cores is similar to the present-day atmospheric composition ; on the other hand the CO$_2$ contents measured were highly variable and generally much higher than the present-day CO$_2$ concentration of the atmosphere, suggesting that these CO$_2$ measurements from ice cores were not representative of the atmospheric concentrations. The hope to extract from ice core a reliable record of atmospheric CO$_2$ became virtually tenuous until as it appeared that the gas extraction method used was critical : in particular, the ice is easily contaminated by carbonate dust during sample preparation and this contamination introduces an excess of CO$_2$ when gases are extracted using a procedure involving ice melting.

Consequently, dry extraction methods based on the extraction of the gas from the air bubbles by crushing the ice at temperature well below the freezing point began to be systematically developed a few years ago. Their use provides, with a high degree of confidence, measurements of the atmospheric CO_2 changes during the past. A first spectacular information was obtained using this method when two different laboratories indicated from several ice cores that the atmospheric CO_2 content was much lower during the Last Glacial Maximum (about 18,000 years BP) than during the Holocene period (1, 2). We present in this paper the recent results that we have obtained in the frame of the EC Climatology Programme. These results concern the evolution of the atmospheric CO_2 during the last centuries and during the period centred on the transition between the Last Ice Age and the Holocene. They have been obtained using the analytical procedure (dry extraction of the gas and CO_2 analysis by gas chromatography) described in reference 3.

2. Sampling sites

The choice of the sampling site is important. As pointed out above the site must be cold enough : the best conditions are found in areas where the mean annual temperature is ≤ − 30°C. Furthermore, the atmospheric CO_2 is isolated more or less progressively from the atmosphere during the transformation of the snow into ice, depending on the rate of air bubble closure and on the mixing of the air in the firn layers where the snow-ice transformation occurs. This last parameter is unfortunately poorly known but, by assuming that the air is well mixed, the CO_2 measured in ice samples could represent an average atmospheric concentration over a time interval varying between (4) :
1) a few tens of years in the areas where the mean annual temperature is about − 30°C and the snow accumulation rate of the order of a few tens of cm of water equivalent per year, and
2) several hundred of years in the most central parts of the ice sheets where accumulation rates and temperatures are extremely low.
The first category of sites is appropriate when looking at the CO_2 record over the last centuries because of the better time resolution and the second category provides the longest time records.
Good sites may be found in Antarctica, Greenland or in some very high altitude glaciers but ice drilling in such areas is difficult and expensive and very few ice cores are available. Furthermore, the quality of the ice cores may be critical (especially the presence of fractures, see reference 5). The results presented here were obtained from the study of two ice cores taken in East Antarctica :
− a 203 m long core drilled in 1980-81 at the Station D 57 (68° 11' S, 137° 33' E, mean annual temperature of − 32°C, accumulation rate of about 40 cm of water equivalent per year), and
− a 905 m long core drilled in 1977-78 at Dome C (74° 39' S, 124° 10' E), mean annual temperature of − 54°C, accumulation rate of about 3 cm of water equivalent per year).

3. The last centuries

The D 57 ice core provides the first opportunity to reveal the CO_2 record over the last few centuries from ice core analysis. This study, that we summarize below, is discussed in details in reference 6.
The dating of the D 57 CO_2 record is unfortunately difficult to obtain with the information available and we have used two independent approaches to establish this dating (see Fig. 1a and 1b). The important differences (from 50 to 220 years) obtained in the mean age of the atmospheric gas for the levels measured in CO_2 between the two approaches demonstrate the large uncertainties in the dating but because of the occurence in the core

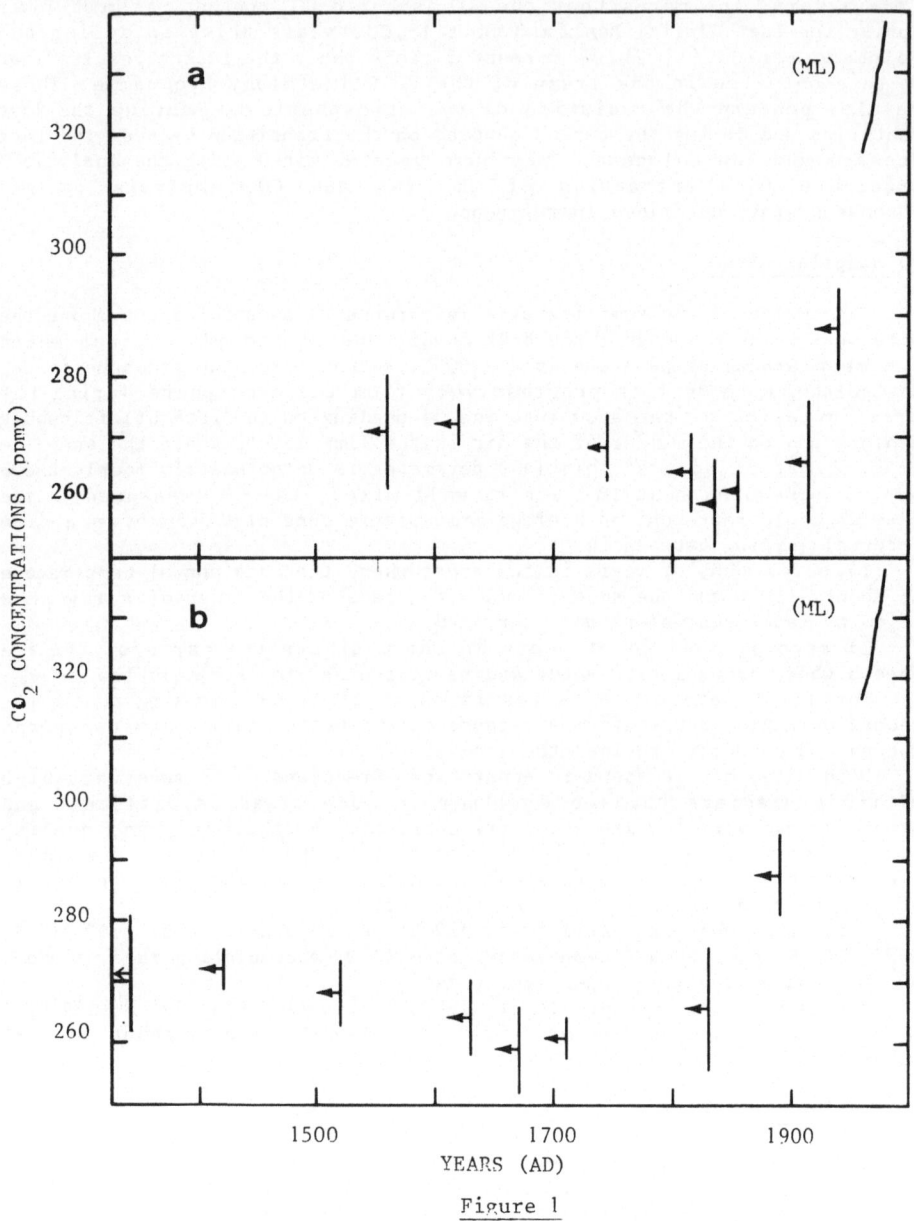

Figure 1

(see the Figure captions below the Acknowledgements)

of a sulfate horizon attributed to the Tambora event (1815 AD) by Delmas and Legrand (personal communication), we have much more confidence, for the youngest part of the CO_2 record, in the time scale shown in Figure 1b.

Whatever the dating uncertainties are, the results (Figure 1) indicate that the atmospheric CO_2 concentration during the last few centuries may have been as low as 260 ppmv[*]. This value is much lower than the concentration of about 295 ppmv calculated for the pre-industrial level by extrapolating the observed atmospheric CO_2 variations measured continuously since 1958 and assuming that fossil fuels were the only sources (see reference 7). The difference between the low concentration observed from the D 57 ice core and the result of the calculation suggests strongly that another significant source besides the fossil fuels has been acting. Taking into account the ages plotted in Figure 1b that we favor at least for the youngest part of the record, this other source may well be the agricultural "revolution" which began to seriously modify the biosphere during the last centuries. The D 57 CO_2 record reveals also that a mean decrease of the order of 10 ppmv may have occurred during the few centuries preceding 1800 AD (cf. Figure 1). This fluctuation could be linked with climate induced changes in the biospheric and/or oceanic CO_2 reservoirs and could reflect the effect of the Little Ice Age on the carbon cycle. It suggests that the so-called "pre-industrial" CO_2 level was not constant over the few centuries preceding the important anthropogenic perturbation of the various CO_2 pools. Furthermore, the D 57 record was the first ice core study to clearly indicate the increase in atmospheric CO_2 due to the anthropogenic influence (see Figure 1).

4. The Ice Age -Holocene transition

Previous CO_2 results from ice core studies (see in particular references 1, 2 and 5) suggest that the atmospheric CO_2 concentrations over the last 40,000 yrs were (i) minimum and as low as about 200 ppmv during the Last Glacial Maximum, and (ii) increasing between the Last Glacial Maximum and the Holocene to reach an average value of about 270 ppmv for the Holocene period.

We decided, consequently, to study in detail how the CO_2 was changing between these two well marked climatic periods. Our recent results obtained from the Dome C ice core (8, 9 and Figure 2) are in agreement with the general trend previously measured on the same core (2) and the detailed study of the Ice Age-Holocene transition indicate, as shown in Figure 2, that :

1) The increase of the CO_2 concentrations associated with the climatic change betwen the Last Glacial Maximum and the Holocene begins before or simultaneously with the climatic warming in Antarctica which is revealed by the large shift observed in the isotopic composition of the ice ;

2) This CO_2 increase may have occurred in two steps ;

3) The CO_2 concentrations fluctuate abruptly and widely near the end of the Ice Age - Holocene transition.

[*] At this stage of the discussion it must be pointed out that it is not possible to disregard the possibility of a systematic error on the CO_2 measurements which would be less than + 15 ppmv (6). Even if we have to correct the measured values by as much as + 15 ppmv the discussion below remains valid.

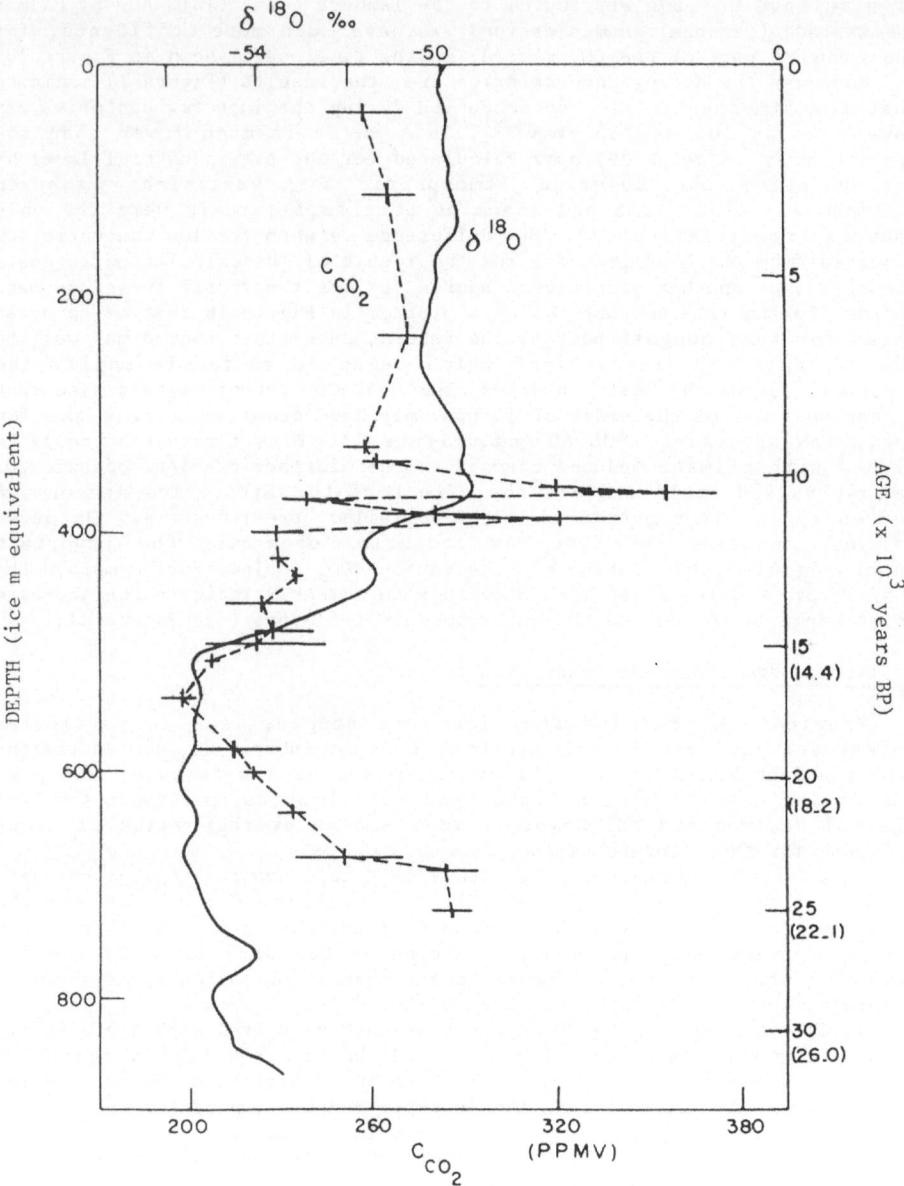

Figure 2

(see the Figure captions below the Acknowledgements)

It is premature to discuss in detail the atmospheric significance and the climatic interpretation of these recent results. They nevertheless suggest strongly a close interaction between CO_2 and climate.

5. Conclusion

Antarctic ice data indicate that the atmospheric CO_2 may have played a significant role in the climate evolution over the last 40,000 years. Nevertheless, these data suggest that similar relative variations in the atmospheric CO_2 have taken place near the Ice Age - Holocene transition in a few thousands years (see Figure 2) and during the most recent period in about 100-150 years (see Figure 1), although the global temperature change has been quite obviously much larger during the transition than during the recent period. Possible reasons include the probable role of other climatic forcing and the fact that we are, for the recent time, in a transient period.

Another interesting information obtained from these ice core measurements is the confirmation that the "pre-industrial" atmospheric CO_2 concentrations were likely significantly lower than the 290 ppmv figure adopted previously in modelling the evolution of the atmospheric CO_2 during the present period and the corresponding climatic response.

Although the CO_2 record from ice core provides currently the most direct evidence of the CO_2 changes in the atmosphere during the last 40,000 years, more studies are needed to reach firm conclusions about the atmospheric significance of the abrupt and wide CO_2 fluctuations observed near the end of the Ice Age - Holocene transition and about the interpretation of the timing of the CO_2 increase at the end of the Last Glacial Maximum in terms of worldwide CO_2 - climate interaction. Further efforts are also needed to obtain more accurate CO_2 determinations and to reduce the uncertainties on the age of the gas when looking at the record over the last few centuries. New CO_2 records from ice cores taken in different sites are necessary to confirm the first results and to cover the whole last climatic cycle.

ACKNOWLEDGEMENTS

This work was supported by the Commission of the European Communities, the Programme Interdisciplinaire de Recherche sur l'Environnement, Terres Australes et Antarctiques Françaises, Expéditions Polaires Françaises and the US National Science Foundation.

FIGURE CAPTIONS

Figure 1. D 57 ice core : CO_2 concentrations of the air bubbles trapped in the ice during the last few centuries (6). The mean ages of the gas have been obtained using two different approaches (a and b). The arrows indicate qualitatively that the mean ages may be older if the air in the firn layers is not well mixed. The atmospheric record observed at Mauna Loa since 1958 AD (10) is also plotted.

Figure 2. Dome C ice core : CO_2 concentrations of the air bubbles trapped in the ice (from reference 9) and the isotopic composition ($\delta^{18}O$) of the ice (from reference 11) versus depth and time.

REFERENCES

1. DELMAS, R.J., ASCENCIO, J.M. and LEGRAND, M. (1980). Polar ice evidence that atmospheric CO_2 20,000 yr BP was 50 % of present. Nature, 284, 155-157.
2. NEFTEL, A., OESCHGER, H., SCHWANDER, J., STAUFFER, B. and ZUMBRUNN, R. (1982). Ice core sample measurements give atmospheric CO_2 content during the past 40,000 yr. Nature, 295, 220-223.
3. BARNOLA, J.M., RAYNAUD, D., NEFTEL, A. and OESCHGER, H. (1983). Comparison of CO_2 measurements by two laboratories on air from bubbles in polar ice. Nature, 303, 410-412.
4. SCHWANDER, J. and STAUFFER, B. (1984). Age diffference between polar ice and the air trapped in its bubbles. Nature, 311, 45-47.
5. LORIUS, C. and RAYNAUD, D. (1983). Record of past atmospheric CO_2 from tree-ring and ice core studies. In : Carbon dioxide : Current Views and Developments in Energy/Climate Research, W. Bach et al. (eds), D. Reidel Publishing Company, 145-176.
6. RAYNAUD, D. and BARNOLA, J.M. (in press). An Antarctic ice core reveals atmospheric CO_2 variations over the past few centuries. Nature.
7. W.M.O. (1983). Report of the WMO (CAS) meeting of experts on the CO_2 concentrations from pre-industrial times to I.G.Y. WMO project on research and monitoring of atmospheric CO_2 (WMO-Geneva), Report N° 10
8. RAYNAUD, D. and BARNOLA,J.M. (1984). The CO_2 record in ice cores : a reconstruction of the atmospheric evolution between 18 ka BP and 1850 AD (Abstract). Annals of Glaciology, 5, 224.
9. BARNOLA, J.M. (1984). Etude des variations passées du CO_2 atmosphérique à partir de l'analyse de l'air piégé dans la glace. Thèse de 3e cycle, Université Scientifique et Médicale de Grenoble.
10. KEELING, C.D., BACASTOW, R.B. and WHORF, T.P. (1982). Measurements of the concentration of carbon dioxide at Mauna Loa Observatory, Hawaii. In : Carbon Dioxide Review : 1982, W.C. Clark (ed.), Oxford University Press, 377-385.
11. LORIUS, C., MERLIVAT, L., JOUZEL, J. and POURCHET, M. (1979). A 30,000 -yr isotope climatic record from Antarctic ice. Nature, 280, 644-648.

ATMOSPHERIC CO2 CONCENTRATION IN SUBANTARCTIC COUNTRIES

A. Gaudry,[1] J.M. Ascencio-Parvy[2] and G. Lambert[3]
Centre des Faibles Radioactivités, Laboratoire mixte CNRS-CEA, BP1
91190 - Gif-sur-Yvette, France
[1] Also ORSTOM/TAAF, Paris, France
[2] Accidentally deceased
[3] Also Université de Picardie, Saint-Quentin, France

Summary

The atmospheric CO2 concentrations at Amsterdam Island (37° 41'S;77° 47'E) showed three kinds of variations from October 1980 to November 1983. The first one, attributed to local effects of the photosynthesis of the island vegetation, does not seem considerably to affect the monthly mean CO2 concentrations. The second one, attributed to long range transport phenomena, is correlated with variations of the atmospheric radon concentrations. Therefore the existence of regional CO2 sources and sinks in the Southern Hemisphere can be felt at a very large distance. Moreover, an average increase rate of 0.11 ppm/month for a 3-year period has shown a considerable variability within each year: 0.06-0.14 ppm/month in 1981, 0.04-0.12 in 1982 and 0.13-0.25 in 1983. The natural evasion of oceanic CO2 following the 1982/83 El Nino event could account for a part of the anomalous change.

1-Introduction

The knowledge of the whole carbon cycle is necessary to explain the atmospheric CO2 variations in the past and to forecast its future level, more especially in connexion with human activities. The long-term evolution of the atmospheric concentration of CO2 has been modeled by considering the exchanges between three main reservoirs of carbon: oceans, continental biomass and atmosphere. Those exchanges should explain seasonal effects as well as irregular space and time variations.A good knowledge of these exchanges should imply to sight the main sources and sinks of atmospheric CO2.

One possible strategy consists in the determination of the geographical gradients of the atmospheric concentration and of their seasonal changes (1). It has been shown by Komhyr and Machta (2) that in the Southern Hemisphere the average differences of concentration, between polar and tropical latitudes could be of 2 ppm or even less. Consequently the accuracy required for CO2 monitoring should be as close as possible to 0.1 ppm. Such an accuracy requires, in addition to solving problems of metrology and standardization, avoiding any interference due to local effects. Now it has been shown at several stations that the CO2 concentration variations reach sometimes more than 10 ppm under the influence of local sources and sinks from human or vegetal origin (3), as well as from the volcanic activity (4). For this reason, we have selected the Amsterdam Island, Terres Australes et Antarctiques Françaises, to establish in that part of the Southern Hemisphere a station for the study of the background pollution, as free as possible from local perturbations.

2-Experimental Procedures and Results

Since April 1980, CO_2 has been continuously monitored at Amsterdam Island, 37 ° 41'S and 77° 47'E (fig1).There is no active volcanism and no regular marine or air traffic corridors in the vicinity of the island. The population is limited to about 30 people living at the scientific base Martin de Vivies (fig2). The vegetation is scarce. The surroundings of the sampling point at Pointe Benedicte, 2.5 km westwards from Martin de Vivies, are mainly composed of graminea. Therefore this site looks particularly well adapted to a background station.

The air is collected on top of an 8m-high aluminium tower set up at several tens of meters from a cliff, generally directly exposed to winds blowing from the sea.

The CO_2 concentration is measured using a non-dispersive infrared gas analyser (Hartmann-Braun/Uras 2T). The standardization was previously performed by using CO_2 in N2 standard gases provided by the Scripps Institution of Oceanography (SIO) and expressed in the 1974 WMO/N2 scale. Recently, CO_2 in air standard gases produced by "L'Air Liquide" and certified by the SIO in the WMO Mole Fraction Scale, enabled us to determine all our data in this scale.

Despite its geographical isolation and scarce vegetation, the site might still be affected by local CO_2 contamination. Thus, very strict selective criteria were imposed to define a marine sector : wind direction between 300 ° and 020° and velocities greater than 4m per second.

Two different shapes of records have been obtained and are shown in fig 3. The first kind consists in a very steady curve and corresponds to either winds blowing from the marine sector and faster than 4 m per second or strong winds faster than 10m per second. The second kind, obtained in weak winds, for which the direction was not well defined, or for air masses having rather slowly crossed over the island, presents significant fluctuations.

In agreement with the preceding remarks, it can be observed that, as far as marine air masses are concerned, no diurnal variation of CO_2 concentration occurs (fig 4a). In contrast, by land sector winds, a minimum of the CO_2 concentration can be observed at the middle of the day (fig 4b), which could be attributed to the photosynthesis of the island vegetation.

3-Long Range Transport of CO_2 Variations

As shown above, it has been possible to select meteorological conditions for which any local interference can be ruled out: winds stronger than 4m per second blowing from a 260°-60° sector. However, even in these conditions, a significant CO_2 variability is still observed,which can be related to changes in the meteorological parameters (fig 5).

During the short period between April 20 and September 30, 1980, not less than 29 meteorological fronts of air masses were registered at Amsterdam Island. They are listed in Table I. In 8 of these 29, the CO_2

- 248 -

variations do not stand out from the noise. However, in the other 21 cases the large scale meteorological perturbations were associated with significant CO2 variations, also listed in Table I, with their plus or minus sign amplitudes (in ppm).

A correct interpretation of the phenomenon could be made easier by a correlative study of the atmospheric Rn222 concentration. Indeed, this radioactive gas is known to be a good tracer of continental air masses and has been extensively used at the French subantarctic stations (5,6,7).It was shown that significant increases of Rn222 concentration are associated with meteorological fronts, during several days, and can be observed successively at Crozet, Kerguelen, and Amsterdam islands(the positions of these islands are shown in Figure 1). Hence, the existence of significant peaks of Rn222, that is larger than 1.5 picocuries per m3, is mentioned in Table I.

One typical examples of the progression of a radon peak from west to east is given in Figure 6, where the CO2 concentration at Amsterdam Island is also indicated. A clear correlation between CO2 and Rn222 concentrations can be seen in this figure. However,such a correlation seems to be sometimes positive and sometimes negative.

The results gathered in Table I can be summarized as follows : out of over 29 meteorological fronts, 5 are neither accompanied with a radon event, nor a CO2 event, 7 are correlated simultaneously with a radon peak higher than 4 pCi per m3, and with a CO2 concentration increase, 14 are correlated with a CO2 concentration decrease, 8 of these events being accompanied with a small, but significant radon peak, and 3 correspond to a small peak of radon, without a significant CO2 concentration variation.

The existence of peaks of radon concentration has been interpreted as an advection of air from South Africa. Therefore, it seems very likely that the CO2 variations should have the same origin.

It is well known that, owing to the low vertical eddy diffusion of the air during the night, the radon concentration over a continental area is then higher than during the day. The same diurnal variation is also observed in the case of CO2 and is attributed to photosynthesis. This similitude has been shown by Dorr et al.(8), who found, in Europa, a good correlation between Rn222 and CO2 concentrations with a diurnal pattern.

Actually, it can be observed in Table I, that all Rn222 peaks reaching or exceeding 4 pCi/m3 are correlated with CO2 concentration increases. Conversely the CO2 concentration decreases are systematically obtained when the Rn222 concentrations are smaller than 2.5 pCi/m3 or even at its background level.

To summarize, it seems that we can observe at Amsterdam Island, with similar frequencies, three different kinds of non-local variations of the atmospheric CO2 concentration, of the order of + or -1 ppm:

i. CO2 concentration increases associated with Rn222 peaks superior or equal to 4 pCi/m3, which are interpreted as advections of night air masses from South Africa.

ii. CO2 concentration decreases associated with small Rn222 peaks between 1.5 and 2.5 pCi/m3 interpreted as advections of day air masses from South Africa.

iii. CO2 concentration decreases non-associated with any significant Rn222 variation that possibly could be attributed to an absorption process by the sea water.

In conclusion, it seems that the concentrations of atmospheric trace gases, like Rn222 and CO2 are rather well preserved along trajectories longer than 5000 km. Therefore, the existence of CO2 regional sources and sinks in the Southern Hemisphere can be left at a very large distance. It involves CO2 variations of the order of 0.5 to 2.5 ppm during 1 to 3 days in a remote station.(9)

4- YEAR TO YEAR CO2 VARIATIONS

The CO2 monthly mean values obtained from hourly data, either with or without selection are shown in Table II. The CO2 concentration in the marine sector increased between October 1980 and November 1983 from 337.55 to 341.61 ppm, (341.74 in September 1983) corresponding to an average increase rate of 0.11 ppm/month. No clear seasonal effect appears, in contrast with the results at Mauna Loa and South Pole (10,11). The increase rate was not constant though, as shown in Fig 7a.

The changes of the yearly mean value of the monthly increase rate were determined by calculating linear regressions for 11-month periods (5 months preceding and following the considered month) from March 1981 to June 1983. Such a simple calculation was possible because of the lack of seasonal variation, in contrast with the situation in most of other WMO background stations. The results are shown in Fig 7b. These variations are characterized by three periods. The first one, from March 1981 to December 1981, corresponds to an irregular increase rate between 0.06 and 0.13 ppm/month, averaging 0.09 ppm/month. A second period, from January to August 1982 shows a much lower CO2 increase rate, within 0.04 and 0.08, averaging 0.05 ppm/month. The last period, up to the end of the curve B displays an anomalously high CO2 increase rate, between 0.12 and 0.25 ppm/month.

Such changes in the increase rate of the CO2 concentration have been already observed in the past and very generally related to changes in the sea surface temperature (SST) of the Pacific Ocean (12,13,14,15). More recently Gammon et al. (16), found that El Nino Southern Oscillation events of 1972, 1976 and 1982 were able to induce slower than normal atmospheric CO2 increases during the first year, followed by very rapid CO2 increases in the following year. That is precisely what we observe here.

Gill and Rasmusson (17) described a SST anomaly in the Pacific Ocean, off the Peruvian coast, beginning in March 1982, but showing a first important peak only in December 1982, perfectly correlated to the resumption of the CO2 increase, also observed in Table II. In the same way the high monthly CO2 increase rates measured in June and July 1983 in our data follow immediately the highest peak of the SST anomaly observed in June 1983 (17).

The first idea is therefore to attribute fast atmospheric increases of CO2 directly to a stronger outgassing from anomalously warm surface waters. MacIntyre (18) showed that, assuming an equilibrium of CO2 between the whole atmosphere and subtropical surface-ocean waters, as well as a constant alkalinity, the variation of the atmospheric CO2 (dp) was related to a SST variation (dT) by $dp/dT=1.54$ ppm/°C for a 100m-deep mixed-layer. Bacastow (13) confirmed such an order of magnitude ($dp/dT=1/0.93=1.08$ ppm/°C), for a 75m-deep mixed-layer, by considering a negligible time lag for attaining the thermodynamical equilibrium and for a mean "Revelle factor" of 10.In fact, it is more realistic to take here a "Revelle factor" of 9 for warm waters (19). Moreover, the depth of sea water containing the same amount of total CO2 as the atmosphere was 84 m in 1983, instead of 76 m in 1966 as in Bacastow's paper. These corrections lead to a figure of about 1.4 ppm/°C. However, it was pointed out by Machta(20) and Zimen et al (21) that the oceanic surface layer exchanging Carbon with the atmosphere could be thicker than the wind-mixed layer. Therefore, there is a large uncertainty on the value of the coefficient dp/dT.

An average SST anomaly during the 1982-1983 El Nino event can be calculated from the NOAA SST anomaly maps (22). Table III shows the increases of the CO_2 atmospheric concentration calculated from quarterly values of the SST anomaly area. The three different figures 1.08, 1.4 and 1.54 ppm/°C were used with both the following hypotheses : either the Southern Hemisphere was considered separately, or the calculations were conducted for the whole atmosphere and oceans. In this last case, the atmospheric CO_2 increases are smaller than 0.2 ppm and therefore the SST anomaly influence could account for a maximum of 1/3 of the CO_2 increases actually observed. This conclusion is not so clear when the Southern Hemisphere is considered separately. In effect, the order of magnitude of the CO_2 amount possibly injected into the atmosphere varyied from figures comparable to the preceding case (0.13 to 0.30 ppm) in June/August 1982 and from March to November 1983, to values significantly higher (0.32 to 0.50 ppm) from September 1982 to February 1983. It is also worth while to remember that our study did not take account of possible changes in the total carbon of the superficial waters. Finally, a modeling of the changes in the atmospheric CO_2, directly related to the occurrence of a strong El Nino, will only be possible in the future from very accurate measurements of the physical and chemical properties of the superficial sea water in the involved region. Moreover, it is also necessary to consider the possible existence of other sources and sinks of CO_2, indirectly related to the El Nino.

The global CO_2 increase, generally observed, of the order of 0.11 ppm/month, is essentially attributed to anthropogenic injections. There is no reason to assume that the input in 1983 was significantly different than in 1981 and 1982. The sources and sinks, above mentioned, could be rather a small change in the growing rate of the vegetation, possibly related to meteorological anomalies, generated in turn by the El Nino event. Rasmusson and Wallace (23) described some meteorological anomalies, such as droughts in South Africa, Australia and Indonesia, during each El Nino events, and particularly from December 1982 to February 1983. Such meteorological effects might affect the CO_2 cycle in the Southern Hemisphere.(24)

Acknowledgements

This programme has been supported by the Administration of the Territoire des Terres Australes et Antarctiques Françaises, the Commission des Communautés Européennes, and the Programme Interdisciplinaire de la Recherche en Matière d'Environnement du CNRS. We thank B.Ardouin, A.Jegou, and the staff in charge of the experiment at Amsterdam Island for their technical assistance.

References

(1)-Fraser P.J., G.I. Pearman, and P. Hyson, The Global Distribution of Atmospheric Carbon Dioxide ,2. A Review of Provisional Background Observations, 1978-1980, J. Geophys. Res. ,88 ,3591-3598, 1983.
(2)-Komhyr W.D., and L. Machta, Preliminary results and interpretation of the NOAA Flask Programme, WMO/ICSU/UNEP Conference on Analysis

and Interpretation of Atmospheric CO2 Data, Bern, Switzerland, 1981.

(3)-Lowe D.C., P.R. Guenther, and C.D. Keeling, The concentration of atmospheric Carbon Dioxide at Baring Head, New Zealand, Tellus, 31, 58-67, 1979.

(4)-Miller J.M., and J.F.S. Chin, Short-term disturbances in the Carbon Dioxide record at Mauna Loa Observatory, Geophys. Res. Lett.,5 , 669-671, 1978.

(5)-Lambert G., G. Polian, and D. Taupin, Existence of periodicity in radon concentrations and in large scale circulation at lower altitudes between 40 and 70 S, J. Geophys. Res. , 75, 2341-2345, 1970.

(6)-Lambert G.,B. Ardouin, G. Polian, and J. Sanak, Natural radioactivity balance in the atmosphere of the Southern Hemisphere, in The Natural Radiation Environment II, ed. J.A.S. Adams, W.M. Lowder, and T. F. Gesell, Conf. 720 805-P2, National Technical Information Service, U.S. Dept. of Commerce, Springfield, Va., 1972.

(7)-Polian G., Transports atmosphériques dans l' Hémisphère Sud et bilan global du radon 222, Thesis, Université de Paris, 1984.

(8)-Dorr H., B. Kromer, I. Levin, K.O. Munnich, and H.J. Volpp, CO2 and Radon 222 as tracers for atmospheric transport, J. Geophys. Res., 88, 1309-1313, 1983.

(9)-Gaudry A, J.M.Ascensio-Parvy and G.Lambert,Preliminary study of CO2 variations at Amsterdam Island (Terres Australes et Antarctiques Francaises),J.Geophys.Res.88,1323-1329,1983.

(10)-Keeling C.D., R.B.Bacastow, A.E.Bainbridge, C.E.Ekdahl, P.R.Guenther, L.S.Waterman and J.F.S.Chin, Atmospheric Carbon Dioxide variations at Mauna Loa observatory, Hawaii, Tellus, 28,538-551, 1976a.

(11)-Keeling C.D., J.A.Adams, A.Jr.Ekdahl and P.Guenther, Atmospheric Carbon Dioxide variations at the South Pole, Tellus, 28, 552-564, 1976b.

(12)-Newell R.E.,A.R.Navato and J.Hsiung, Long-term global sea surface temperature fluctuations and their possible influence on atmospheric CO2 concentrations, Pure Appl. Geophys., 116, 351-371, 1978.

(13)-Bacastow R., Dip in the atmospheric CO2 level during the mid-1960's, J. Geophys. Res., 84, 3108-3114, 1979.

(14)-Wong C.S.and K.G.Pettit, Global-scale secular CO2 trends and seasonal changes at Canadian CO2 stations : ocean weather station P, Sable Island and Alert., J. Geophys. Res., in press.

(15)-Schnell R.C., J.M.Harris and J.A.Schroeder, A relationship between Pacific Ocean temperatures and atmospheric CO2 concentrations at Point Barrow and Mauna Loa, WMO/UNEP/ICSU Conf. on Analysis and interpretation of CO2 data, Bern September 1981.

(16)-Gammon R.H., W.D.Komhyr, L.S.Waterman, T.Conway and K.Thoning, Estimating the natural variation in atmospheric CO2 since 1860 from interanual changes in tropospheric temperature and the history of Major El Nino events. Abstr. Chapman Conf. Natural Variations in Carbon Dioxide and the Carbon Cycle. Tarpon Springs, Florida, 1984.

(17)-Gill A.E., Rasmusson E.M., The 1982-1983 climate anomaly in the equatorial Pacific, Nature , 306, 17, 229-234, 1983.

(18)-MacIntyre F.,On the temperature coefficient of pCO2 in seawater, Climatic Change, 1, 349-354, 1978.

(19)-Broecker W.S., T.Takahashi, H.J.Simpson, T.H.Peng, Fate of fossil fuel Carbon Dioxide and the global Carbon budget, Science, 206, 409-418, 1979.

(20)-Machta L., Prediction in the atmosphere, in Carbon and the Biosphere, Woodwell and Pecan,ed.,CONF-720510, National Technical Information Service, U.S. Department of Commerce, Springfield, Virginia 22151, 51-85,1973.
(21)-Zimen K.E.,P.Offermann and G.Hartmann,Source Functions of CO2 and Future CO2 Burden in the Atmosphere, Z.Naturforsch.32a,1544-1554 ,1977.
(22)-Arkin P.A.,J.D.Kopman,R.W.Reynolds, 1982-1983 El Nino/Southern Oscillation Event Quick Look Atlas,NOAA/National Weather Service,Nov 1983.
(23)-Rasmusson E.M. and J.M.Wallace, Meteorological aspects of the El Nino/Southern Oscillation, Science,222, 1195-1202,1983.
(24)-Ascensio-Parvy J.M.,A.Gaudry and G.Lambert,Year to year CO2 variations at Amsterdam Island in 1980-83,Geophys.Res.Let.,in press,1984.

Table I. Correlative variations of CO2 and radon 222 Concentrations with the passage of meteorological fronts registered at Amsterdam Island.

Date (year : 1980)	CO2 event amplitude (ppm)	Radon event (pCi / m3)
April 20	-0.8	0
April 24	-2	0
April 27-28	-2.5	0
May 12	0	1.8
June 2	+0.7	4
June 6-9	-0.8	1.6
June 13	0	0
June 15-17	+1.3	4
June 20	-1.3	0
July 1-2	-0.7	1.6
July 4-5	-1.3	0
July 8	0	0
July 10-11	0	0
July 12	0	0
July 18	0	2
July 25-27	+0.65	4.5
July 29-31	+0.7	8.5
August 2-3	0	0
August 4-5	+0.5	5
August 12-13	+0.9	6.5
August 17-18	+0.85	4
August 21-23	-1.1	1.6
August 25-26	-0.6	1.9
August 29-31	0	2
September 2-3	-1.1	2.5
September 7-8	-0.7	1.5
September 15-16	-1	0
September 17-21	-1.6	2
September 23-26	-1.1	1.6

Table II. Monthly mean CO2 concentration at Amsterdam Island (1981 WMO Mole Fraction Scale); A-all data; B-data by marine sector; C-Difference with the preceding month (marine sector).

Data		A	B	C
Year	Month			
1980	10	337.98	337.55	
	11	337.92	337.55	0.00
	12	338.22	337.72	+0.17
1981	01	338.21	338.12	+0.40
	02	338.54	338.27	+0.15
	03	338.20	338.23	-0.04
	04	338.13	338.09	-0.14
	05	338.10	338.33	+0.24
	06	338.15	338.03	-0.30
	07	338.37	338.17	+0.14
	08	338.73	338.74	+0.57
	09	338.60	338.43	-0.31
	10	338.67	338.64	+0.21
	11	338.90	338.81	+0.17
	12	339.01	338.89	+0.08
1982	01	339.35	339.28	+0.39
	02	339.78	339.52	+0.24
	03	339.40	339.31	-0.21
	04	339.24	339.00	-0.31
	05	339.23	338.86	-0.14
	06	339.28	338.97	+0.11
	07	339.50	339.05	+0.08
	08	340.06	339.81	+0.76
	09	340.06	339.70	-0.11
	10	339.92	339.48	-0.22
	11	339.89	339.51	+0.03
	12	339.98	339.54	+0.03
1983	01	340.19	339.82	+0.28
	02	340.35	340.12	+0.37
	03	340.67	340.21	+0.02
	04	340.76	340.50	+0.29
	05	340.73	340.53	+0.03
	06	341.11	341.04	+0.51
	07	341.70	341.65	+0.61
	08	341.87	341.72	+0.07
	09	341.91	341.74	+0.02
	10	341.90	341.61	-0.13
	11	341.91	341.61	0.00

Table III. Atmospheric CO_2 ascribed to the 1982-1983 El Nino. (1)-Area (10 exp(6) km2) showing a temperature elevation; (2)-Mean SST anomaly (C); (3)-CO2 increase (ppm) for dp/dT = 1.08 ppm/°C; (4)-CO2 increase (ppm) for dp/dT = 1.4 ppm/°C; (5)-CO2 increase (ppm) for dp/dT = 1.54 ppm/°C

Southern Hemisphere					Date	Both Hemispheres				
(1)	(2)	(3)	(4)	(5)		(1)	(2)	(3)	(4)	(5)
14.8	1.04	0.16	0.20	0.23	June to Aug.1982	21.2	1.03	.065	.08	.09
20	1.57	0.32	0.42	0.45	Sept.to Nov.1982	32.5	1.51	0.14	0.18	0.20
24.8	1.35	0.35	0.45	0.50	Dec.82 to Feb.83	35.7	1.36	0.14	0.18	0.20
13.9	1.30	0.19	0.25	0.27	March to May 1983	19.5	1.32	0.08	0.10	0.11
11.4	1.73	0.21	0.27	0.30	June to Aug. 1983	16.8	1.64	0.08	0.10	0.11
9.8	1.07	0.10	0.13	0.14	Sept. to Nov.1983	9.9	1.07	.035	.045	0.05

Fig. 1. French islands in Subantarctic Area

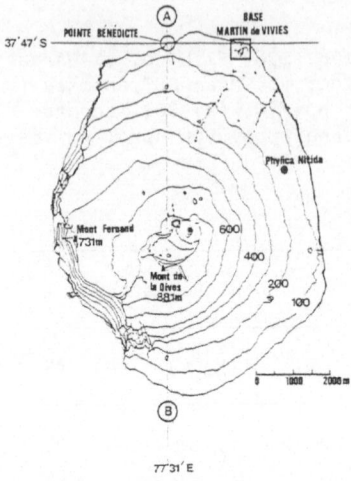

Fig. 2. Amsterdam Island map

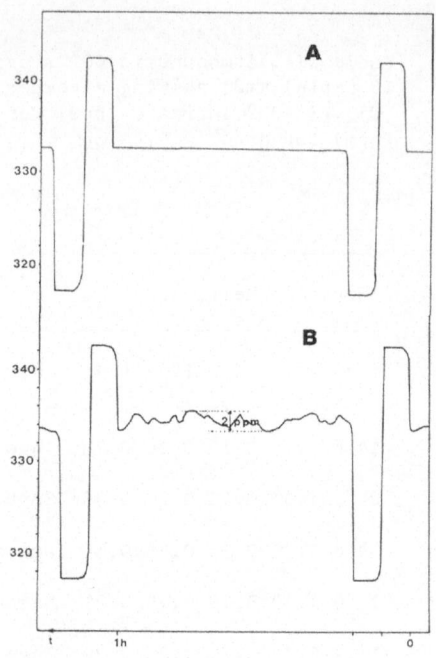

Fig. 3. One-hour records of atmospheric CO2
a- by a 13 m/s wind from the 300
b- by 1 to 2 m/s winds from the 100

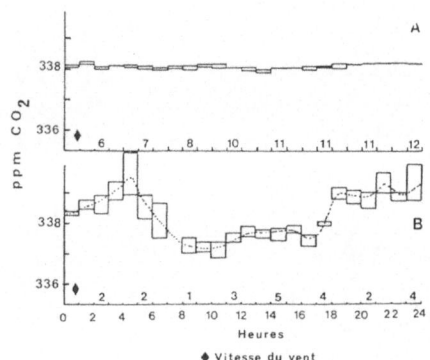

♦ Vitesse du vent

Fig. 4. One-day evolutions of CO2 in 1980
a- marine sector; b- land sector

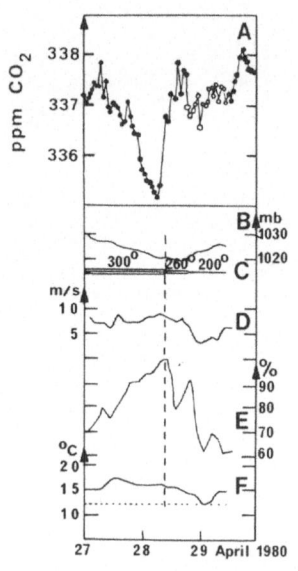

Fig. 5

a- CO2 decrease;   b- barometric pressure
c- wind direction;   d- wind velocity
e- humidity;   f- air temperature

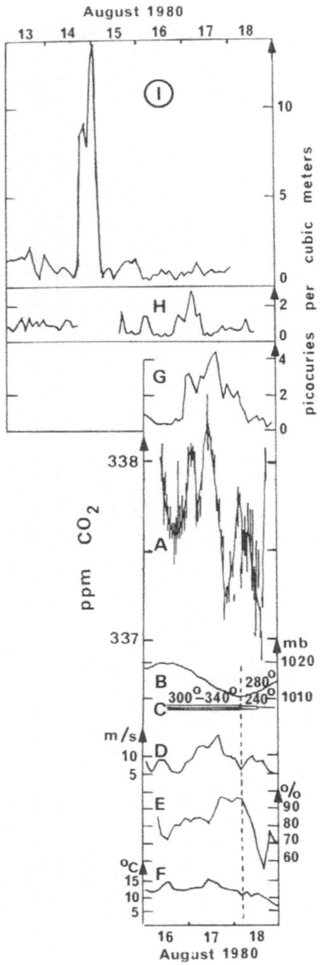

Figure 6. Positive correlation between radon 222 (G) and CO2(a) at Amsterdam Island;
b- Barometric pressure; c- Wind rotation; d- Wind velocity
e- Humidity; f- Air temperature
The radon events are observed at Kerguelen (H) and Crozet (I) with respective time lags of 12 and 60 hours, by reference to Amsterdam.

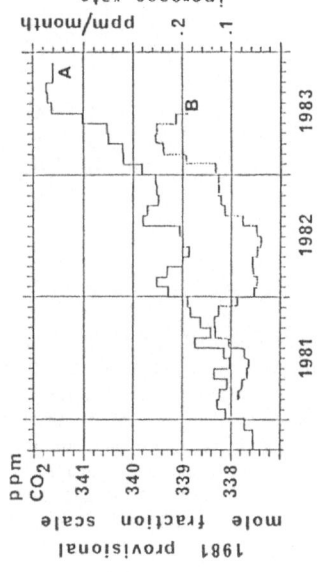

Figure 7. a- 3-year trend of atmospheric CO2 at Amsterdam Island
b- Monthly increase rate averaged over 11 months.

THE INFLUENCE OF MAN-MADE SO_2 AND PARTICLE EMISSION ON THE GLOBAL AEROSOL CONCENTRATION AND THE OPTICAL PROPERTIES OF THE ATMOSPHERE

H. Hinzpeter
Max-Planck-Institute for Meteorology
Hamburg, FRG

Summary

Using a two-dimensional grid point model and zonal mean values of temperature, humidity, wind field and exchange coefficients the zonal distribution of aerosol particle concentration and SO_2 concentration are calculated. This is done for the natural sources of aerosol particles, H_2S and DMS and for the man-made emission of aerosol particles and SO_2 as estimated for natural emission conditions, for the present man-made emission and for twice the amount of present man-made emission.

All sources and sinks, including gas-to-particle conversion are used. The calculated particle concentration, roughly divided in three size classes, is then used to determine the number of cloud droplets in stratiform clouds. Using radiation routines the albedo of the stratus clouds and the resulting planetary albedo has been calculated. The result shows that due to the present man-made emission the albedo is increased by about 1% in the region from 30-70° north. The southern hemisphere is not influenced.

Due to the increased aerosol particle concentration the reduction of the short wave net flux at the earth's surface turns out to have the same value as the increase of the long wave net flux by doubling the CO_2 concentration.

1.1. Introduction

The often discussed influence of mans activity on climate is that of a possible warming of the atmosphere by increasing CO_2 concentration. But generally the discussion neglects the influence of variing cloudiness with increasing CO_2. We know that the increase in coverage by the lower stratus clouds by 4% only would compensate the effect of doubling the CO_2 concentration due to enhanced reflection of solar radiation. But not only the amount of clouds, also the change of their optical properties, could influence albedo and therefore climate. Such a change could be created by a modified number of cloud condensation particles due to an increased aerosol particle concentration. The variation of the aerosol concentration due to man-made emission of SO_2 and aerosol particles is therefore estimated and the resulting change of the albedo is calculated.

1.2.   The model

        Our zonally averaged two-dimensional model uses the climatological data of different autors: temperature (northern hemisphere, Defant, southern hemisphere, Newell), humidity (northern hemisphere, Oort, southern hemisphere Sasamori), wind field (Newell), exchange coefficient (Gidel et al.).
        The grid point model has a high resolution in the industrial belt of the northern hemisphere of 2° between 33° and 57° N, while in the other regions the resolution is 10°. The model has in the 19 vertical levels 10 of which are in the lowest kilometer. For the numerical integration a mixed forward backward method is used (Richtmeyer and Morton, 1967). Starting with Budykos (1963) evaporation data the model has been tested by a computation of the latitude distribution of the mean precipitation for summer and winter. Fig. 1 demonstrates the test results which agree favourably with reality.

1.3.   The SO_2 and the aerosol particle balance

        The aerosol particle spectrum is characterized by only three size classes: nucleus-mode (0.01 - 0.1 μm), accumulation-mode (0.1 - 1 μm), and coarse-mode (1 - 10 μm). To determine the concentrations the sources of H_2S, SO_2, $(CH_3)_2S$, DMS, and natural and man-made particle emission are considered. As an example the natural H_2S and the man-made SO_2-emission are shown in fig. 2. The oxidation of SO_2 in droplets and also in the gaseous phase to H_2SO_4 and the subsequent gas to particle conversion by homogeneous condensation has been taken into account.
        The complex oxidation of HSO_3 in droplets is parameterized and the following is assumed
-        The ion spectrum of the droplets is not influenced by the cloud condensation particle and other trace gases possibly solved.
-        Even at large relative humidity aerosol particles do not react with SO_2; aerosol particles are considered as dry particles.
-        Cirrus clouds are not taken into account. All other clouds are considered as pure water clouds.
        For the gas phase SO_2-oxidation to H_2SO_4 the reaction with OH radicals is considered to be the most important one in the troposphere. Crutzen's OH concentrations were taken. The sulfur acid droplets are created by homogeneous condensation and the production rate is computed as a function of relative humidity and of partial pressure of gaseous sulfur acid.
        The change of the size distribution by coagulation within in the classes and between the particles of different classes is considered but the latter has only a small effect due to the rough description of the spectrum by three classes only. Nevertheless, the particle concentration in the classes is considerably reduced due to coagulation. The washout process is calculated (McMahon and Denison, Radke et al.) but turns out to be also a minor effect.
        The rainout process is important and is calculated in the following way: The coverage by As, Ac; Ci, Cs, Cc; St, Sc; Ns; Cu, and Cb is taken from climatological studies (Telegadas and London, Sasamori et al.), and it is assumed that the total coverage is the sum of the coverage by each cloud type given above. For the different clouds the following mean vertical velocity is used:

Cb	100 cm/s^{-1}
Ns	5 cm/s^{-1}
Cu	65 cm/s^{-1}
St, Sc	1 cm/s^{-1}
As, Ac	1 cm/s^{-1}
Ci, Cs, Cc	---

Cb and Ns are considered as raining clouds, and only in the region between 30°N and 30°S also Cu is considered as a raining cloud.

The following scheme (Fig. 3) - based mainly on formulas given by Twomey - determines the numbers of cloud condensation nuclei. The procedure starts with a reasonable value for k, the Twomey-exponent, and a typical vertical velocity v for the cloud type. C(N) is the number of CCN for S = 1. For the Junge distribution $C(N) = N/B^3$. The coefficient B represents the chemical composition of the particles which are assumed to consist of amoniumsulfid. As a first step the supersaturation S is determined. Assuming a Junge distribution the radius r_k of the smallest particle which is involved in the condensation process is calculated (Junge and McMahon). Then the number of the cloud condensation nuclei CCN is determined in a third step. CCN, C(N) and S than determine the exponent k. With the such determined k the procedure is started again to determine S and so on. After 4-6 iterations we get stable results for k and CCN.

Due to Twomey the mean radius of the cloud droplets is $\bar{r}$ $CCN^{-1/3}$. The amount of precpitation and the droplet volume then give the number of particles that are removed by the rainout process.

The sedimentation velocity is calculated as usual and for the deposition velocity the following values were used: SO_2 0.6 cm/s^{-1}, nucleus-mode 0.1 cm^{-1}, accumulation-mode 0.02 cm^{-1}, coarse-mode 0.2 cm^{-1} (Garland).

1.4.   The aerosol concentration for different szenarios

For winter (December, January, February) and summer (June, July, August), the aerosol concentration and SO_2 concentration is calculated. After 70 model days a nearly stationary state is reached. For each of the two seasons the aerosol concentration has been calculated for the following conditions:
- Only natural emission of H_2S, DMS and aerosol particles
- natural emission and man-made emission of SO_2 and aerosol particles
- natural emission, man-made emission of aerosol particle and doubled man-made SO_2-emission
- natural emission and doubled man-made emission of aerosol particles and SO_2-emission.

As a by-product the deposited volume of the aerosol particles and the deposited sulphur mass both depending on latitude have been determined. The height and latitude dependence of concentrations of SO_2 and of the three aerosol size classes are given in Figs. 4 - 8.

Fig. 4 shows the concentration of the three particle classes due to natural emission only. The coarse-mode is restricted to the lowest km, the accumulation-mode to the lower troposphere and only the nucleus-mode shows considerable concentrations in the whole troposphere but of course with a great decrease of concentration in greater heights. Due to the small SO_2 concentration - shown in Fig. 5 - the gas-to-particle conversion is not very efficient and therefore there is only a small difference between summer and winter.

Considering man-made emission the SO_2 concentration in the lowest km is about 10-20 times greater then for natural emissions only (Fig. 6). Due to the efficient gas-to-particle conversion in the winter season near to the surface of the industrial belt the nucleus-mode is 10 times greater then without man-made emission. The change of the concentration of the other size classes is remarkably smaller, the coarse-mode is nearly unchanged.

A doubling of the man-made emission shows again for the nucleus-mode an increase of several 100% for the lower troposphere of the industrial belt. The concentrations of the other modes increases also by several 100% in the middle and higher troposphere of the northern hemisphere but as the particle concentration is small the concentrations of that modes still remain small. Again the contrast between summer and winter conditions is remarkable.

1.5.   Change of the albedo

For the winter season the above described concentration of aerosol particles is then used to calculate the droplet concentration for the different cloud types using the procedure sketched in Fig. 3. For Cb, Ns, and Cu this is inconsistent because we have used their precipitation to remove particles from the atmosphere, but calculations show that due to their large optical thickness the albedo is not very sensitive to the number of cloud condensation nuclei CCN. St and Sc turn out to be most sensitive to a change of the aerosol concentration.

For the calculation of short wave radiation fluxes the   -Eddington method (Joseph et al., 1976) has been used. For the aerosol particles the complex index of refraction as given by McClatchy is taken. Cloud droplets are assumed to be pure water spheres. The results of short wave net fluxex at the upper and lower boundaries of the atmosphere are presented in table 1.

In that connection some modifications were discussed.

1.   It is assumed that the relative particle spectrum is discribed by a Junge spectrum with   = 3.5 independent of height and latitude. The change of the planetary albedo is than mainly restricted to the industrial belt and the albedo of that region is increased by about 1-2%.

2.   Due to the sinks of the aerosol particles the particle spectrum changes with increasing distance from the source region. The relative number of small particles and therefore   will increase. Considering this variation of the relative spectra the maximal albedo increase is shifted to lower northern latitudes and the mean change is then reduced to 0.5 - 1%. The approximation of the particle concentration in the three classes by a Junge spectrum is a very rough one. Also the influence of the coalescence of cloud droplets to greater ones which do not contribute to the rain but re-evaporate is neglected. Therefore the difference between both results should be considered as a sensitivety test, but the variation of the albedo with changing particle spectrum will probably be smaller than given by the two examples.

In an additional run of the radiation scheme it has been assumed that 20% of the particles of the nuclei-mode are soot particles. The results of the albedo change for the cloudless part of the atmosphere, for the stratus clouds and for the planetary albedo are given in the Figs. 9 - 11.

For the short wave net flux at the earth surface the aerosol effect can be compared with the change due to CO_2-doubling. The increase of the aerosol particle concentration due to man-made emission as presently estimated results in a decrease of the short wave net flux at the earth surface. The amount of that change of the short wave net flux has about the same magnitude as the increase of the long wave net flux due to CO_2 doubling. This means that both changes nearly compensate each other.

REFERENCES

1.   BUDYKO, M.A. (1963). Atlas of heat balance of the earth's surface. Glavnaia Geofizicheskaia Observatoriia, Moscow, USSR.
2.   CRUTZEN, P.J. (1982). Private communication
3.   DEFANT, F. (1975). Private communication.
4.   GARLAND, J.A. (1982). in: Georgii, H.W., and Pankrath, J. (Ed.). Deposition of atmospheric pollutants. Reidel publishing company, Dordrecht, Boston, London, 9-16.
5.   GEORGII, H.-W. (1982). The atmospheric sulphur-budget. in: Georgii, H.-W., and Jaeschke, W. (Ed.): Chemistry of the unpolluted and polluted troposphere. Reidel Publ. Company, London, 295-324.

6. GIDEL, L.T., CRUTZEN, P.J., and FISHMAN, J. (1983). A two dimensional photochemical model of the atmosphere. 1. Chlorocarbon emissions and their effect on stratospheric ozone. J. of Geophys. Res., 88, 6622-6640.

7. JAEGER, L. (1976). Monatskarten des Niederschlages für die ganze Erde. Berichte des Deutschen Wetterdienstes Nr. 139 (Band 18), Offenbach/Main.

8. JOSEPH, J.H., WISCOMBE, W.J., and WEINMAN, J.A. (1976). The delta-Eddington approximation for radiative transfer. J. of Atmos. Sci., 33, 2452-2459.

9. JUNGE, C.E., and McLAREN (1971). Relationship of cloud nuclei spectra to aerosol size distribution and composition. J. of. Atmos. Sci., 28, 382-390.

10. LASKUS, L., and LAHMANN, E. (1977). Korngrößenverteilungen von Stäuben im Rauchgas von Kraftwerken und in atmosphärischer Luft. Staub, Reinhaltung der Luft, 37, Nr. 4, 136-140.

11. McCLATCHEY, R.A., BOLLE, H.J., KONDRATYEV, K.Y. (1980). A cloudless standard atmosphere for radiation computations. Radiation Commission, IAMAP, Oct. 1980.

12. McMAHON, T.A., and DENISON, P.J. (1979). Empirical atmospheric deposition parameters - a survey. Atm. Env., 13, 571-585.

13. MÖLLER, F. (1951). Vierteljahreskarten des Niederschlags für die ganze Erde. Petermanns Geograph. Mitt., 95, 1-7.

14. NEWELL, E.E., KIDSON, J.W., VINCENT, D.G., and BOER, G.J. (1972): The general circulation of the tropical atmosphere and interaction with extratropical latitudes. Vol. I, MIT Press, Cambridge, Mass./USA.

15. OORT, A.H., and RASMUSSON, E.M. (1971). Atmospheric circulation statistics. NOAS Professional paper 5, US Department of Commerce.

16. RADKE, L.F., HOBBS, P.V., and ELTGROTH, M.W. (1980). Scavenging of aerosol particles by precipitation. J. of Appl. Met., 19, 715-722.

17. SASAMORI, T., LONDON, J., and HOYT, D.V. (1972). Radiation budget of the southern hemisphere. Met. Monogr., 35, 9-23.

18. TELEGADAS, K., LONDON, J. (1954). A physical model of the northern hemisphere troposphere for winter and summer. Scientific report No. 1, New York University, College of engineering research division.

19. TWOMEY, S. (1959). The nuclei of natural cloud formation. Part II: the supersaturation in natural clouds and the variation of cloud droplet concentration. Geof. Pura e Appl., 43, 243-249.

20. UMWELTBRIEF 20 (1980): Bericht über die Auswirkungen von Luftverunreinigungen auf das globale Klima. Hrsg.: Der Bundesminister des Innern, Bonn.

Table 1: SHORTWAVE NET FLUX F_n AND NET FLUX DIFFERENCE ΔF_n [Wm^{-2}], AVERAGED

OVER THE NORTHERN HEMISPHERE,

IN COMPARISON TO THE CO_2-EFFECT. CLOUDS: Ns, St

SCENARIO	LEVEL	F_n a	b	$a \cdot \Delta F_n$	b
NATURAL EMISSIONS	GROUND	81.34	77.42	---	---
	TOP	120.27	117.59	---	---
NATURAL PLUS MAN-MADE	GROUND	80.46	76.08	-0.89	-1.34
EMISSIONS WITHOUT SOOT	TOP	119.48	116.80	-0.79	-0.79
NATURAL PLUS MAN-MADE	GROUND	80.27	76.04	-1.07	-1.38
EMISSIONS WITH 20% SOOT	TOP	119.62	116.86	-0.65	-0.73
2 . CO_2	GROUND			1.2	
(RAMANATHAN, 1981)	TOP			4.0	

a. CONSTANT PARTICLE SIZE DISTRIBUTION
b. HEIGHT-LATITUDE DEPENDENT SIZE DISTRIBUTION

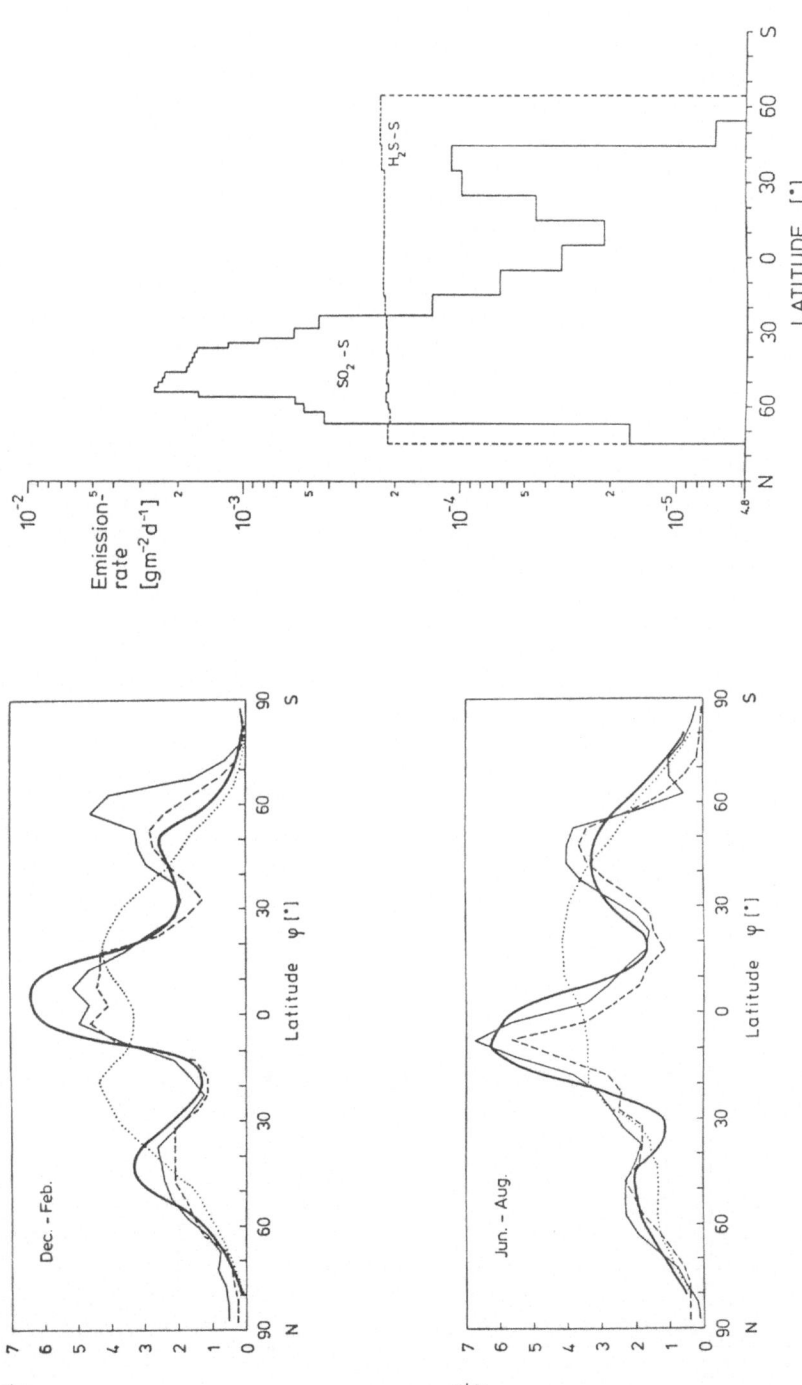

Fig. 1: Calculated and observed precipitation distribution for winter and summer

... Evaporation (Budyko, 1963)
—— Observed precipitation distribution (Jaeger, 1976)
— — Observed precipitation distribution (Möller, 1951)
=== Calculated precipitation distribution (this model)

Fig. 2. Distribution of the emitted H_2S-S (H_2S + DMS) and SO_2 as a function of latitude

- 263 -

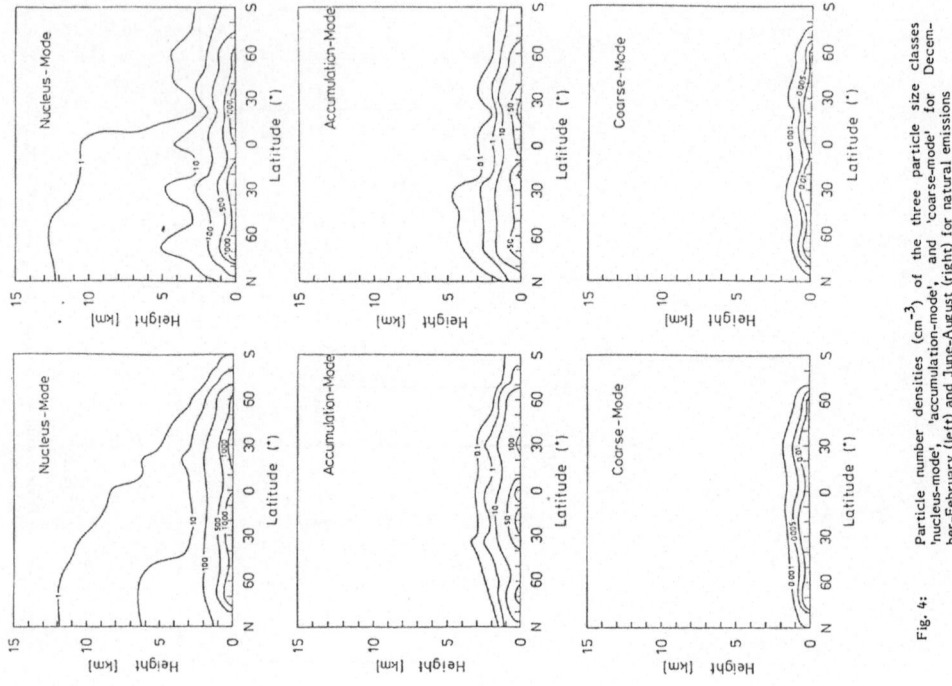

Fig. 4: Particle number densities (cm⁻³) of the three particle size classes 'nucleus-mode', 'accumulation-mode', and 'coarse-mode' for December-February (left) and June-August (right) for natural emissions

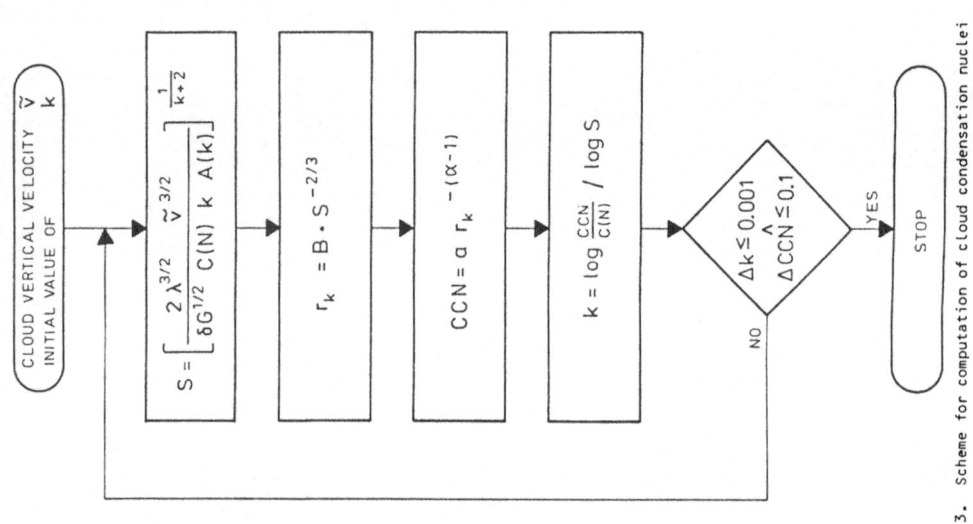

Fig. 3. Scheme for computation of cloud condensation nuclei

$$S = \left[\frac{2\lambda^{3/2}}{\delta G^{1/2}} \frac{\tilde{v}^{3/2}}{C(N) \, k \, A(k)} \right]^{\frac{1}{k+2}}$$

$$r_k = B \cdot S^{-2/3}$$

$$CCN = a \, r_k^{-(\alpha-1)}$$

$$k = \log \frac{CCN}{C(N)} \Big/ \log S$$

CLOUD VERTICAL VELOCITY $\tilde{v}$
INITIAL VALUE OF k

$\Delta k \leq 0.001$
$\wedge$
$\Delta CCN \leq 0.1$

NO

YES

STOP

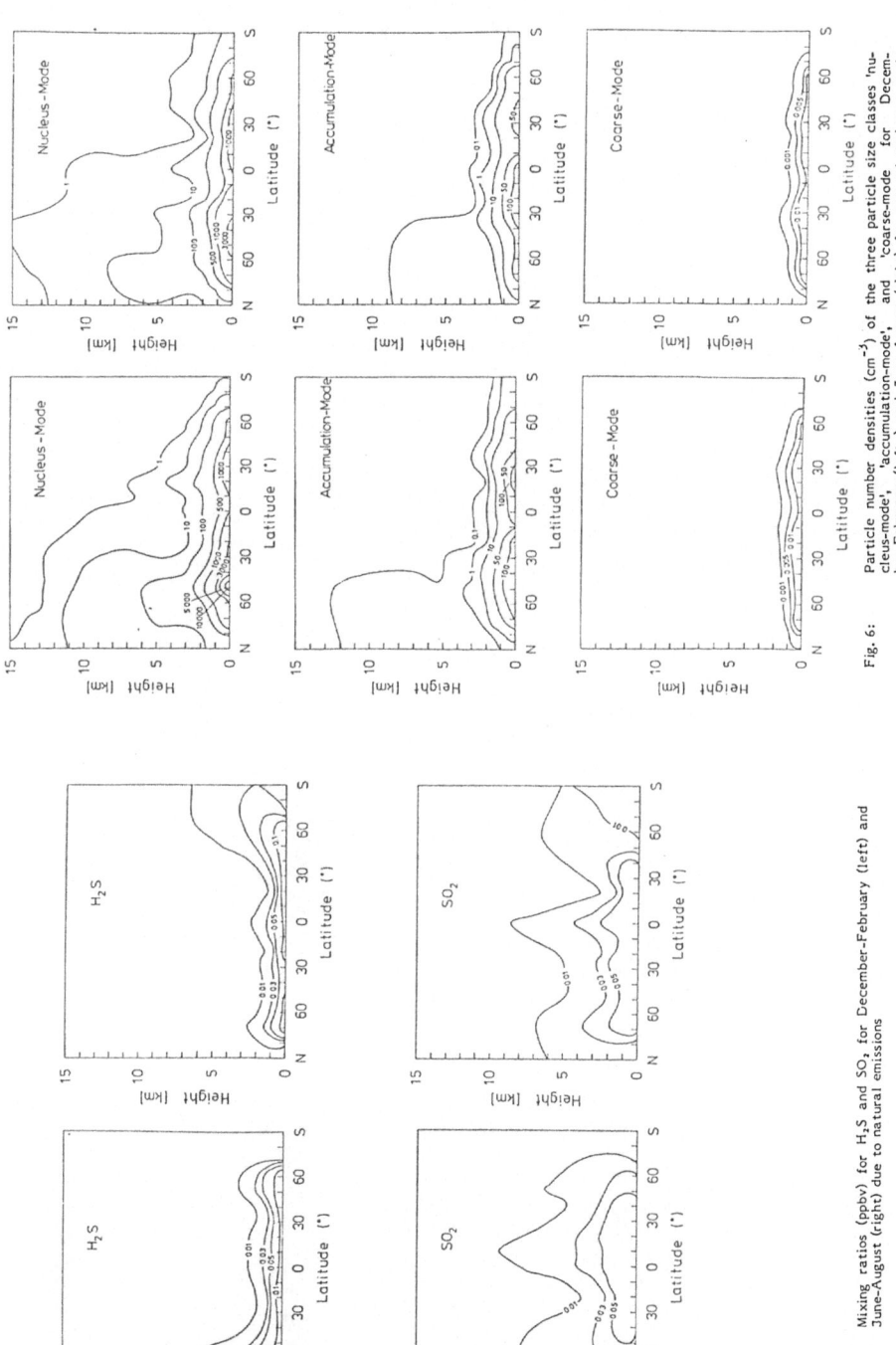

Fig. 6: Particle number densities (cm⁻²) of the three particle size classes 'nucleus-mode', 'accumulation-mode', and 'coarse-mode' for December-February (left) and June-August (right) for natural and man-made emissions.

Fig. 5: Mixing ratios (ppbv) for H₂S and SO₂ for December-February (left) and June-August (right) due to natural emissions

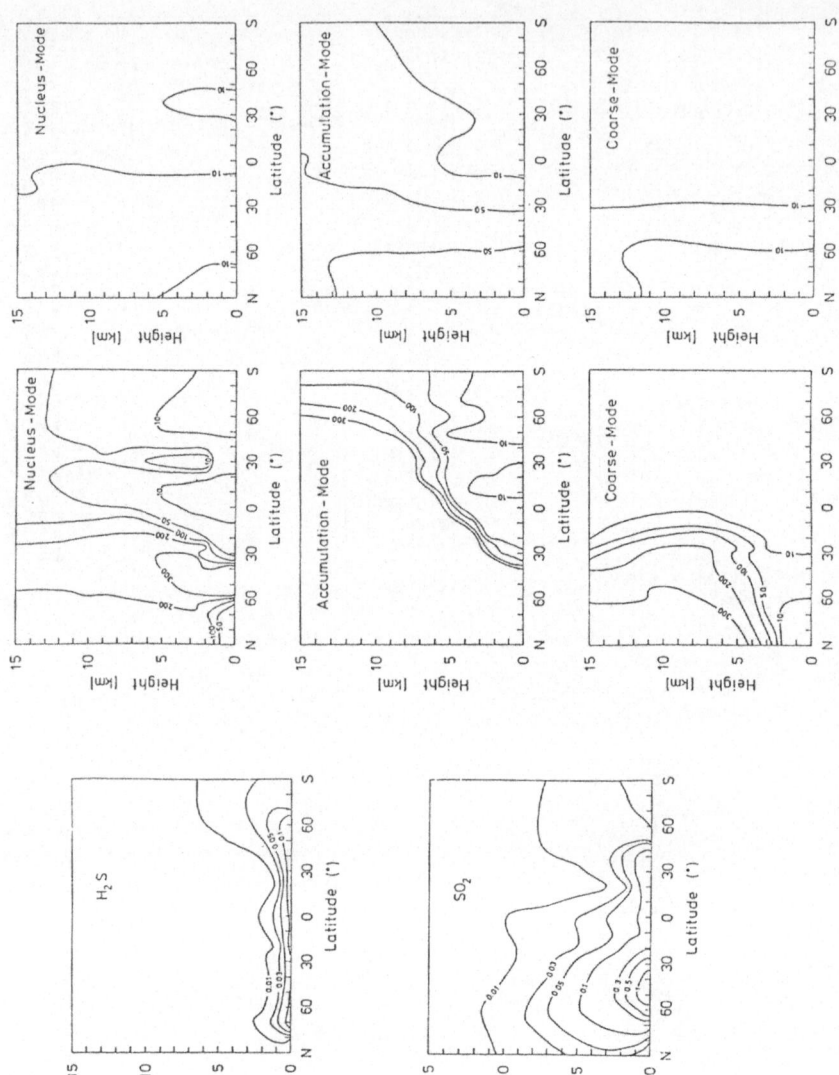

Fig. 3: Relative increase of the particle number density (%) between natural plus present man-made emissions and natural plus twofold man-made emissions. December-February (left), June-August (right)

Fig. 7: Mixing ratios (ppbv) for H₂S and SO₂ for December-February (left) and June-August (right) for natural and man-made emissions

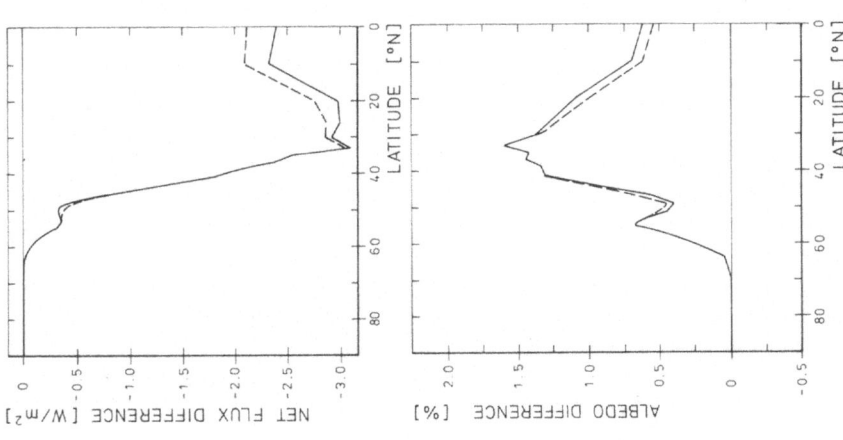

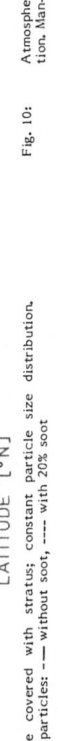

Fig. 10: Atmosphere covered with stratus; height-latitude dependent size distribution. Man-made particles: —— without soot, --- with 20% soot

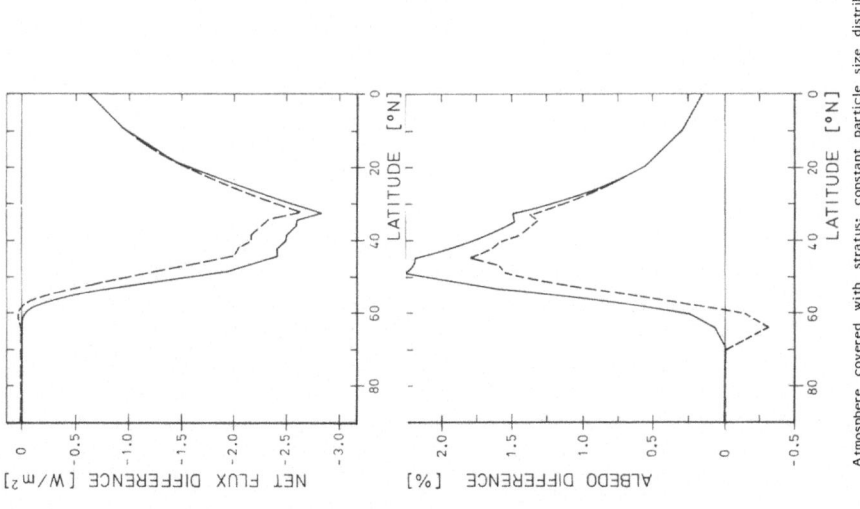

Fig. 9: Atmosphere covered with stratus; constant particle size distribution. Man-made particles: —— without soot, --- with 20% soot

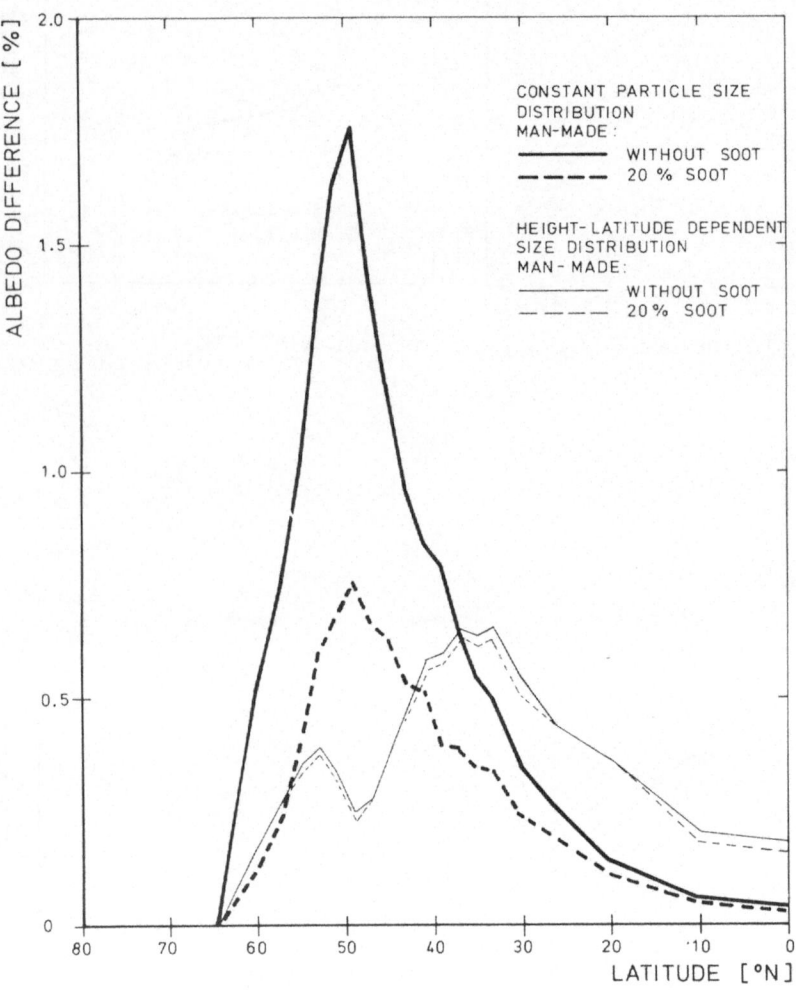

Fig. 11.   Change in albedo for mean cloud cover

C H A P T E R I V : C L I M A T I C I M P A C T S

- Socio-economic impacts of a climatic change due to a doubling of atmospheric CO_2 content

- The use of crop models in assessing likely effects of variation in climate

- The effects of elevated carbon dioxide levels on the growth of crop plants : an attempt to predict the consequences for grassland and maize production in Europe

- Agroclimatic classification of central Italy

SOCIO-ECONOMIC IMPACTS OF A CLIMATIC CHANGE DUE TO A DOUBLING OF
ATMOSPHERIC CO$_2$ CONTENT

CH. SCHNELL
Dornier System, Friedrichshafen, FRG

Summary

During the last 10 years, several major studies have attempted to
assess the possible effects of a defined climatic change. The study
here reported is the second European investigation to address the
problem of potential effects of increasing atmospheric CO$_2$ levels to
specific economic sectors. It was one task of the study to assess and
select methods and data available and necessary for impact analysis.
In a second step a case study for the EC area was performed to demonstrate methods and their usefulness resp. limits of applicability.
The starting point for all impact analyses is provided by a climatic
change scenario. The climatic change has often been treated as being
uniform over a large geographic area, and as acting by means of increases or decreases in annual values of temperature and precipitation. By means of selected general circulation models in this investigation the scenarios were given with spatial and temporal resolutions,
in order to look for the regional differences and effects dependent on
annual cycle for instance in the agricultural sector - a first attempt
to incorporate the regional and seasonal components of a climatic
change in the impact analysis.

1.1 Introduction

Two recent conferences in Austria (1) and West Germany (2) have
dealt intensively with different aspects of climatic impact analysis. In
both conferences emphasis was placed upon the role of impact investigations
with particular reference to the role they can play in assessing the
effects of a CO$_2$-induced climatic change on natural and managed resources
and human activities.
The state of discussion can be summarized as follows:

- In general the need for impact analysis in accepted. The value of the
  results depends on the attainable degree of confidence. This is again
  dependent on the quality of input data (climatic change scenarios) as
  well as on the performance of the methods and tools applied for impact
  assessment (e.g. crop-weather models).

- Until now there is a limited performance of models and methods to
  simulate the climatic response to a given perturbation like increasing
  CO$_2$ concentrations and to derive climatic change scenarios with desir-

able spatial and temporal resolution. However, this cannot lead to a cancellation of impact analysis development. Instead a permanent and intensive dialogue between climatologists, modellers and impact analysts should be established in order to assure improvements and adaptations of the whole set of instruments for impact analysis.

- There is special need for the selection and improvement of instruments and tools for impact assessment. This holds true for most of the relevant impact areas like the agricultural, water and energy resources sector.
  However, the simple stock-taking of improved tools useful for impact analysis is not sufficient: a methodology for the application of such tools has to be developed as well. Despite of the uncertainties of climatic change scenarios and the deficiencies of impact assessment tools there is a need to perform case studies. But, again, as in the case of climatic change scenarios it is of major importance not to interprete the derived results as reliable predictions and incontrovertible facts: the experimental character of model application and scenario construction and the interpretation as qualitative description of perspectives and possible future effects must be emphasized.

The study to be reported here on "Socio-economic impacts of climatic changes due to a doubling of atmospheric CO_2 content" (3) started in 1981 within the EC climatology programme. It addressed the whole range of questions mentioned above: Analysis and assessment of availability and quality of methods, data and results for use in impact analysis within the whole chain of causes and effects from Socio-economic development/Energy consumption/Fossil fuel consumption to CO_2-emissions/Development of atmospheric CO_2 concentration to Climatic response/Construction of climatic change scenarios to Impact analysis in the agricultural, water and energy resources sector and description of socio-economic consequences.
A case study was performed with selected (best available) models for impact estimation on the basis of two different climatic change scenarios for the European study area (20°W - 20°E, 30°N to 70°N).
The case study was the second one addressed to the problem of potential effects of increasing CO_2 levels on specific European economic sectors including a first attempt to incorporate regional and seasonal components of a climatic change in the impact analysis.
The study was performed in close cooperation between Dornier System, Friedrichshafen and the Research Center for Applied Climatology, Münster (Prof. Bach and associates). Results of the investigations are given in the following.

2. FROM CO_2-EMISSIONS TO CLIMATIC CHANGE SCENARIOS

The development of global energy, fossil fuel consumption and of the closely related CO_2-emissions depends on several influencing factors, which are partially dependent on political decisions and measures. Different assumptions are leading to different projections and scenarios, ranging from the IIASA 36 TW-high scenario (total global energy consumption) with a 69 % portion of fossil fuels to the Lovins 5,2 TW efficiency scenario with only a 18 % portion of fossil fuels in the year 2030.

Computation of the resulting different values and time dependencies of CO_2-emissions and simulation of the resulting CO_2-concentration in ambient air lead to the conclusion, that a doubling of CO_2-level is likely to occur within a period of 50 to 100 years from today (detailed discussion and references in (3)).

This well-known result should not be interpreted in the sense that abundant time for measures and political (re)actions is still available. It rather shall illustrate one problem of impact analysis: it is nearly impossible to draw a realistic picture of the future technical and economic scene (e.g. in the year 2060), in which changing climatic parameters will cause physical and economical consequences. This problem is enlarged when considering the fact, that there will be a lot of reaction and adaptation processes to slowly changing climatic conditions. Therefore, the impact analysis will first of all give an estimation of effects resulting under status-quo conditions in a stepwise changed climate. The interpretation of such results must include the consideraton of possible future trends and chances for reaction and adaptation.

In our study the question of future climatic changes was related to the perturbing factor: doubling of the atmospheric CO_2-concentration. The climatic change scenarios - understood as a set of plausible and selfconsistent patterns, showing the structure and magnitude of a climatic change resulting from a CO_2 doubling - were intended to provide zonal, regional and seasonal distributions. From the different techniques for the construction of scenarios: utilization of paleoclimatic data, utilization of instrumental data, application of numerical models only the last method is considered to be capable to reflect a doubled CO_2-situation. The objective to dispose of regional climatic change patterns requires the application of 3-dimensional general circulation models (GCM), the resolution of which depends on the distribution of grid points for the calculation procedure. The order of magnitude of changes as well as the ability of the models to describe temporal effects and annual cycles depends strongly on the way, a GCM simulates the behaviour of the ocean. Table I gives a survey of different GCMs and their performance with respect to criteria for their use in impact analysis. It is obvious, that from all GCMs, with which $2xCO_2$ experiments have been performed, (only) two meet the given criteria - even with some remaining restrictions:

- the BMO model has a 5-layer atmospheric GCM and a climatological ocean with prescribed sea surface temperature, upon which a latitudinally unvarying +2 K increment is superimposed. The ocean does not interact with the atmospheric GCM. The spatial resolution is 330 km (4).

- the GISS model consists of a 9-layer atmospheric GCM coupled with an interactive, mixed-layer ocean. The low spatial resolution is related to the coarse grid point distribution of 8° x 10° (latitude x longitude) (5).

Both modelers, Mitchell and Hansen, placed all necessary data and results to the disposal of Bach et.al. to construct and assess the respective climate and climatic change scenarios. Results of other GCM experiments as well as the results of instrumental data evaluations completed the data basis for comparison purposes etc.

The data of all GCM experiments have been evaluated by Bach et.al. transferring them, after space and time filtering procedures, to a reference grid (4° x 5°), comparing control run data with measured data and test-

TABLE 1: CRITERIA FOR THE USE OF CLIMATE MODELS IN IMPACT ANALYSIS

| MODEL/INSTITUTE | OCEAN MODEL | | | | SPATIAL RESOLUTION | TEMPORAL RESOLUTION | TYPE OF CO_2 PERTURBATION | |
	CLIMATO-LOGICAL MODEL	MODIFIED CLIMATO-LOGICAL MODEL	SWAMP MODEL	MIXED LAYER MODEL	GRID LAT. x LONG.	ANNUAL MEAN/CYCLE	$2xCO_2$	$4xCO_2$
MANABE AND STOUF-FER (1980); MANA-BE, WETHERALD, STOUFFER (1981); GDFL				o	4,5x5 -	o	o	
GATES, COOK, SCHLESINGER (1981); OSU	o				4x5	o	o	o
SCHLESINGER (1983); OSU			o		4x5	o	o	
WASHINGTON AND MEEHL (1983); NCAR			o		-	o	o	
MITCHELL (1983A); BMO	o	●			330 km	o ●	o ●	
MITCHELL (1983B); BMO		o			330 km	o		o
HANSEN ET. AL. (1983); GISS				●	8x10	●	●	

BMO: BRITISH METEOLOGICAL OFFICE, BRACKNELL, UK
GFDL: GEOPHYSICAL FLUID DYNAMICS LABORATORY, PRINCETON, USA
GISS: GODDARD INSTITUTE FOR SPACE STUDIES, NEW YORK, USA
NCAR: NATIONAL CENTER FOR ATMOSPHERIC RESEARCH, BOULDER, USA
OSU: OREGON STATE UNIVERSITY, CORVALLIS, USA

ing the statistical significance of change data. The evaluation was mainly related to temperature and precipitation data, zonal averaged and regional distributions for annual mean values as well as for seasonal mean values have been plotted.

Values of monthly means also for additional climatic parameters have been calculated as far as the subsequent impact analysis made this necessary.

As results of these evaluations in Fig. 1 the longitudinally averaged changes in temperature and precipitation rate for the BMO and GISS experiments as annual means and for winter and summer are given. For annual mean temperature as well as for seasonal distributions the response of the GISS

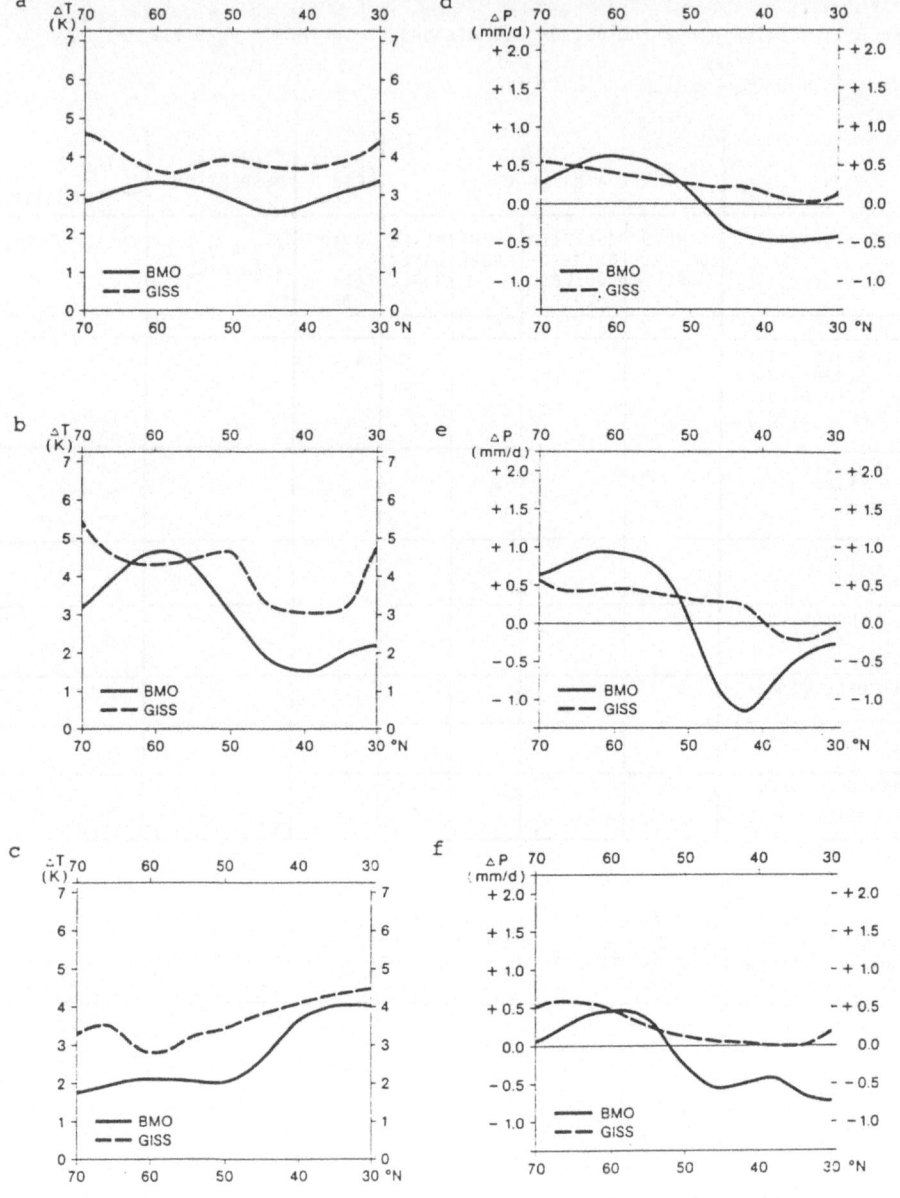

Fig. 1   Longitudinally-averaged (20°W-20°E) changes in temperature (a-c)
and precipitation rate (d-f) for the BMO and GISS 2xCO$_2$ experiments a,d: annual mean; b,e: winter; c,f: summer

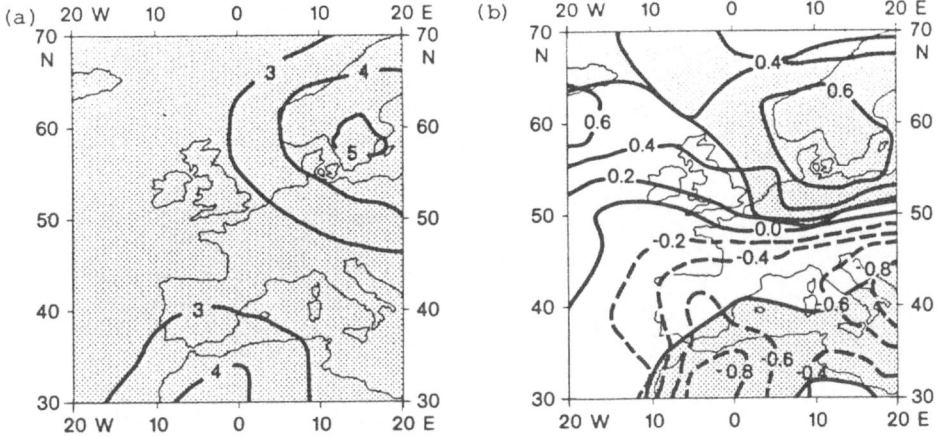

Fig. 2: Regional distribution of the temperature (K) (a) and precipitation rate (mm/day) (b) change for the BMO 2xCO$_2$ experiment for annual means.

model is generally greater (mean difference ca. 1 K). Regarding the changes of precipitation rate, the BMO experiment shows an increase north and a decrease south of about 50°N for all distributions, while the GISS experiment results in increases at nearly all latitudes.

Regional patterns are given in the following only for annual means: Fig. 2 shows temperature and precipitation rate changes for the BMO experiment, Fig. 3 the respective GISS results.

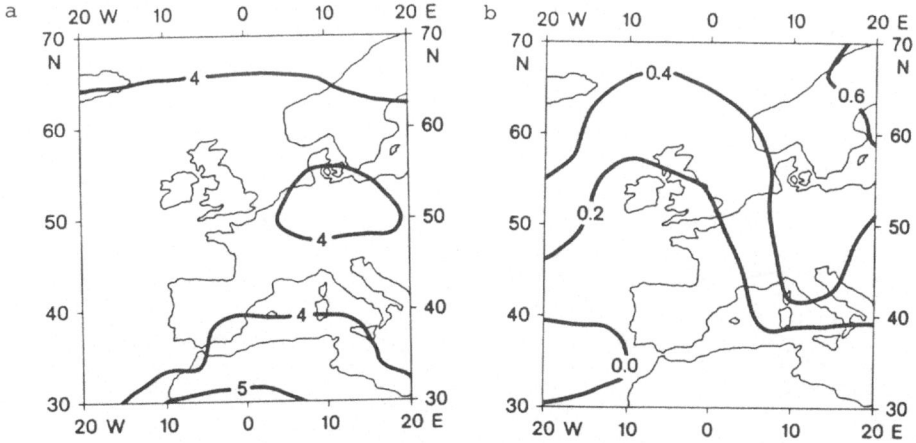

Fig. 3: Regional distribution of the temperture (K) (a) and precipitation rate (mm/day) (b) change for the GISS 2xCO$_2$ experiment for annual means.

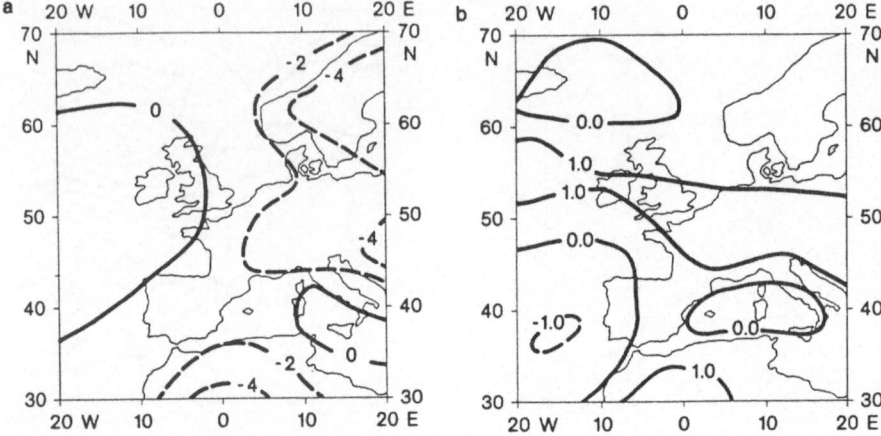

Fig. 4:   Regional distribution of the differences between the BMO control
           experiment and the measured climate
           a: temperature (K)
           b: precipitation rate (mm/day)

        Before discussing differences or conformities between these regional
patterns, it is necesary to examine the model performances expressed in
terms of their ability to describe the regional (and seasonal) features of
the present-day climate. Figures 4 and 5 are showing temperature and preci-

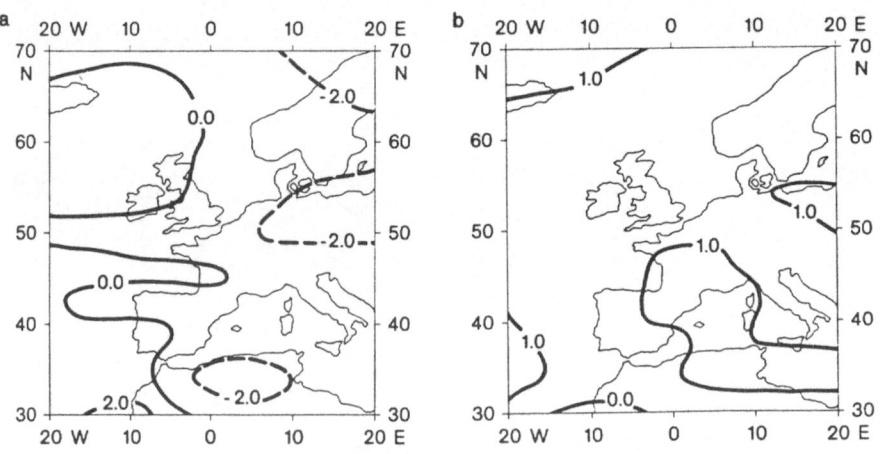

Fig. 5:   Regional distribution of the differences between the GISS control
           experiment and the measured climate
           a: temperature (K)
           b: precipitation rate (mm/day)

pitation rate anomalies – i.e. the differences between control run results and observed data. Even if one could state, that – taking into account the regional variations of observed data and considering the coarse resolution of the models (esp. GISS) – the differences are not too bad, it is obvious that the anomalies are in the same range or even greater than the respective change values themselves.

An analysis of seasonal control run data compared with measured data shows also significant differences.

These results could suggest the conclusion, that the uncertainties of results should prevent from any utilization for subsequent impact assessment at all. But there are also several reasons for another conclusion:

– utilization of relative change values instead of absolute values
– conformities of GCM results – e.g. sign and order of magnitude of global or zonal temperature changes; regional structures of changes expressed as sign of deviation from regional mean change values (M) as shown in Fig. 6.

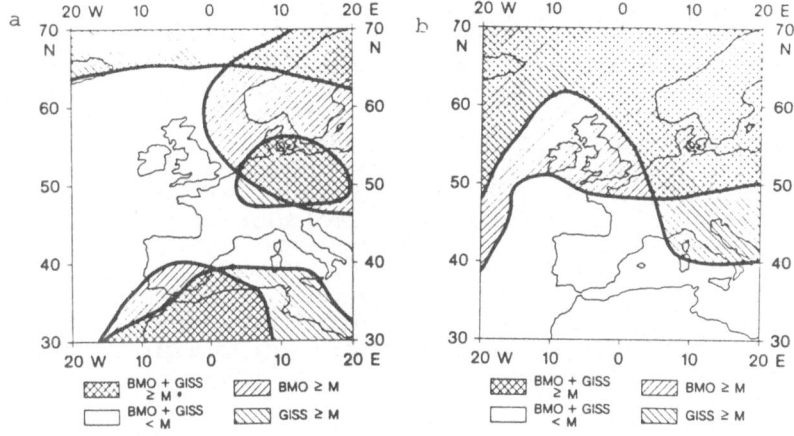

Fig. 6: Pattern analysis for the temperature (a) and precipitation rate (b) distributions of the GISS and BMO 2xCO_2 experiments for anual means

But – of course – one should realize the consequences of the evident limitations of GCMs to simulate the fine-scale change patterns: the climatic change scenarios cannot be seen – as noted above – as a prognosis of future developments, even its interpretation as description of most probable change effects has to be doubted. But it represents certainly a pattern of possible climatic response and insofar it can be used to look for risks – risk expressed in terms of climatic changes as well as in terms of possible effects evaluated in the subsequent impact analysis.

The differences between the BMO and the GISS experiment results suggest not to combine or to average the two sets of values to a "most probable" scenario, but to use the different results as two scenarios of possible developments. This was the decision we have taken for the present study.

3.   IMPACT ANALYSIS

The impact analysis of the reported study considered the effect of a CO_2 doubling on the sectors of agriculture, energy and water resources in Europe. Here shortly some aspects of the methodological considerations, selection of available models and application experiences from the case study will be reported.

3.1 Agricultural Sector

An evaluation of previous impact analyses shows, that no former attempt was made to give a detailed assessment of the response of the European agricultural sector to a doubling of atmospheric CO_2 levels. The fundamental basis of such an assessment requires, beneath an adequate climatic change scenario, a structural framework which is capable of identifying and quantifying the complex relationships between climate and crop yields. A reliable quantitative estimation of future effects must include the consideration of anthropogenic activities and technological advances.

There are direct influences of climate on crop yields acting via the pathway of plant processes (e.g. photosynthesis, respiration, transpiration, water uptake) during the various phenological stages. The most important influencing climatic factors here are temperature, precipitation and radiation expressed also in terms of length of growing season or number of growing degree days and variability of temperature and precipitation. There are critical phenological stages of the different crops with respect to temperature or water requirements and tolerances. As a direct—but not climatic effect the influence of the increased CO_2 concentration should be considered.

Indirect climatic influences on crop yield – that means that climate has the potential to influence different factors which in turn have the capability to affect crop yields – are given via biological factors as pests (parasitic plant diseases, insects) and weeds, or via application of chemical technologies, mechanical technologies and agricultural management strategies.

Without discussing the different influences in detail it is obvious that there must be a variety of complex and by no means simple, linear and additive relationships.

There are two principal ways to represent the dependencies of crop yields on climatic influences by use of different model types: empirical/statistical models and simulation models.

Almost all impact studies with comparable objectives had concentrated exclusively on the empirical/statistical approach. For use in our case study one model of each type was chosen applying the following selection criteria: availability, simple model structure, climatic input data given or derivable from the GCM-results, applicability to the whole study area.

The empirical/statistical model chosen was developed by Hanus for predicting yields of winter wheat and is described in detail in an investigation conducted by Hanus for the European Community (6). It consists of a series of linear regression equations for each of 42 European meteorological stations, using only data for the seven months from January to July. The model expresses empirical relationship (of 20-30 years experience) between nationally-averaged winter wheat yields and the predictor variables mean monthly temperature, total monthly precipitation, mean monthly maximum and minimum temperatures (and time for the extrapolation of a linear yield trend). In the case study the temperature and precipitation change data as interpolated from the BMO and GISS grid point results for the locations of the 42 meteorological stations were used directly in the Hanus-model. Minimum and maximum temperatures were estimated from observed data assuming that the same changes as for mean values would occur for the perturbed situation.

The second model employed in the impact analysis is a (simple) simulation model. It was developed by Briggs during the course of the European Ecological Mapping Project (7). It is of the yield indicator type and was designed for the evaluation of the "biomass potential" of the EC. The potential above-ground biomass production by a standard crop (in this case a mixed species grass sward) under constant management procedures is calculated from climatic and soil characteristics, as is roughly described in Fig. 7. In a short assessment the restricted comparibility of growth pattern of grass with that of different crops must be pointed out as a disadvantage, the consideration of edaphic conditions (of soil characteristics) beneath the climatic information is marked as one advantage.

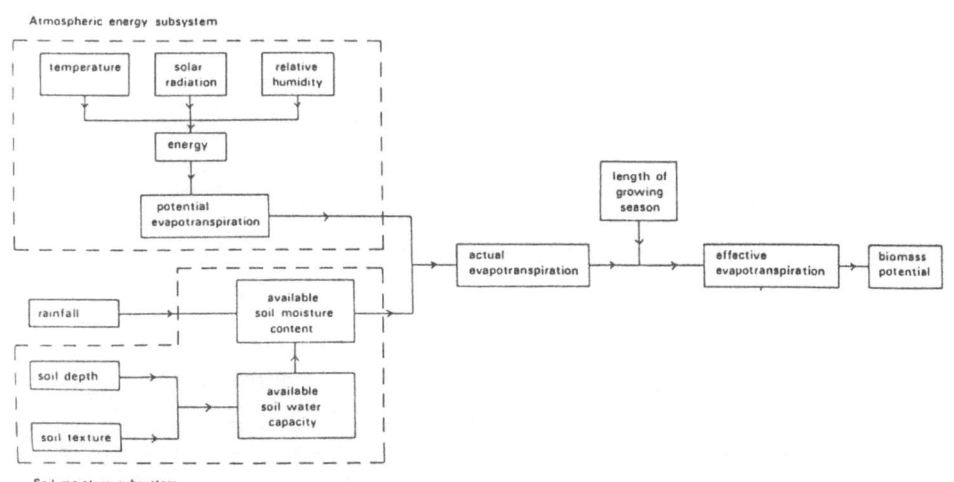

Fig. 7: Chief components of the Briggs model (after Briggs, 1983).

The input data for the Briggs model were delivered by the GCM results resp. (as in the case of net solar radiation) derived from other parameters. The different parameters were interpolated from the GCM grid points to the 233 meteorological stations used by Briggs.

Because the Briggs model operates with absolute values the change values resulting from the BMO and GISS experiments were added to the observed climatic data (BIODATA). The resulting composite meteorological data sets were then employed to calculate new absolute values of biomass potential.

TABLE II: PERCENTAGE YIELD CHANGE ESTIMATES

Country	HANUS MODEL		BRIGGS MODEL	
	BMO mean monthly yield changes as percentage of 1975-79 average yield	GISS mean monthly yield changes as percentage of 1975-79 average yield	Change in biomass potential index BIO.GISS	Change in biomass potential index BIO.BMO
Ireland	- 1.25%	- 2.5%	+ 16	+ 11
Great Britain	-	-	+ 20	+ 17
Denmark	+ 2.7%	+ 0.2%	+ 31	+ 36
Netherlands	+ 0.2%	0.0	+ 20	+ 17
Belgium	- 1.5%	- 1.1%	+ 18	+ 15
Luxemburg	+ 1.0%	+ 1.0%	+ 26	+ 15
France	- 1.4%	- 1.8%	+ 15	+ 2
West Germany	- 0.2%	- 1.1%	+ 27	+ 19
Italy	0.0	- 0.4%	+ 19	- 5
Greece	-	-	+ 11	- 38

The results of applying the two models with the different BMO and GISS data sets are given in Tab. II.

It can easyly be seen, that there are some effects within the Hanus model, which cannot be understood or explained by physical reasons. At first there are the unexpected low values of yield change predictions which seem to be serious underestimates. As a second example the substantial differences in precipitation rate changes e.g. for Italy between the BMO and GISS experiments are not reflected in any way in the results. An explanation may be, that in the range of empirical experience there could have been multicollinearity effects between the variables which could have caused, that the precipitation coefficients are often several orders of magnitude smaller than the temperature coefficients thus underestimating precipitation change influences. In addition, the linear regression model may reach its limits, because substantial changes of temperature and precipitation could cause non-linear effects or even exceed of critical requirement/tolerance limits.

The few examples illustrate, why simple linear regression approaches on an empirical basis may be fundamentally unsuitable for making credible assessments of the impacts of substantial climatic changes on crop yields.

One result of the Briggs model calculation is immediately apparent: The BIO.GISS scenario (GISS climatic change values added to the BIODATA set of observed data) predicts a positive average change in biomass potential for all 10 EC countries. In contrast the BIO.BMO scenario forecasts that Italy and Greece will experience decreases in biomass production, and that France in particular will show a far smaller increase than in the BIO.GISS scenario. These results appear to reflect the BMO model's predicted decrease in precipitation rate over southern Europe.

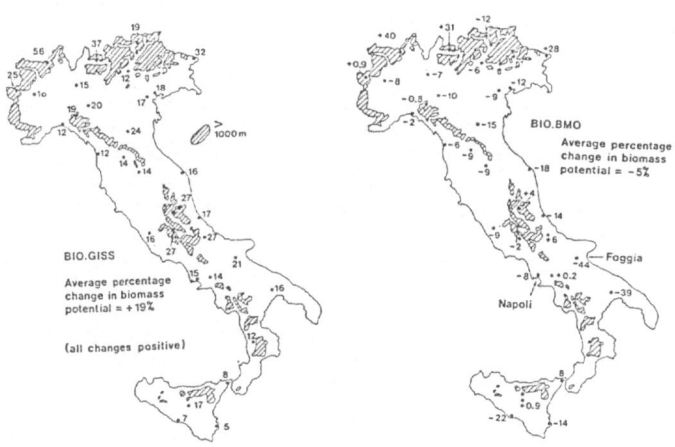

Fig. 8:   The Briggs model: calculation of percentage changes in the average biomass potential of 35 Italian stations, using meteorological information from the BIO.BMO and BIO.GISS scenarios.

Precipitation is not the only factor of importance in the explanation of percentage changes in biomass potential. In Fig. 8 the percentage changes are given at the level of individual stations in Italy, and it is evident, that even in the BIO.BMO scenario not all stations exhibit decreases. Those stations located within the boundary of the 1000 m contour line sometimes show increases (or only small decreases). This correlation between station elevation and percentage change in biomass potential is also to be found in the BIO.GISS scenario. A possible explanation is that here the temperature increase with the resulting amelioration of low-temperature stresses in winter and the extension of growing season length is effective beneath the precipitation rate changes in the different scenarios.

Looking at the order of magnitude of predicted changes of biomass potential, the regional distributions in relation to the climatic change scenario patterns and the evidence of considering temperature and moisture stress effects allows the statement, that a simple simulation model may provide the impact analyst with results which are more based on physical reality than the results of linear regression models.

Supplementary to the above estimations and evaluations human response to possible positive and negative impacts in the agricultural sector and the economic relevance of possible yield changes in the order of the BIO. BMO percentage changes in biomass potential have been discussed in our study.

3.2 Water Resources Sector

Water is both a vital element in the life of man and nature and a basic resource of human activities. World climate is capable of changes, as was demonstrated with the BMO and GISS scenarios, and it is probable that climatic changes will also affect water availability and water management in a way, that could lead to shortages in certain regions.

Problems to meet regional water demands may result from changes in climatic and natural conditions, as well as human activities may be of major influence: concentration of population and industries may result in high local water demands, which cannot always be met by local resources; sealing of surface by cities and transportation networks or intensification of land use affect immediately regional water budgets; water pollution by human activities is directly limiting the utilization of water.

To the extend that these conflicts concern local or regional areas, they can be compensated for a large degree by water budgeting and technical measures such as redistribution, storage and upgrading - but evidently at the expense of considerable economic means.

The trend to increasing water demands and limitations in water availability makes it even more probable, that climatic changes will be of importance because of their capability to mitigate or to amplify existing conflict potentials.

In our investigation the interdepencies between the climatic and hydrologic system, the derivation of water availability, the conditions of regional catchment areas and drainage basins with their influences on water budgets and water supply and as affected by human activities have been discussed in detail, in order to select feasible methods and applicable models for the impact analysis.

As a result it must be stated, that a comprehensive simulation of climatic change impacts up to the description of effects on regional supply conditions is far away from the capabilities of today available instruments. Studies of water demand and demand perspectives on an regional scale are not available and it is recommended, to exclude these aspects also in the next step of development. For the determination of water availability on a regional scale two ways have been discussed. The first can be signed as a small-scale investigation of climate/water resources relationships for single catchment areas. This way allows detailed examination and results, which are restricted to the special area conditions. For this type of investigation a large amount of data is required and flexible models for application under different conditions, which also allow an easy transfer of results to comparable regional situations, are not available. Therefore it is recommended, to look here in a next step for improved instruments using hydrogeological experiences as well as statistical/empirical methods for the developmet of a transferable or at least adaptable water budget model, which can be applied to the given variety of regional conditions in the EC. In the present investigation it was decided to apply an indicator

concept for the whole study area by using the modified Briggs model developed for evaluation of biomass potential.

In the modified model the part of rainfall, which remains after
evapotranspiration and soil moisture replenishment is interpreted as "Runoff potential". Beneath the simplification of this approach of course all
neglections and restrictions by the different assumptions of the Briggs
model are given again, especially the Uniform coverage with a

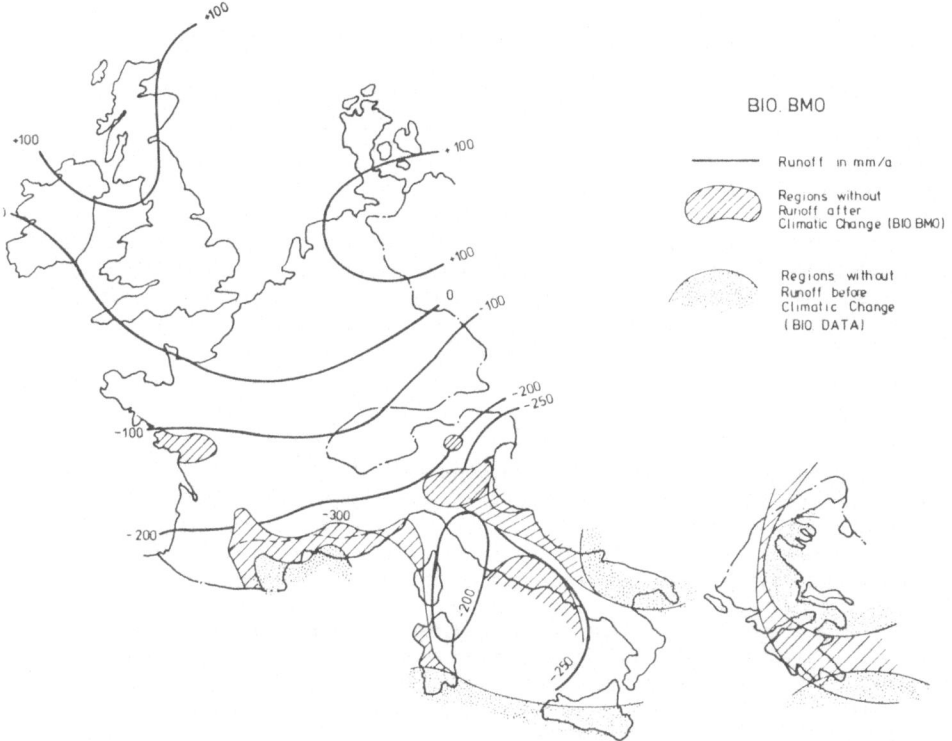

Fig. 9:   Regional runoff difference between BIODATA and BIO.BMO

grass sward, neglection of interception losses, percolation processes and
topographical differences. Especially the neglect of interception is leading to calculated runoff values higher than actually given. By comparing
measured data with Briggs results it could be shown that the diffference
lies in the order of 25 to 50 %.

The results of the run off calculations are shown in Fig. 9 as
changes of the BIO.BMO case compared with the BIODATA baseline. The runoff
changes range from -300 mm/a to ≥ +100 mm/a, compared with the smaller
range of -100 mm/a to +50 mm/a in the BIO.GISS case. The distribution pattern for the BMO scenario exhibits a marked north-south component. Areas
with no runoff increase substantially.

In our investigation a more detailed discussion of results including
testing of statistical significance or comparing runoff calculations with
water availability estimations by Walter-Lieth-diagrams is given.

Here the presentation of selected results should only demonstrate a
first step with yet imperfect tools in the direction of answering questions
about water resources future under changed climatic conditions.

3.3 Energy Resources Sector

The data sets of BIODATA and BIO.BMO and BIO.GISS scenarios have been
used in a case study to look at the effects on heating and cooling degree
day totals and solar radiation. These investigations as well as discussion
of available methods and experiences were performed by J. Jäger.

The temperture changes caused by the CO_2 increase in the models
to a reduction in heating degree day total by 30 - 40 % at the six stations
(1 in Denmark, 2 in F.R. Germany and 3 in Italy) studied here. The model
temperature changes also led to considerable increases in the cooling degree day totals, but at all stations except the most southern location, the
cooling degree day total remained smaller than the heating degree day
total. Thus it seems that the CO_2-induced temperature changes would lead to
energy savings due to reduced heating energy demand, but there can also be
demand increases due to air-conditioning requirements. The model results
suggest a decrease or no change in the amount of solar radiation in the
case of a CO_2-doubling. Only in the GISS results for Rome is there a solar
radiation increase in the winter months.

4. CONCLUSIONS

Assessing the results of the preceeding chapters it must be emphasized, that the investigation of the "Socio-economic Inmacts of Climatic
Change" must be viewed as a pilot-study. In having performed such a pilot
study, a basis is formed for assessment of the most important steps of
impact analysis and of the most promising ways of improvement and qualification of methods.

As mentioned above, the most important prerequisite for a successfull
impact analysis is the ability to construct climatic change scenarios as
probable, not only possible perspectives of climatic response to a given
perturbation. The scenario should be presented with reasonable spatial and
temporal resolution. Preliminary condition for a high degree of confidence
is a good presentation of the todays climates features by the models in the
undisturbed case.

The most important steps of impact analysis development must be in
the next future the improvement and validation of models for direct influences of climatic changes or indirect effects via biological factors
on physical properties like crop-yields or water availability. Modelling
of these indirect effects (if not included in statistical/empirical models),
of economic consequences and of technical trends and adaptation strategies
is of less priority.

In the following a list of indications where and how to improve
instruments for impact analysis is given :

Improvement of Climatic Change Scenarios

- Development and application of improved GCMs-Introduction of mixed-
layer variable-depth ocean models or of oceanic GCMs

- Intercomparison of new results from different GCMs and instrumental data approaches

- Improvement of Spatial Resolution
  - o Application of higher grid point density
  - o Employing of model hierarchies - coupling of GCMs with regional models
  - o Improvement of interpolation techniques using instrumental data

- Improvement of Temporal Resolution
  - o Use of short time data of extended GCM control run and perturbation integrations
  - o Use of instrumental data informations about annual cylce statistics and climatic variability

Improvement of Impact Analysis Techniques

- Improvement of Crop-Weather Models
  - o Formulation of simple simulation models for different crops
  - o Development of models for the forestry sector

  - o Improvement of statistical/empirical models by incorporation of phenological information
  - o Validation of different models with uniform data sets

- Improvement of Impact Analysis Techniques in the Water Resources Sector
  - o Development of simple water budget models
  - o Empirical/statistical analysis of water budgets on a regional scale.

REFERENCES

1.   Villach Study Conference, 19-23 September (1983):
        The Sensitivitiy of Ecosystems and Society to Climatic Change:
        Possible Impacts of CO_2 Increase in the Atmosphere.
        UNEP/WMO/ICSU
2.   Bad Sooden Allendorf, Oct. 1983:
        Second West German Climatology Conference
3.   H. MEINL; W. BACH; B.D. SANTER; H.J. JUNG; H. KNOTTENBERG;
        J. JÄGER; G. MARR and G. SCHWIEREN; 1984. The Socio-Economic
        Impacts of Climatic Change. Report to the Commission of the
        European Communities, Contract No. 83319.
4.   J.F.B. MITCHELL; 1983. The Seasonal Response of a General
        Circulation Model to Changes in CO_2 and Sea Surface Temperature.
        Quarterly Journaly of the Royal Meteorological Society, 109,
        113-152.
5.   J. HANSEN; G. RUSSEL; P. RIND; P. STONE; A. LACIS; S. LEBEDEFF;
        R. RUEDY and L. TRAVIS; 1983. Efficient Three-Dimensional Global
        Models for Climate Studies: Models I and II. Monthly Weather
        Review, 110, 609-662.

6.   D.J. BRIGGS; 1983. Biomass Potential of the European Community.
        Report for the Environment and Consumer Protection Service of the
        European Community, Sheffield, U.K.
7.   H. HANUS; 1978. Forecasting of Crop Yields from Meteorological
        Data in the EC Countries. Agricultural Statistical Studies, No.
        21, Statistical Office of the European Communities.

THE USE OF CROP MODELS IN ASSESSING LIKELY EFFECTS OF VARIATION IN CLIMATE

W. DAY
Physiology and Environmental Physics Department
Rothamsted Experimental Station
Harpenden, Herts. U.K.

Summary

Crop models are currently being developed at many research centres to describe how weather conditions and agronomic practice affect the various components of crop growth and yield. The development of these models requires continuing experimental study and analysis in order to define the major environmental influences and how they contribute to yield variation, but, when appropriately tested, such models do provide a tool for investigating the possible consequences of climate change for agriculture.

.1 Introduction

The topics covered already in this meeting have identified changes in climate on a variety of time scales. The consequences of the rising level of CO_2 concentration are probably the topic of most current interest, and the predicted climate changes are on a time scale that can have direct effect on present agricultural planning. The predicted changes in temperature and precipitation can each influence crop production independently of the direct effects of the CO_2 concentration. Crop models offer a means of assessing the potential effects.

The type of crop model with which I am concerned is the mechanistic model or, in its most complex form, the crop simulation model. Mechanistic models can be quite simple when a single environmental variable is considered to limit yield: for example, Penman's (10) model of the irrigation response of crop yield determines yield loss solely in terms of maximum potential soil moisture deficit in the growing season. But in many practical circumstances there is a wide range of possible factors that can limit yields. To take these many factors into account, comprehensive crop simulation models are being developed, defining how each factor influences the component processes of crop growth and integrating these processes to determine yield. Models at all levels from simple to comprehensive can all contribute to our understanding of how crops respond to the climate. In my use of the term crop model, I would discount purely statistical models that relate crop yield to an ad hoc assembly of weather variables. Though such models can be valuable in specific studies and can highlight where weather seems to have a significant influence, by themselves they provide little insight on the causes of such influences. Without a proper understanding, extrapolation to future changed climates is dangerous.

For any mechanistic model to be useful, we must have confidence in the performance of the crop model in a wide range of conditions. This paper presents some experimental approaches that contribute to the development of sound models, and identifies some of the ways climate can influence crops.

.2 Practical approaches
 Crop models are built on a set of physiological and physical rules. Our recent experimental programme with spring barley has investigated some

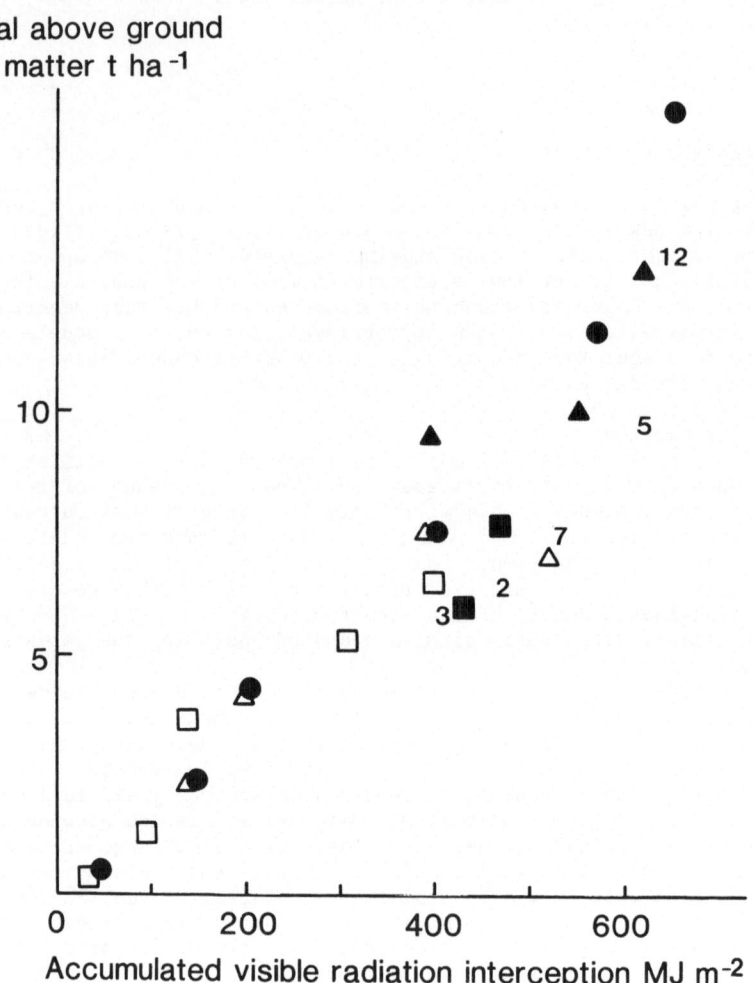

Fig. 1. The relationship between dry matter growth and light interception for samples taken during the growing season from five drought treatments: no drought, ●, treatment no.12; late drought, △, 7; mid-season drought, 5, ▲; early drought, ■, 2; and continuous drought, □,3. The numbers indicate the relation for samples from each treatment at final harvest. For further treatment details see (4).

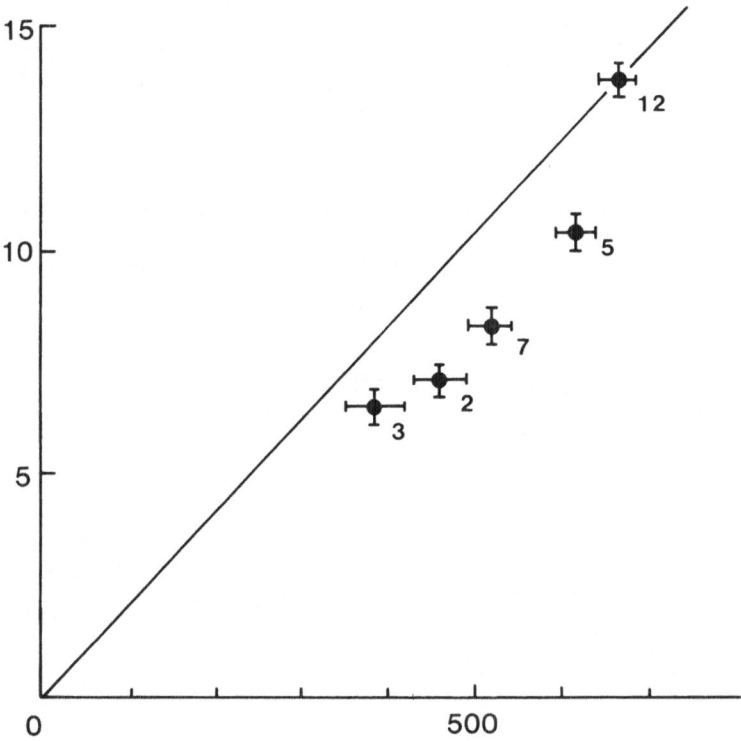

Fig. 2. The final harvest figures from Fig. 1 together with estimates of
standard errors. The line indicates the values of final dry
matter that would be expected if the effect of drought was
entirely through decreasing light interception.

of these rules, and results from this programme give an insight into how
models of crop response to water shortage may be constructed.

Crop growth can be analysed in terms of the solar energy intercepted
by the crop's foliage, and the efficiency with which that energy is
converted into dry matter. In many circumstances this conversion
efficiency is quite constant, and Fig. 1 shows data from a range of
drought treatments (for further details, see (4, 8)). A large part of
the variation in growth due to drought is associated with decreases in
radiation interception: leaves grow less and die earlier under water
stress. More precise data on dry matter at final harvest (Fig. 2) shows
that there are other contributions: these result from changes (Fig. 3)
in photosynthetic rate (9) and respiration (8) and from stomatal closure
(3). From these results it is clear that a model of crop response to
water shortage needs primarily to take into account effects on leaf area,

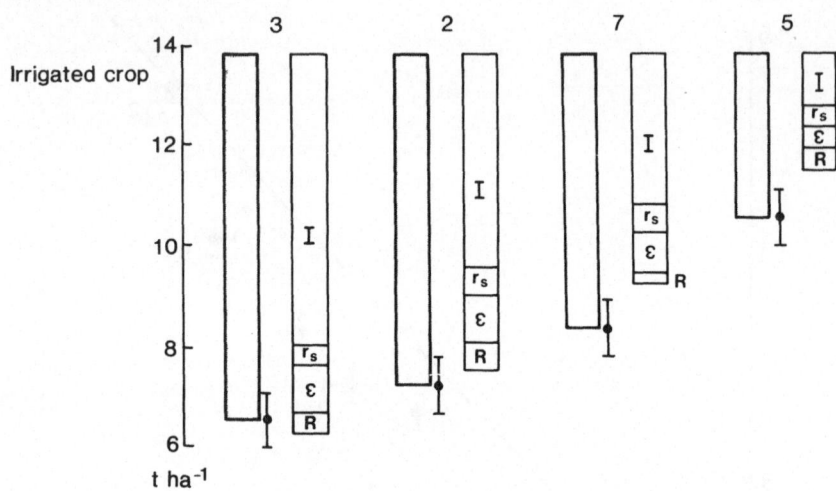

Decrease in total dry matter yield

Fig. 3. The shaded bars show the decreases in total dry matter production below that of the fully irrigated treatment 12 for the four treatments defined in Fig. 1. The adjacent bars show the effect on dry matter production expected from differences in light interception (I), stomatal resistance (r_s), photosynthetic efficiency (ε) and respiration (R), following the analyses of Legg et al. (8) and Parkinson and Day (9).

but that in some circumstances (e.g. when drought occurs towards end of leaf production; treatments 5 and 7, Fig. 3) photosynthetic responses can be equally important.

An alternative approach to the modelling of crop growth can be based on the efficiency of water use. In irrigated agriculture, the water requirement of crops has long been appreciated as a determinant of yield. If the use of water by a crop is constrained, e.g. by drought, then yield is limited, generally in proportion to the amount of water actually used (Fig. 4). The differences between seasons apparent in Fig. 4 can be resolved by taking into account the atmospheric saturation vapour pressure deficit (Fig. 5) following the analysis of Bierhuizen and Slatyer (1). This analysis is more appropriately done in terms of transpiration rather than total water use, which includes evaporation from the soil surface, and Tanner and Sinclair (11) have shown that it can be expected to apply to all crops in a similar manner. Thus the relation between growth and water use is modified by atmospheric humidity, and also by climate variables that affect the rate at which the crop foliage grows to cover the ground, and decrease soil surface evaporation. Again a crop model can be proposed that is based on the definition of crop transpiration, and incorporates the factors that affect transpiration e.g. leaf growth and senescence, soil water availability and rooting depth and density.

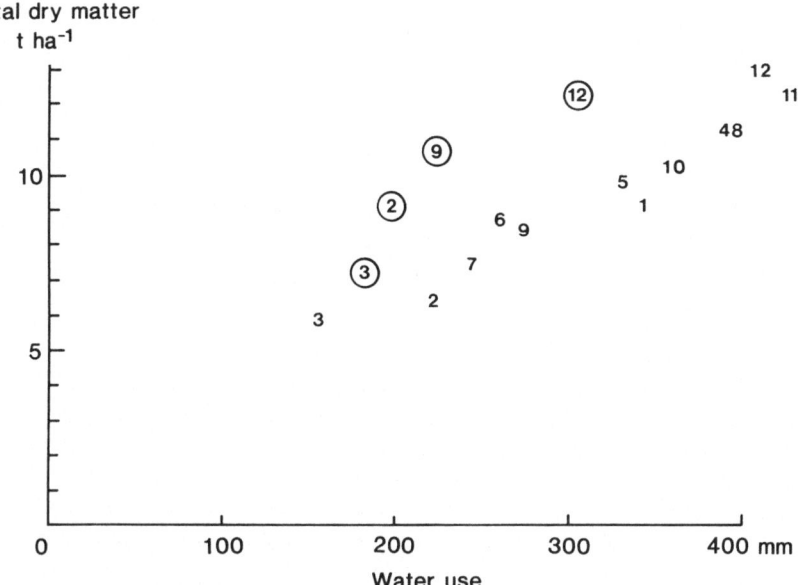

Fig. 4. The relation between dry matter production and total water use
for twelve drought treatments in 1976 and four in 1979.

.3 Opportunities for use of crop models in relation to climate variation

Our researches and other practical studies throughout the world
provide much of the information necessary to formulate crop models, and
many such models are being used (examples for wheat can be found in
(2)). There are many ways in which such models can contribute to the
interpretation of potential changes in climate.

Changes in temperature have their greatest influence on crop
development, higher temperatures shortening the growing season. For
overall crop growth, a shorter season gives less opportunity for the crop
to intercept solar energy whilst a longer season can, in some climates,
expose the crop to increasing probability of significant drought.
Temperature changes can also shift key physiological stages of crop
development; for example for winter cereals, more rapid development
leading to an earlier switch from vegetative to reproductive growth could
expose the crop to damage by frosts, as the reproductive plant is often
more sensitive. An understanding of such responses to climate change can
influence agronomy through choice of sowing date and cultivar.

Changes in radiation, associated with changes in cloudiness, have
their greatest influence on the rate of dry matter production, and hence
directly on yield. If climate change brings cooler temperatures and
increased cloudiness, the effect of the former in extending the growing
season will to some extent counter decreased growth rates.

Changes in rainfall, and in evaporative demand, can have a number of
influences. Most commonly considered is the effect on crop growth via

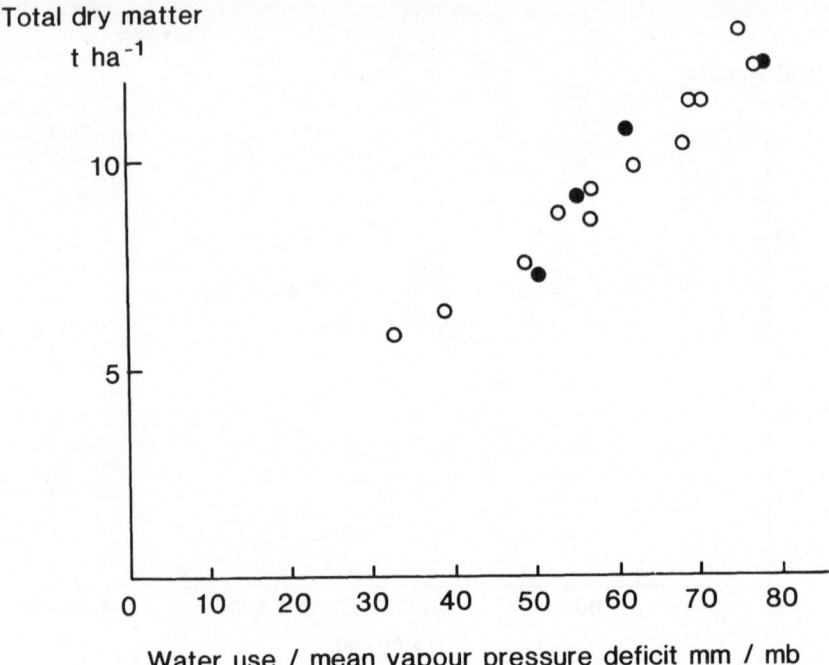

Total dry matter t ha^{-1}

Water use / mean vapour pressure deficit mm / mb

Fig. 5. The relation between dry matter production and the ratio of water use to mean saturation vapour pressure deficit for the same experiments shown in Fig. 4 (1976, o; 1979, ●).

soil water shortage. The effect on yield is greatly dependent on soil type, particularly the soil water holding capacity, and analysis of the consequence of a shift to a drier climate would need to consider the distribution of soil types through the region of interest, and the distribution of crop species of different rooting habit amongst these different soils. Strategies for maximising yield with irrigation, predicting irrigation demand and optimizing the use of limited or expensive irrigation water can all be assessed using crop models (5). But changes in rainfall can also have a major influence on soil trafficability, and this can modify production both via changes in planting dates and cultivation practices, and via damage to crops during subsequent farming operations. When soil properties that influence trafficability and damage can be suitably quantified, then crop growth models can be used to assess the consequences for crop production.

Increasing CO_2 concentration can directly increase growth and yield by increasing photosynthesis. Experimental (6) and modelling (7) studies have been made that quantify the possible effects. And for this particular aspect of climate change, statistical models of yield in terms of current climate variables can offer no insight.

The understanding of crop growth responses to climate that can be derived from crop models is potentially of great value to assessment of the impacts of climate change. The models still require that

experimentation and analysis continue but already provide a sound basis
for that assessment.

REFERENCES

1. BIERHUIZEN, J.F. and SLATYER, R.O. (1965). Effect of atmospheric
   concentration of water vapour and CO_2 in determining transpiration
   - photosynthesis relationships of cotton leaves. Agricultural
   Meteorology, 2, 247-258.
2. DAY, W. and ATKIN, R.K. (eds.) (1985). Wheat Growth and
   Modelling. Plenum Press. (in press).
3. DAY, W., LAWLOR, D.W. and LEGG, B.J. (1981). The effect of drought
   on barley: soil and plant water relations. Journal of Agricultural
   Science, Cambridge, 96, 61-77.
4. DAY, W., LEGG, B.J., FRENCH, B.K., JOHNSTON, A.E., LAWLOR, D.W. and
   JEFFERS, W.de C. (1978). A drought experiment using mobile
   shelters: the effect of drought on barley yield, water use and
   nutrient uptake. Journal of Agricultural Science, Cambridge, 91,
   599-623.
5. FISCHER, R.A. (1985). The role of crop simulation models in wheat
   agronomy. In: Wheat Growth and Modelling. Day, W. and Atkin, R.K.
   (eds.), Plenum Press. (in press).
6. GIFFORD, R.M., and EVANS, L.T. (1981). Photosynthesis, carbon
   partitioning and yield. Annual Review of Plant Physiology, 32,
   485.
7. GOUDRIAAN, J., VAN LAAR, H.H., VAN KEULEN, H. and LOUWERSE, W.
   (1985). Photosynthesis, CO_2 and plant production. In: Wheat
   Growth and Modelling. Day, W. and Atkin, R.K. (eds.), Plenum
   Press. (in press).
8. LEGG, B.J., DAY, W., LAWLOR, D.W. and PARKINSON, K.J. (1979). The
   effects of drought on barley growth: models and measurements showing
   the relative importance of leaf area and photosynthetic rate.
   Journal of Agricultural Science, 92, 703-716.
9. PARKINSON, K.J. and DAY, W. (1983). The influence of water stress on
   photosynthesis in a barley crop. Marcelle, R., Clijsters, H. and van
   Pousse, M. (eds.) In: Effects of Stress on Photosynthesis, M.
   Nijhoff/Dr. W. Junk, Published The Hague - Boston - London,
   pp.65-74.
10. PENMAN, H.L. (1971). Irrigation at Woburn. VII. Rothamsted
    Experimental Station Report for 1970, Part 2, pp.147-170.
11. TANNER, C.B. and SINCLAIR, T.R. (1983). Efficient water use in crop
    production: Research or re-search? In: Limitations to Efficient
    Water Use in Crop Production, Taylor, H.M., Jordan, W.R. and
    Sinclair, T.R. (eds.), American Society of Agronomy.

THE EFFECTS OF ELEVATED CARBON DIOXIDE LEVELS ON THE GROWTH OF CROP
PLANTS: AN ATTEMPT TO PREDICT THE CONSEQUENCES FOR GRASSLAND AND
MAIZE PRODUCTION IN EUROPE

M.B. JONES
Botany Department, Trinity College, Dublin 2, Ireland

Summary

The projected increasing levels of atmospheric carbon dioxide (CO_2) has
the potential for raising rates of photosynthesis in most but not all
plants. Doubling the present atmospheric CO_2 concentration could lead
to a 30-50% increase in photosynthesis in plants with C3 photosynthesis
but C4 plants are unlikely to be affected. Other environmental factors
such as light and temperature show considerable interaction with elevated CO_2, such that the maximum response occurs at high light levels and
the temperature optimum for photosynthesis is likely to rise. Apart
from the direct effects of CO_2 on photosynthesis the stomata on plant
leaves will tend to close in response to elevated CO_2 with the effect
that water use efficiency of plants will increase. It is predicted
that C4 plants will show greater benefit than C3 plants in this respect.
    The influence of elevated CO_2 levels on crop yield is difficult to
predict but ryegrass with C3 photosynthesis and maize with C4 photosynthesis are considered to be useful models. They are both used extensively for forage production in Europe. It is suggested that the use
of mechanistic models of crop growth could help in understanding the
complex of responses to elevated CO_2 but experimental work is required
to provide inputs for the model and for subsequent validation.

1 Introduction

    The growth and final yield of crop plants is ultimately dependent upon
the amount of carbon they are able to assimilate in the process of photosynthesis during their growing season. It has long been recognised that the
rate of photosynthesis in many plants in strong sunlight is limited by the
availability of carbon dioxide (CO_2) in the air (34) and this has been put
to practical use by feeding CO_2 to plants grown in greenhouses to give up
to several-fold increases in yield (6,13). However, despite the fact that
atmospheric CO_2 levels are currently increasing at 1.5 ppm per annum (14),
for practical purposes the availability of CO_2 has been considered as constant and the tendency has been to emphasise the interception of light by
the crop as the major limiting factor for growth. There have been several
demonstrations recently that the rate of dry matter production of crops
early in the growing season is proportional to the amount of light they
intercept (4,31) and in some cases the final yield of a crop depends on the
total amount of light intercepted during the growing season (3,50). In the

longer term though, there is little doubt that we will see a doubling of the present level of atmospheric CO_2 of approximately 340 ppm within the next 100 years (14) and this will have a significant positive effect on photosynthesis.

Not all plants respond by increasing photosynthesis equally for a given increase in CO_2 concentration. This is because, although the ultimate pathway for CO_2 assimilation in all plants is via the enzyme ribulose biphosphate carboxylase-oxygenase (Rubisco) and the so-called Calvin Cycle, some of them have an additional cycle which in effect makes them less CO_2 limited (57). In fact, the latter plants have a mechanism for concentrating CO_2 at the site of the Rubisco enzyme and therefore making the reaction much more efficient, principally by overcoming oxygen inhibition of the reaction (45). The two groups of plants are referred to as C3 (those without the concentrating mechanism) and C4 (those with the concentrating mechanism) because the first products of CO_2 fixation are 3-carbon or 4-carbon compounds respectively. Europe's natural vegetation consists primarily of C3 species with most C4's occurring in warmer drier climates (9) while most major crops are also C3, apart from maize. The effect of the different types of metabolism is that, at present levels of atmospheric CO_2, the C4 plants have higher rates of photosynthesis at high light levels than C3 plants because the latter are CO_2 limited and the former are not. However the effect of increasing the amount of CO_2 available to these C3 and C4 plants is to raise the rate for C3 plants while it has little effect on C4 plants (see Figures 1 and 2). Even in C3 plants the increase in photosynthesis is not always proportional to increases in atmospheric CO_2 levels because increasing CO_2 tends to close the stomatal pores on the leaf surface which are the primary pathway for diffusion of CO_2 into the leaf.

There have been many demonstrations that a short term doubling of the atmospheric CO_2 concentration leads to a 30 to 50% increase in photosynthesis by C3 plants due primarily to elimination of oxygen inhibition (12). Further, Kimball and Idso (27) have pooled data for a large number of CO_2 enrichment experiments and have predicted a 33% increase in marketable yield of crops with a doubling of present CO_2 concentrations. However, much of this data is from experiments that were not carried out under strictly controlled environmental conditions and these increases in yield cannot be wholly attributed to the direct effect of CO_2 on photosynthesis. Also, most experiments have examined the response of plants to relatively short term exposures to elevated CO_2. There is relatively little known about plants grown through their total growth cycle in high CO_2 concentrations, especially when exposed to the normal stresses in a natural environment such as drought, temperature extremes or nutrient shortage (30). The direct effects of elevated CO_2 levels on photosynthesis may be relatively simple to predict but the final yield of an agricultural crop will also be influenced by less direct effects of high CO_2 on processes such as transpiration, leaf expansion, nutrient uptake and plant phenology (e.g. time of flowering). Kramer (28) has made the following tentative generalisation about the response of plants to enhanced CO_2:

(i)   There are large variations in the response of different species
(ii) The response is greater in C3 than C4 plants
(iii) The response is greater in indeterminate plants such as ryegrass than in determinate plants such as maize
(iv) The largest growth response is in the seedling stage.

In the rest of this article I wish to review some of the evidence for the effects of elevated CO_2 on crop plant physiology and growth. It will be clear that we are dealing with a complex of responses and I will conclude by considering how we can handle this information in a mechanistic model of crop growth. Two important European crops, perennial ryegrass which is a C3 plant and forage maize which is a C4 plant, will be used as examples of how we might predict the relative performance of crop plants when CO_2 concentrations are double their present levels. It should be emphasised at this point that we are simply considering responses to elevated CO_2 levels and not the climatic changes, such as increased ambient temperatures and changes in patterns of precipitation, which will probably accompany the increase in CO_2 concentration (17,27).

2 Direct effects of carbon dioxide on photosynthesis

At moderate or high light intensities and moderate temperatures the rate of CO_2 fixation by C3 plants approximately doubles with a doubling of the present CO_2 concentration (Figure 1). However the concentration of CO_2 at the chloroplasts, where CO_2 fixation occurs, is lower than in the atmosphere because of the resistance to diffusion of CO_2 through the stomata; the more closed the stomata, the higher the resistance and the larger the difference in concentration between atmospheric (C_a) and internal or intercellular (C_i) CO_2 concentration. In both C3 and C4 plants the relationship between photosynthetic rate and internal CO_2 concentration is linear at low CO_2 levels but with C4 plants showing a steeper slope. At higher internal CO_2 levels saturation occurs abruptly in C4 plants at about 140 ppm but much more slowly in C3 plants (38,45,53)(Figure 2). It has recently been suggested (61) that when the ambient CO_2 concentration increases, the stomata respond in such a way as to maintain the ratio of intercellular to ambient CO_2 (C_i/C_a) at 0.6 to 0.8 in C3 plants and at 0.3 to 0.5 in C4 plants. This means that in order to establish the steeper gradient the stomata are closed more in C4 plants compared with C3 plants and consequently the amount of water lost in transpiration will be reduced. The implications of this are considered in section 4.

3 Interaction of increased atmospheric carbon dioxide with light and temperature

The amount of enhancement of photosynthesis in C3 plants at elevated atmospheric CO_2 concentrations is influenced by other environmental factors such as light and temperature. Increased temperatures appear to have the effect of maximising the enhancement of photosynthesis in C3 plants while low temperatures reduce the response. The reason for this is that the increased level of CO_2 overcomes the normal decline in affinity of the enzyme Rubisco for CO_2 with rise in temperature (2). In general the optimum temperature for photosynthesis in C4 plants is higher than that for C3 plants at present levels of atmospheric CO_2, but elevated CO_2 shifts the optimum temperature for C3 plants closer to the C4 temperature optimum (41)(Figure 3). The implication of this response for the performance of C3 crops in warm climates could be significant as even at present the optimum temperature for photosynthesis in these plants is often below ambient temperatures at midday (59).

The maximum enhancement of photosynthesis in C3 plants at elevated CO_2 levels occurs at high light levels but there is still a significant increase at much lower light levels (Figure 1). This explains why CO_2 enrichment in greenhouses is still effective even during the winter months when light levels are low (19). The reason for the enhancement at low light levels is an increase in the efficiency of use of light by the photosynthetic mechanism, which is reflected in the steeper slope of the relationship between photosynthesis and light levels at low light, and a decrease in the light compensation point (the light intensity at which photosynthesis is counter-

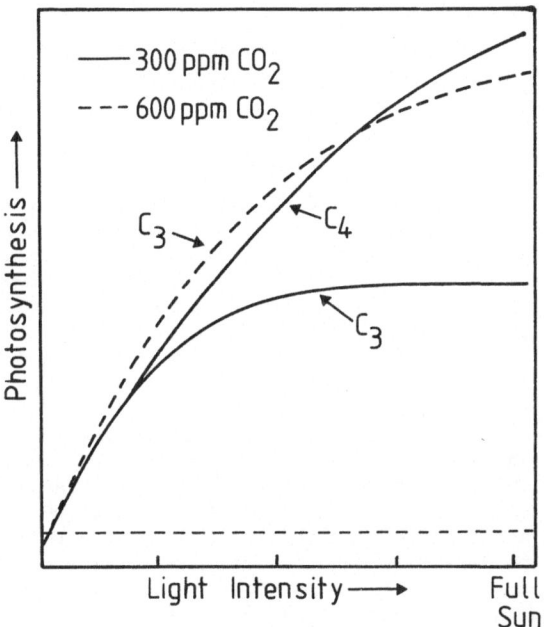

Figure 1.   The generalised response of photosynthesis to light in C3 and C4 plants.

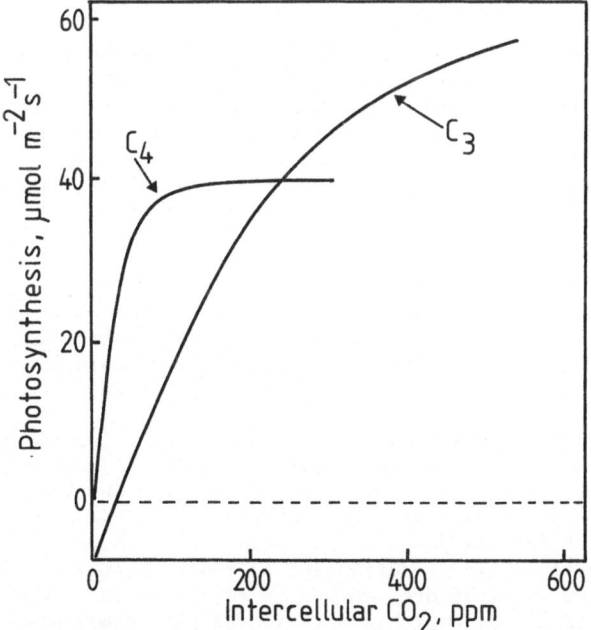

Figure 2.   The response of photosynthesis to intercellular CO_2 in C3 and C4 plants (from 45).

balanced by respiration). The significance of this to crop plants is that leaves in a plant canopy are exposed to a whole range of light levels due to mutual shading by other leaves, with the result that leaves nearer the base of the canopy are normally exposed to relatively low light levels. It is to be expected that even these leaves will show increased photosynthesis at elevated CO_2 concentrations.

4 Effect of carbon dioxide on stomata and water use efficiency

Stomata are ventilating pores on the surface of leaves formed by a pair of specialised cells, the guard cells, which can open and close to control gas exchange (58). In effect the stomata impose a resistance to diffusion of CO_2 into the leaf for photosynthesis and similarly a resistance to diffusion of water vapour out of the leaf in transpiration. When the ambient CO_2 concentration increases stomata tend to close, thus increasing resistance (20), but the mechanism by which this response occurs is still unknown (32). Guard cells probably sense the CO_2 concentration in the intercellular spaces of the leaf because if the stomata are closed they do not open if the CO_2 concentration outside the leaf is reduced (32). The magnitude of the response of stomata to CO_2 concentration is extremely variable between species and it also depends on the pretreatment of the plant, particularly its history of water stress (32). The variability is illustrated by the observation that in many cases the stomata seem to respond in such a way that, as mentioned above, the ratio of leaf internal CO_2 concentration (C_i) to atmospheric CO_2 concentration (C_a) remains constant, but in other cases the stomata may close even more to maintain a constant C_i while in others they do not respond at all (45). The effect of all of these responses is to alter the ratio of CO_2 uptake in photosynthesis to water loss in transpiration; the photosynthetic water use efficiency (WUE). Water use efficiency can also be expressed in terms of plant biomass gained per amount of water lost in transpiration over a period of time. Consequently, if WUE is increased then productivity is also increased when water supply limits growth.

At present levels of atmospheric CO_2 concentration the WUE of C4 plants is about twice that of C3 plants (45) because C4 plants are more effective at taking up CO_2 from the atmosphere. In C3 plants the direct enhancement of photosynthesis by elevated CO_2 concentrations increases WUE. However, even if the stomata close in response to increasing CO_2 levels and the rate of photosynthesis is not increased there is still a substantial rise in WUE because transpiration decreases more than photosynthesis as the stomata close (47). This is illustrated in Figure 4 where the response of WUE to increasing atmospheric CO_2 levels in C3 plants is presented for conditions where the ratio C_i/C_a is constant (photosynthetic rate increases) or C_i remains constant (photosynthetic rate is constant). More interestingly though, it can be shown (Figure 4) that the WUE of C4 plants also increases at elevated CO_2 levels and that the absolute improvement in WUE of C4 plants may be even greater than for C3 plants. However, recent work has cast some doubt on this prediction (37).

It is probable therefore that as atmospheric CO_2 levels increase the amount of plant biomass produced per unit volume of water lost will increase substantially. As a consequence, where the availability of water now limits yields then crop productivity should increase. This could be significant even in cool temperate parts of northern Europe, because it has been estimated that water deficits sufficient to reduce growth of ryegrass develop after only two weeks of dry weather in summer in the U.K. (26). More indirect effects of reduced transpiration on plants may also become apparent because high transpiration rates can lead to a fall in plant water poten-

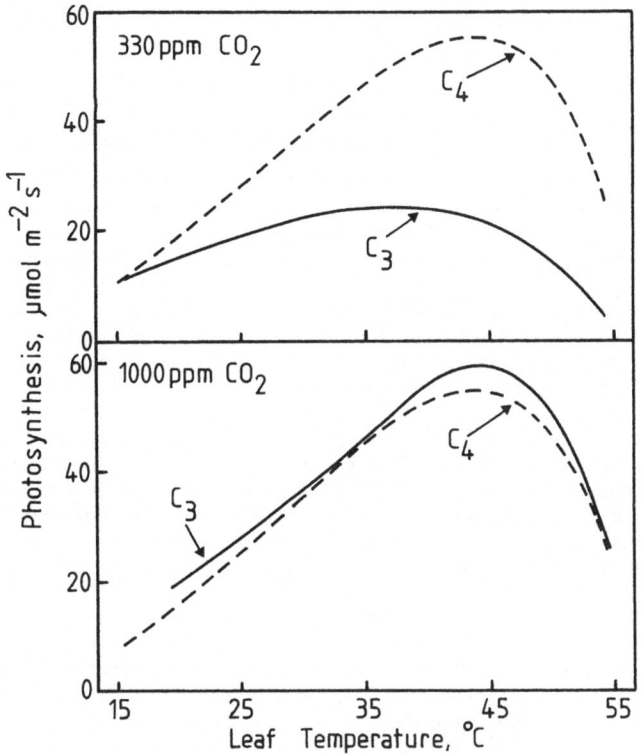

Figure 3.   The effect of temperature on photosynthesis in C3 and C4 plants
at 330 ppm and 1000 ppm atmospheric CO_2 (from 41).

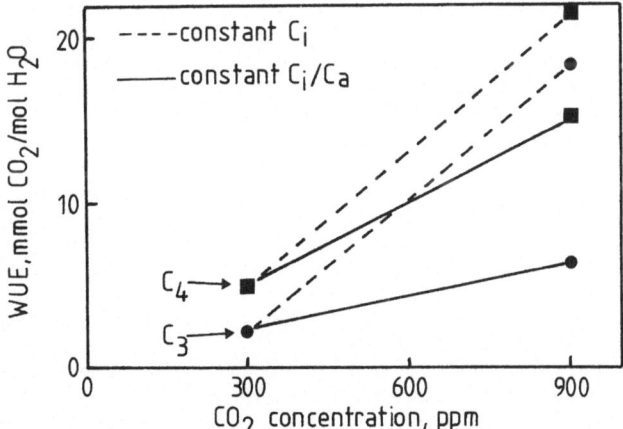

Figure 4.   The response of water use efficiency to CO_2 in C3 and C4 plants
assuming either a constant C_i of 240 ppm in C3 and 120 ppm in C4
plants, or a constant C_i/C_a of 0.72 in C3 plants and 0.36 in C4
plants (from 45).

tials, which can reduce growth with even moderate falls (23). Water potentials are an indication of the 'water status' or the level of 'water stress' of the plants and it has been demonstrated recently that wheat and peas develop water stress more slowly when grown at elevated CO_2 levels (42,43,52).

5 Consequences of increased leaf photosynthesis and improved water status for the whole plant

The growth of plants depends on the export of products of photosynthesis (assimilates) from the 'source' leaves to the regions of growth, the so-called 'sinks'. Photosynthetic products are stored in the form of starch in the leaves and subsequently transported in the phloem to the sinks primarily in the form of sucrose (36). The capacity to load sucrose into the sieve tubes of the phloem does not appear to limit export of photosynthate from the leaves, as more sucrose is exported from leaves that have their photosynthetic activity increased by high light levels (21). However the increased accumulation of starch in chloroplasts of plants grown at high CO_2 is commonly observed (8,22,55), and this may lead to a fall in photosynthetic rate after a relatively short term enhancement at high CO_2 levels (1, 46). The cause of the decline is unclear but it is unlikely to be due to a simple direct effect of mechanical distortion of chloroplasts by starch grains. In general, the long term effect of high photosynthetic rates on plants seems to be variable but it may be related to their ability to generate new sinks for the assimilates (45) so that determinate and indeterminate species may respond differently.

The partitioning of assimilates to different parts of the plant does not appear to be affected by increased atmospheric CO_2, so that the root/shoot ratio and the harvest index (economic yield/total biomass yield) are not altered (15,51). However the increased total allocation of assimilates to leaves at high CO_2 concentrations is particularly important for whole plant and canopy development. This is because during the early stages of growth the rate of biomass production is dependent on the amount of leaf area the canopy has available to intercept light for photosynthesis (see Section 1). Also, at this time the canopy development is exponential so that any effect of elevated CO_2 on leaf expansion is amplified over time so that a larger canopy results from small increases in the rate of leaf expansion during early stages of growth. It is therefore possible that the effect of elevated CO_2 levels on leaf area has a greater consequence for crop production than its direct effect on leaf photosynthesis (35). However, when the canopy becomes fully closed and intercepts virtually all the incoming solar radiation then growth is much more dependent on the photosynthetic capacity of the leaves which form the canopy.

The improved water status of the plant under elevated CO_2 (Section 4) may also increase the rate of leaf expansion and canopy development because leaf expansion is very sensitive to water stress (23). This may explain the increased growth of C4 maize exposed to high CO_2 levels observed by Rogers et al. (48) despite the fact that leaf photosynthetic rate is not increased. They also found that maize plants growing at 520 and 910 ppm did not undergo the wilting which inhibits leaf expansion that was observed in control plants on hot summer afternoons.

Growth of plants in elevated CO_2 has been observed to change leaf morphology, which in turn may have an effect on leaf photosynthesis. In general, leaves of plants grown for extended periods at high atmospheric CO_2 are thicker and have a higher weight per unit area (specific leaf weight) than those grown at normal CO_2 levels (16,22,55). In some cases the leaf tissue develops an additional layer of palisade cells to the normal two layers (55), but the contribution of this third palisade layer to the photosynthetic

capacity of the leaves has not been investigated. The higher specific leaf
weight is partly structural material and partly stored carbohydrate (starch)
which appears to accumulate most at lower temperatures (22). These effects
on leaf thickness and specific leaf weight seem to be limited to C3 plants
as no similar effects on C4 plants have been demonstrated (11,55).

6 Interaction of elevated atmospheric carbon dioxide with nutrient supply

The potential for more rapid growth and increased dry matter production
of plants at elevated atmospheric CO_2 concentrations can only be achieved
if sufficient nutrients are available to sustain this growth. Under intensive agricultural management the growth of crops is no longer limited by
the availability of nutrients because of the application of large amounts
of fertiliser. However, the higher growth rates in CO_2 enriched atmospheres
will presumably result in a faster depletion of mineral nutrients in the
soil with the result that high CO_2 levels may lead to the development of
nutrient stress in crop plants. This effect will be even more marked in
natural communities which even at present are often nutrient limited (54).
Some attention has been given recently to the interaction between atmospheric CO_2 levels and mineral nutrition, primarily nitrogen availability
(16,51,52,60). In all cases the enhancement of yield at elevated CO_2 is
reduced under nitrogen shortage but there is still some stimulation of
growth. This continued response to CO_2 while nitrogen is limiting may be
due to either improved uptake of nitrogen from the soil or increased nitrogen efficiency. Goudriaan and de Ruiter (16) suggest that both effects occur and it has certainly been demonstrated that C3 plants have a much lower
nitrogen content under elevated CO_2 (60). This reduced nitrogen content
alters the carbon/nitrogen ratio of the plant material but the influence of
this on the quality of crops as an animal feed, especially digestibility,
has not been investigated.

There is even less information on the limiting effect of other nutrients such as phosphorus and sulphur at high atmospheric CO_2. However, it
has been suggested (16) that for a range of crop plants the interaction of
CO_2 with shortage of phosphorus appears to be simply governed by the law of
limiting factors because, unlike nitrogen, the biochemical role of phosphorus cannot be reduced at elevated CO_2.

7 Interaction of elevated atmospheric carbon dioxide with phenology

The largest stimulation of growth by CO_2 appears to occur in seedling
plants. For example, Mauney et al. (35) reported higher relative growth
rates in the seedling stage of soybean, cotton, sorghum and sunflower at
630 ppm CO_2 than at normal atmospheric levels but the effect was not continued during later development, while Neales and Nicholls (39) found a large
CO_2/plant age interaction in wheat.

It has also been found that elevated CO_2 levels can alter the balance
between growth and reproductive development. This effect can be manifested
in either or both of two ways. First, flowering may be determined by the
attainment of a certain minimum amount of vegetative growth and increased
growth at high CO_2 would reduce the time necessary to reach this threshold.
Secondly, the enhanced CO_2 concentration may have a direct effect on the
flowering process. Marc and Gifford (33) have recently investigated floral
initiation in wheat, sunflower and sorghum at high CO_2 levels and found
that there are marked differences in the response of these species. In
wheat and sunflower, floral initiation was advanced up to three days, but
while there was an equivalent increase in dry weight in wheat this did not
occur in sunflower. In sorghum, the time of floral initiation did not
change but the period for inflorescence development was significantly re-

duced. Clearly there are marked differences between species in the effect of CO_2 on their phenology and further research is required to elucidate these.

8 Crop/Weed interactions at elevated atmospheric CO_2

Weeds can be extremely damaging to crop plant growth and development because they compete directly with the crops for the same limited environmental resources upon which growth depends (54). If crop/weed competition is affected by increased atmospheric CO_2 levels then this could have a significant effect on crop yields. In particular, if the weed and crop have the different photosynthetic pathways, either C3 or C4, then as we have seen they could respond differently to increased CO_2 levels.

In general, increases in global atmospheric CO_2 should increase the impact of C3 weeds in C4 crops and decrease the problems of C4 weeds in C3 crops (44). In Europe, the only important C4 crop is maize and it is to be expected that its competitive relationship with its weeds would be strongly reduced. However the research in this area to date suggests that the relative reponses of crops and weeds will also vary according to other factors such as species, growth habit, seed size and nutrient availability (54).

9 Model Crops

As we have seen, the response of crop plants to increasing atmospheric CO_2 is difficult to predict not only because of the differences of response between species but also because CO_2 has both direct effects on the process of photosynthesis and indirect effects which relate to plant water and nutrient fluxes. If we are to attempt to predict the effects of elevated CO_2 on crop production in Europe then it is important to select economically significant examples and study their response to CO_2. For this reason we have chosen two crops which are grown extensively for forage production in Europe; viz. perennial ryegrass and maize. At the same time as being important economically they are also interesting physiologically, as ryegrass is C3 and maize is C4 and we have already seen that they might be expected to respond differently to elevated CO_2. These then are our model crops.

Grasses, amongst which perennial ryegrass (Lolium perenne) is the most important, are arguably one of the most important crops in Europe. A wet maritime climate, like that in the Republic of Ireland, is the most favourable for grass growth and there are more than 4.8 million ha of grassland in the country. However it is still important in other European countries, for example grassland occupies more than half the agricultural land in France and in the Netherlands grassland products meet 80% of the total nutritional requirements of cattle in the summer and 50% in the winter (40). Grass is immensely variable in yield, quality and nutrient content; so that yield from cut, non-irrigated swards of S23 ryegrass are between 5 and 18 t ha^{-1} yr^{-1} but Cooper (10) estimated the potential annual yield in Britain to be 29 t ha^{-1}.

The expansion of the maize (Zea mays) crop has been one of the most significant developments in European agriculture in the last 30 years. Initially it was grown predominantly in southern France and parts of Italy but now almost five million hectares are grown in the European Community, of which two million is for forage, and it spreads into northern Europe as far as Scotland and Ireland (5). Maize originates from subtropical regions and temperature strongly influences its growth and productivity (7). Seeds germinate slowly below 10°C and growth only starts when soil temperatures reach this level. Even during the summer daily maximum temperatures are nearly always below the optimum for photosynthesis (30-33°C)(7) and growth is terminated by low temperatures in autumn. Despite these temperature

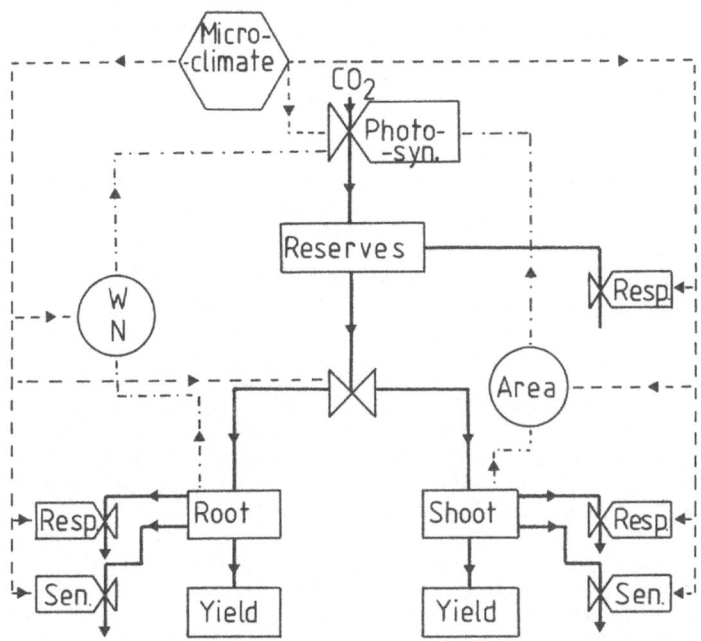

Figure 5. Plant physiological processes affected by the environment.
— is flow of carbon, --- is effects of environment; -.-.
is feedback, W is water supply, N is nutrients, Resp. is
respiration, Sen. is senescence (adapted from 29).

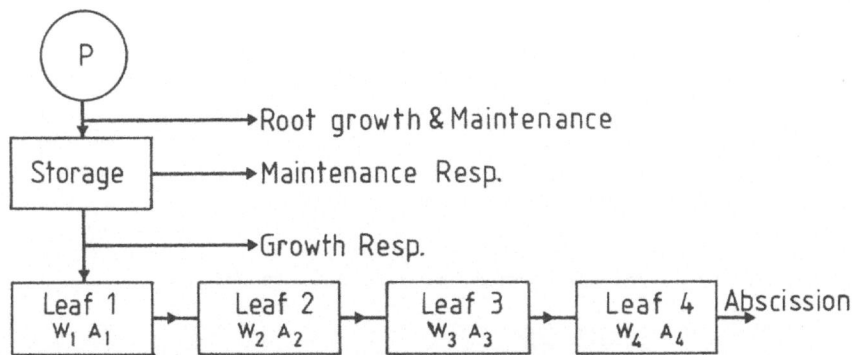

Figure 6. Schemic representation of Johnson and Thornley model (24). P
is photosynthesis, Resp. is respiration, W is leaf weight, A
is leaf area.

limitations maize gives yields in northern and central Europe which are comparable to ryegrass. In addition, maize silage is easier to make than grass silage and the crop can be harvested over a longer period of time because the period of optimum digestibility is greater. Grass and maize are therefore direct alternatives for forage production in most of France, in southern Britain and in the Netherlands (40).

10 Crop Models
The growth and yield of crop plants depends upon many factors amongst which photosynthesis is contributory but not of overriding importance. The other factors include nitrogen metabolism, translocation, partitioning of assimilates to plant parts and leaf expansion; many of which we have seen are influenced by atmospheric CO_2 concentration. It is therefore difficult to predict the yield of plants which are responding to a whole complex of factors. However one way to relate these interacting factors is through a mechanistic model which attempts to predict or simulate plant growth using quantitative data from physiological measurements. Figure 5 shows a conceptual model which includes plant physiological processes affected by the environment including levels of atmospheric CO_2. Simulation models should aim to express the effects of the environment (weather) on each plant process in a way that reflects the mechanism involved. However, when individual processes are not understood modellers have used other options which include fixed functions, measured values and comprehensive statistical descriptions (29). Legg (29) has reviewed the features of several crop simulation models but the two of most direct interest and relevance are BACROS developed by de Wit et al. (62) for maize in Holland and an evolving series for grass (24,25,49). A comprehensive description of these models is beyond the scope of this review but if should be borne in mind that these dynamic simulation models are according to Legg (29) 'still in a primitive stage' and certainly they are restricted in terms of both environmental inputs and physiological processes modelled. For example, the grass model of Johnson and Thornley (24) ignores the influence of nutrient availability and water stress, or it simply assumes that the grass plant has adequate non-limiting supplies of nutrients and water.
In general, the primary processes of photosynthesis and respiration can be adequately modelled (56), as can the light distribution within the canopy (56) but the prediction of expansion of new leaves and their eventual senescence has been more of a problem. In some models it has been assumed that leaf area expansion is a dependent variable calculated from the dry weight of the crop, but Johnson and Thornley (24) have recently treated leaf area as an independent state variable influenced by the assumed level of a storage pool (Figure 6). Using this approach they have been able to adequately model the seasonal pattern of growth of a vegetative grass crop. This has validated the model at the level of its predictions (56) but no attempt has been made so far to incorporate the effects of elevated atmospheric CO_2 concentrations.
Ideally the mechanistic models should be based upon the data obtained from experiments under defined environmental conditions with elevated CO_2 levels both in controlled environments and in the field. The initial procedure should be to determine the parameters for each part of the model from careful experiments in controlled environments.

11 Experimental approach
Measurements made to determine the parameters for mechanistic models of crop growth must of course be on plants growing at elevated CO_2 levels, but also under conditions of light intensity, temperature, humidity, water

availability etc. which reproduce the natural environment in a controlled fashion. These experiments can be done in controlled environment rooms, greenhouses and also in the field using open-top chambers (18). The latter allowing investigations of crops under field conditions but at the same time experiencing controlled elevated CO_2 levels, although results are probably more useful in validating model predictions rather than constructing them.

12 Conclusions

There is substantial evidence that the projected increase in atmospheric CO_2 levels over the next century will have a significant beneficial effect on crop plant production which is independent of the influence of the climatic change that will also occur. However it is possible that some climatic changes, such as increased temperature and lower rainfall, could counteract the benefits derived from higher CO_2. The main problem in predicting the effects seems to be the variation between species in their response to CO_2, so that higher yield in some cases will be due to higher rates of photosynthesis during certain stages of development while in other cases it may be due to higher water use efficiency when availability of water limits growth. A greater understanding of the physiological processes involved should enable us to make predictions for specific crops but more information is required on the response of selected plants to elevated CO_2. In particular, the response of the stomata to elevated CO_2 under a range of environmental conditions is of central importance.

These difficulties in predicting the response of crop plants, grown in monoculture and in non-nutrient limiting conditions, emphasise the problems in predicting the response of natural vegetation to enhanced CO_2 levels. However, if the consequences of increasing CO_2 levels on the world's vegetation is to be assessed then an understanding of effects on natural vegetation must be the long term aim.

REFERENCES

1. AOKI, M. and YABUKI, K. (1977). Studies on the carbon dioxide enrichment for plant growth. VII. Changes in dry matter production and photosynthetic rate of cucumber during carbon dioxide enrichment. Agr. Meteorol. 18, 475–485.
2. BERRY, J.A. and BJORKMAN, O. (1980). Photosynthetic reponse and adaptation to temperature in higher plants. Ann. Rev. Pl. Phys. 31, 491–543.
3. BIERHUIZEN, J.F. et al. (1973). Effects of temperature and radiation on lettuce growing. Neth. J. Agric. Sci. 21, 110–116.
4. BISCOE, P.V. and GALLAGHER, J.N. (1977). Weather, dry matter production and yield. In: Environmental effects on crop physiology. Ed. J.J. Landsberg and C.V. Cutting. pp. 75–100. Academic Press, London.
5. BUNTING, E.S. (1980). History of maize in Europe. In: Production and utilization of the maize crop. Ed. E.S. Bunting. pp. 3–13. The Hereward and Stourdale Press, Ely, Cambs.
6. CALVERT, A. and SLACK, G. (1975). Effects of carbon dioxide enrichment on growth, development and yield in glasshouse tomatoes. I. Responses to controlled conditions. J. Hort. Sci. 50, 61–71.
7. CARR, M.K.V. and HOUGH, M.N. (1978). The influence of climate on maize production in north-western Europe. In: Forage Maize. Ed. E.S. Bunting et al. pp. 15–55. Agricultural Research Council, London.
8. CAVE, G., TOLLEY, L.C. and STRAIN, B.R. (1981). Effect of carbon dioxide enrichment on chlorophyll content, starch content and starch grain

structure in <u>Trifolium</u> <u>subterraneum</u> leaves. Physiol. Plant. 51, 171–174.

9. COLLINS, R.P. and JONES, M.B. (1984). The influence of climatic factors on the distribution of C4 species in Europe. (submitted for publication).

10. COOPER, J.P. (1968). Energy and nutrient conversion in a simulated sward. Report of the Welsh Plant Breeding Station, 1967, pp. 10–11.

11. DOWNTON, W.J.S., BJORKMAN, O. and PIKE, C.S. (1980). Consequences of increased atmospheric concentrations of carbon dioxide for growth and photosynthesis of higher plants. In: Carbon dioxide and climate; Australian Research. Ed. G.I. Pearman. Australian Academy of Science, Canberra.

12. EDWARDS, G. and WALKER, D. (1983). C3, C4: mechanisms, and cellular and environmental regulation, of photosynthesis. Blackwell, Oxford.

13. GARDNER, R. (1965). The response of crops to carbon dioxide enrichment of glasshouse atmosphere. Agric. Prog. 40, 88–93.

14. GATES, D.M. (1983). An overview. In: CO_2 and plants, the response of plants to rising levels of atmospheric carbon dioxide. Ed. E.R. Lemon. pp. 7–20. Westview Press, Colorado.

15. GIFFORD, R.M. (1977). Growth pattern, carbon dioxide exchange and dry weight distribution in wheat growing under differing photosynthetic environments. Aust. J. Plant Physiol. 4, 99–110.

16. GOUDRIAAN, J. and de RUITER, H.E. (1983). Plant growth in response to CO_2 enrichment, at two levels of nitrogen and phosphorus supply. 1. Dry matter, leaf area and development. Neth. J. Agric. Sci. 31, 157–169.

17. GRIBBIN, S. (1984). Meteorology blows hot and cold. New Scientist, 6 Sept. No. 1420, 17–20.

18. HEAGLE, A.S., BODY, D.E. and HECK, W.W. (1973). An open-top field chamber to assess the impact of air pollution on plants. J. Environ. Quality 2, 365–368.

19. HEATH, O.V.S, and MEIDNER, H. (1967). Compensation points and carbon dioxide enrichment for lettuce grown under glass in winter. J. Exp. Bot. 18, 746–751.

20. HEATH, O.V.S. and RUSSELL, J. (1954). Studies in stomatal behaviour. VI. An investigation of the light responses of wheat stomata with the attempted elimination of control by the mesophyll. J. Exp. Bot. 5, 1–15.

21. HO, L.C. (1976). The relationship between the rates of carbon transport and photosynthesis in tomato leaves. J. Exp. Bot. 27, 87–97.

22. HOFSTRA, G. and HESKETH, J.D. (1975). The effects of temperature and CO_2 enrichment on photosynthesis in soybean. In: Environmental and biological control of photosynthesis. Ed. R. Marcelle, Dr. W. Junk, The Hague.

23. HSIAO, T.C. (1973). Plant responses to water stress. Ann. Rev. Pl. Phys. 24, 519–570.

24. JOHNSON, I.R. and THORNLEY, J.H.M. (1983). Vegetative crop growth model incorporating leaf area expansion and senescence, and applied to grass. Plant, Cell & Env. 6, 721–729.

25. JOHNSON, I.R., AMEZIANE, T.E. and THORNLEY, J.H.M. (1983). A model of grass growth. Ann. Bot. 51, 599–609.

26. JONES, M.B., LEAFE, E.L. and STILES, W. (1980). Water stress in field-grown perennial ryegrass. I. Its effects on growth, canopy photosynthesis and transpiration. Ann. Appl. Biol. 96, 87–101.

27. KIMBALL, B.A. and IDSO, S.B. (1983). Increasing atmospheric CO_2: effects on crop yield, water use and climate. Agric. Water Management 7,

55-72.
28. KRAMER, P.J. (1981). Carbon dioxide concentration, photosynthesis, and dry matter production. Bio Science 31, 29-33.
29. LEGG, B.J. (1981). Aerial environment and crop growth. In: Mathematics and plant physiology. Ed. D.A. Rose and D.A. Charles-Edwards. pp. 129-149. Academic Press, London.
30. LEMON, E.R. (1983). Interpretive summary. In: CO_2 and plants, the response of plants to rising levels of atmospheric carbon dioxide. Ed. E.R. Lemon. pp. 1-5. Westview Press, Colorado.
31. LOOMIS, R.S. and GERAKIS, P.A. (1975). Productivity of agricultural ecosystems. In: Photosynthesis and productivity in different environments. Ed. J.P. Cooper. pp. 145-172. Cambridge University Press, Cambridge.
32. MANSFIELD, T.A. (1983). Movements of stomata. Sci. Prog., Oxf. 68, 519-542.
33. MARC, J. and GIFFORD, R.M. (1984). Floral initiation in wheat, sunflower, and sorghum under carbon dioxide enrichment. Can. J. Bot. 62, 9-14.
34. MASKELL, E.J. (1928). Experimental researches on vegetable assimilation and respiration. XVII. The diurnal rhythm of assimilation in leaves of cherry laurel at "limiting" concentration of carbon dioxide. Proc. Roy. Soc. B 102, 467-87.
35. MAUNEY, J.R., FRY, K.E. and GUINN, G. (1978). Relationship of photosynthetic rate to growth and fruiting of cotton, soybean, sorghum and sunflower. Crop Sci. 18, 259-263.
36. MOORBY, J. (1981). Transport systems in plants. Longman, London.
37. MORISON, J.I.L. and GIFFORD, R.M. (1983). Stomatal sensitivity to carbon dioxide and humidity, a comparison of two C3 and two C4 species. Plant Physiol. 71, 789-796.
38. MORISON, J.I.L. and JARVIS, P.G. (1983). Direct and indirect effects of light on stomata. II. In Commelina communis L. Plant, Cell & Env. 6, 103-109.
39. NEALES, T.F. and NICHOLLS, A.O. (1978). Growth responses of young wheat plants to a range of ambient CO_2 levels. Aust. J. Plant Physiol. 5, 45-59.
40. NEDO report (1974). UK farming and the common market: grass and grass products. National Economic Development Office, Millbank, London.
41. OSMOND, C.B., BJORKMAN, O. and ANDERSON, D.J. (1980). Physiological processes in plant ecology: towards a synthesis with Atriplex. Ecological Studies, Vol. 36. Springer-Verlag, Berlin.
42. PAEZ, A., HELLMERS, H. and STRAIN, B.R. (1983). CO_2 enrichment, drought stress and growth of Alaska pea plants (Pisum sativum). Physiol. Plant 58, 161-165.
43. PAEZ, A., HELLMERS, H. and STRAIN, B.R. (1984). Carbon dioxide enrichment and water stress interaction on growth of two tomato cultivars. J. Agric. Sci., Camb. 102, 687-693.
44. PATTERSON, D.T. and FLINT, E.P. (1980). Potential effects of global atmospheric CO_2 enrichment on the growth and competitiveness of C3 and C4 weed and crop plants. Weed Sci. 28, 71-75.
45. PEARCY, R.W. and BJORKMAN, O. (1983). Physiological effects. In: CO_2 and plants, the response of plants to rising levels of atmospheric carbon dioxide. Ed. E.R. Lemon. pp. 65-105. Westview Press, Colorado.
46. RAPER, C.D. and PEEDIN, G.F. (1978). Photosynthetic rate during steady-state growth as influenced by carbon dioxide concentration. Bot. Gaz. 139, 147-149.
47. RASCHKE, K. (1979). Movements of stomata. In: Encyclopedia of Plant

Physiology, New Series No. 7. Ed. W. Haupt and M.E. Feinlieb. pp. 383-441. Springer-Verlag, Berlin.

48. ROGERS, H.H., THOMAS, J.F. and BINGHAM, G.E. (1983). Response of agronomic and forest species to elevated atmospheric carbon dioxide. Science 220, 428-429.

49. SHEEHY, J.E., COBBY, J.M. and RYLE, G.J.A. (1980). The use of a model to investigate the influence of some environmental factors on the growth of perennial ryegrass. Ann. Bot. 46, 343-365.

50. SHIBLES, R.M. and WEBER, C.R. (1966). Interception of solar radiation and dry matter production by various soybean planting patterns. Crop Sci. 6, 55-59.

51. SIONIT, N. (1983). Response of soybean to two levels of mineral nutrition in CO_2-enriched atmosphere. Crop Sci. 23, 329-333.

52. SIONIT, N. et al. (1981). Effects of atmospheric CO_2 concentration and water stress on water relations of wheat. Bot. Gaz. 42, 191-196.

53. SLATYER, R.O. (1970). Comparative photosynthesis, growth and transpiration of two species of Atriplex. Planta 93, 175-189.

54. STRAIN, B.R. and BAZZAZ, F.A. (1983). Terrestrial plant communities. In: CO_2 and plants, the response of plants to rising levels of atmospheric carbon dioxide. Ed. E.R. Lemon. pp. 177-222. Westview Press, Colorado.

55. THOMAS, J.F. and HARVEY, C.N. (1983). Leaf anatomy of four species grown under continuous CO_2 enrichment. Bot. Gaz. 144, 303-309.

56. THORNLEY, J.H.M. (1976). Mathematical models in plant physiology. Academic Press, London.

57. TOLBERT, N.E. and ZELITCH, I. (1983). Carbon metabolism. In: CO_2 and plants, the response of plants to rising levels of atmospheric carbon dioxide. Ed. E.R. Lemon. pp. 21-64. Westview Press, Colorado.

58. WILLMER, C.M. (1983). Stomata. Longman, London.

59. WOLEDGE, J. and DENNIS, W.D. (1982). The effect of temperature on photosynthesis of ryegrass and white clover leaves. Ann. Bot. 50, 25-35.

60. WONG, S.C. (1980). Elevated atmospheric partial pressure of CO_2 and plant growth. 1. Interaction of nitrogen nutrition and photosynthetic capacity in C3 and C4 plants. Oecologia 44, 68-74.

61. WONG, S.C., COWAN, I.R. and FARQHAR, G.D. (1979). Stomatal conductance correlates with photosynthetic capacity. Nature 282, 424-426.

62. de WIT, C.T. et al. (1978). Simulation of assimilation, respiration and transpiration of crops. Simulation Monographs, Pudoc, Wageningen.

AGROCLIMATIC CLASSIFICATION OF CENTRAL ITALY

G.MARACCHI and F.MIGLIETTA
Agronomy Istitute - University of Florence
I.A.T.A. - C.N.R. - Florence

Summary

 We outline the main climatic classifications. Ecological parameters as radiation, temperature and water availability are discussed in order to the impact on crop productivity.

A data base of most relevant meteorological parameters has been built for Central Italy. Classification was performed for winter and summer standard crops on the basis of limiting factors, temperature and water, defining the growing period. For the growing period the productivity index was applied daily and different stations comparated.

1. Introduction

The climatic classification started at the end of the eighteenth centu ry beginning of nineteenth,with the classical studies of Köppen on the climates repartition on the basis of two main parameters as temperature and rainfall. He choosed some arbitrary limits in relation to a general correspondence with the most important ecosystem of the world.

Several other studies were performed based on the same principles and sometimes on other parameters as evaporation or vapour pressure deficit.

All these kind of classification more or less sophisticated are based on the idea of a relation between the annual radiation and heat balance and the annual water availability by means of hidrological balance.

As the life limits are represented by the minimum temperature and aridity, there is a logical over imposition of the values of such indices and the most important ecological association and rain forest, savanna, prairie, steppa and tundra.

This kind of descriptive classification reply to a need of a general knowledge of geography but they don't give any information that allow us to take some decisions on the enviromental management or on agricultural management.

In a more recent period several efforts were made to build up classifications more useful from a point of view of agroclimatology.

The works of Papadakis, De Fina, Thran and Broekuisen, Kimball and Brooks, are examples of a different approach to the climatic features that takes in account the agricultural potentiality as related to the distribution during the year of the main climatic parameters and their conbination in a way that makes the climate of a specific area suitable for the agricultural management.Thesekind of classification were be based on indices having a specific agricultural significance in term of ecological parameters as lenght of growing period, water availability cardinal temperatures, evapotraspiration, lenght of day, minimum temperatures, frost hazard, etc.

This group is characterized by a comparison of ecological parameters with the climate features of a station during all the year.

It represents a progress in the possibility to utilize the information of this kind of classification to plan the agriculture at least at a level of climatic regions. The regional planning is a tool of large interest both in the developed as in the developing countries and these kind of classification could be used at a very large scale . If we take in account the Papadakis classification we find that Italy is divided in four main classes, Subtropical mediterranean, Marine mediterranean, Temperate mediterranean, Continental mediterranean, that means the first area is reliable for citrus and cotton, the second one for citrus and maize, the third one for wheat, oat, maize, rice, the fourth one for wheat, oats , maize. This kind of classification is very useful for having a seeing on the agriculture potentiality all over the world but in term of contribution to solve problems of managers or farmers, is completely useless .

A further improvement was made at the beginning of sixty years with introducing the concept of the model in assessing the potentiality of agricultural agroecosystems .

From the knowledge of the photosynthesis processes, temperature and water relations, some models were built up of the crop growth in relation to the atmosphere and soil characteristics .The possibility was open by the high speed computers of simulating the processes time varying .The utility of this kind of models was at,a first time,more concerning the better understanding of the processes under-lying the production but in the last years the introduction of the personal computer generation available to the farmers because of the low price, opened new possibilities to use the models as operational tools in the fields of farm management.

Surveying the methods of climate classification and the last views on the crop modelling, we performed an intermediate method to assess the agroclimate and potential productivity.

The purpose of such a classification were to discriminate not only large groups of climate but to identify in the same region difference between neighbouring areas with similar characteristics .

Another problems we attempted to face was the possibility to utilize both models based on daily data as on monthly or weekly data .

2 . Ecological parameters

The classification system is based on the main ecological parameters controlling the crop growth as radiation, temperature and water.

From the results of research developed in the last years the crop growth rate is proportional to the quantity of radiation intercepted by the stand.

The photosyntesis by the crop as a whole is much less subject to the light saturation due the fact that the reciprocal sha-ding of the leaves allow to utilize all the radiation impinging on the crop and shape, orientation, inclination and azimuth of leaves could optimize the quantity of light received by the stand as a whole (Fig . 1). For this reason while saturation start to occur for example for the single leaves of maize and wheat at 0.6 and 0. 3 cal $cm^{-2} min^{-1}$ the net assimilation for the stand is increasing linearly until 1 cal $cm^{-2} min^{-1}$. This value is corrisponding to the value of maximum radiation at midday on June 21th in the temperate latitude, on the hour basis .

The experimental relation could be expressed in the following way:

$$\frac{P}{P_{max}} = 1 - \exp^{-kR_g} \qquad (1)$$

where: P = dry matter
          P_{max} = maximum dry matter
          k = 0 .00 25
          R_g = global radiation

So photosynthesis, crop structure and geometry, leaf area index are cardinal parameters to describe the crop growth.

Temperature by means of the effect on photosynthesis and other enzymatic reactions controls the growth both as crop growth rate and phenophase occurrence. In that way the most important clock of life rhitms are connected to the thermal parameter.

Distribution of the plant species and varieties according to the temperature regimes is the most evident feature of sensibility to this factor .

Temperature is dependent from two different conditions, air temperature due to the height on the sea level, to the morphology, to the characteristics of ground cover; the second one is connected with the radiation balance of the crop surface that controls temperature as evapotraspiration.

The relative effect of temperature on growth was assessed in several ways by different authors related to the leaf photosynthesis, single crop production, groups of crop, etc. The main feature we can observe is the steep slope of the function in the first part of the curve decreasing up to the maximum, then when we reach the maximum, we can observe a decreasing process.

So if we interpolate the different equation we obtain the curve in fig . 2

Another effect of temperature is related to the severe constrains due to the low temperature and frost as to the very high temperature .

The frost iniury is variable with species and varieties both for the crops and fruit trees and vegetable . Low temperatures slow the crop growing with possible damages due to the insects and phitopaties, in the fruit trees can reduces the percentage of setting up of flowers .

In the similar way the very high temperature can cause a severe stress due to traspiration coupled with very high radiation increasing the respiratory activity and decreasing photosyntesis until the dehidratation of the tissues.

The period between the phenophases can be computed by means of the temperature summation minus a value corrisponding to the vegetation zero. Correction of temperature summation by radiation or sunshine duration gives a much more accurate estimate of the period necessary to pass from a phenological stage to another.

Water availability and water status of leaves and stems are the third cardinal ecological factor.

The first condition is related to the availability of water in the soil depth explored by roots, the roots depth, the soil texture, the water table height, the phenological stage.

The second condition is related to the transpiration velocity during the day. The water loss by the stomata due to thermal load speed up the translocation from the roots to the leaves. When the water flux in the atmo sphere becomes greater than the water flux in the phloema, the feedback mechanism of regulation of stomata closes the stomata. This takes place very often during the summer days around midday also if the water availability in the roots depth is sufficient to replenish the water reservoir of the crop.

Together with the high value of radiation the factor controlling the traspiration is the air water deficit.

The water consumption changes during the vegetative or ripening periods, most of the crop shows an high ratio between the crop growth rate and the actual evapotraspiration/potential evapotraspiration, during the yield formation the trend is generally reversed. If we consider all the growing period as a whole, the relation between these parameters has a slope close to one.

We can approach the argument of water – productivity relationship from the following point of view. Considering for instance the results of irrigation on crop productivity or of rainfall on primary productivity. (Fig.3) From these kind of data we can arrive to the following function:

$$\frac{P}{P_{max}} = 1 + k \ln \frac{I}{I_{max}} \qquad (2)$$

where: P = dry matter $\qquad P_{max}$ = maximum dry matter

k = coefficient $\qquad I$ = water availability

I_{max} = maximum water availability

in which the coefficient k span between 0.3 and 0.45.

It seems that the processes of productivity both at the crop scale as at ecosystems level abeyes to the same general laws.

As we consider the atmospheric parameters as driving the processes of crop growing, by means of radiative, thermal and hydrological cycle, we must consider the interactions of the meteorological parameters with morphology and landscape. North and south facing slopes, large or narrow valleys, clay or sandy soils, sea or mountains proximity, as the features of macroclimate are elements changing the interaction of meteorological parameters with the land surface, leading to different microclimates in a few kilometers.

Agroclimatic classification must take in consideration this kind of variability or accounting for the ecological reply of crops, fruit trees and forestry.

In order to include in the model these aspect we took in account slope and orientation both on radiative model and on hydrological cycle.

To account for the effect of slope on hydrological balance we start from the data concerning the ratio between rainfall and stream flow during the different seasons and from the slope inclination.

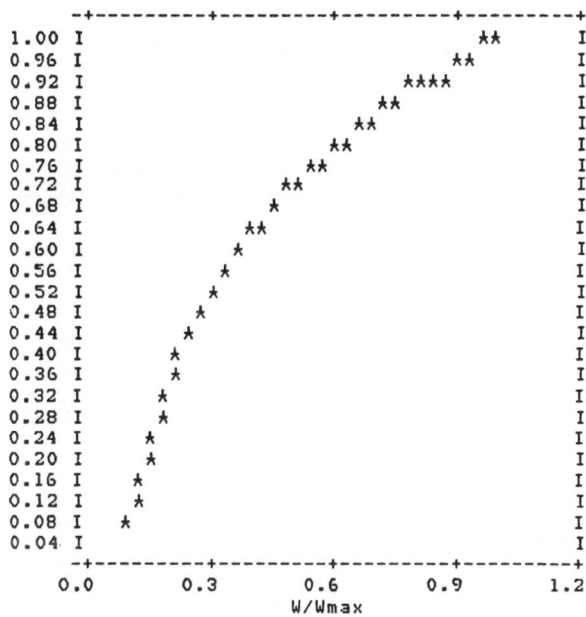

Fig.3 - Relative effect of water availability on dry matter production.

3. Methods

3.1 The computer system

The Institute actually dispose of a DEC VAX 11/750 of 2 Mbytes. The main frame is articulate as follows:
a) Ten pT 100 video terminals
b) Two line printers
c) Two disk units (124 Mbytes and 450 Mbytes)
d) One tape unit TS11 1600 bpi
e) An image processing system consisting in a parallel interface, 512x512, a standard TV camera and a colour video display
f) A graphical station with a digitizer and a high resolution graphic plotter.

3.2 The meteorological data base

Italy is a climatologically non-homogenous country. The main geographic axis is North/South oriented, and a mountain chain divides the Eastern and the Western coasts.

The historical data furnish thermopluviometric information. A number of WMO standard meteorological station are managed by the Aeronautic meteorological office and data are stored in magnetic support. Our Institute is now working on a digitization plan of a large number of thermopluviometric data, according to the goals of C.E.E. - Climatology project for agriculture productivity increase (Table I)

Table I - Meteorological networks of central Italy

Region	Surface Km^2	Thermopluviometer networks	Pluviometer networks	Standard networks
Toscana	22992.38	37	300	13
Umbria	8456.04	12	100	1
Lazio	17202.69	19	180	11
Marche	9693.51	14	120	4

Actually the data of Central Italy reside on magnetic tapes and had been used for drawing this communication.

3.3 Classification method

Multivariate data analysis allows to perform a complete date classification starting from multivariate data matrix of the form

$$X = \begin{vmatrix} x_{1.1} & x_{1.2} & \cdots\cdots\cdots & x_{1.n} \\ \cdots\cdots\cdots\cdots\cdots \\ \cdots\cdots\cdots\cdots\cdots \\ x_{m.1} & x_{m.2} & \cdots\cdots\cdots & x_{m.n} \end{vmatrix}$$

where n = number of data units to classify
       m = number of variables

It is then possible to perform a "cluster analysis" using a wide variety of computing algorithms. In this case the centroid method is used starting from the calculation of an euclidian distance symmetric matrix.

The result of this kind of hierarchical cluster analysis allows to explore the data structure and, finally, to choice the best partitioning of the data set.

All kinds of statistical analysis described here, were performed using the GENSTAT package.

4. Agroclimatic classification

According to the considerations exposed on the general processes driving the crop productivity we performed a simple classification method to apply to the italian agroclimatic classification.

The main section of the method are:
a) the definition of the growing period
b) the computing of an index of productivity related to the growing period.

The definition of the growing period was determined on the basis of:
- thermal conditions
- hydrological balance

We choosed two different treshold of temperature for the winter crops and for the summer crops 5 °C and 10 °C.

For winter crop the parameter, connected with the possibility of seed germination, and initial growing season is depedent on the soil water content after the summer dryness. The computation of soil moisture budget can be done both by daily values of temperature, radiation, saturation deficit, wind to compute potential evapotraspiration and rainfall to compute the positive term of the balance or to allow to use the method for a more large number of stations on the basis of the monthly data with an adequate method of linear interpolation (Fig.4).

The period at which soil moisture can be consider adequate for crop establishing and for techical operations is when the rainfall equals the 0.5 of the ETP.

At that time the growing period will start. If temperature will fall below 5 °C, growing period will stop to start again when temperature will rise up to 5 °C. From this time the growing period last till the value of ETP will be greater than two times the value of the rainfall.

So the final information consists in the number of days useful for vegetation of winter crops.

A further agroclimatic information concern the fruit trees. Period growing period will start for fruit trees when temperature is up to 10 °C and we must compare the period computed in this way with the period computed starting from the last day with minimum temperature 2 °C, until the first day in autumn with the same condition. This period is called the frost free period.

Summer crop growing period is computed starting in spring from the mean temperature of 10 °C until the evapotraspiration equals the rainfall times 0.5. From this time we assume the aridity period start.

Once we computed the number of day of growing period both for the winter and the summer crops we calculate the summation of degree-days for the period and we compare with the figures corrisponding to the crops requirements. In this way we can assess the suitability of the climate for every species and variety.

The second part of classification concern the estimate of productivity of a standard crop having the characteristics of winter or summer crop.

We define a standard crop an hipotetical crop having the most relevant characteristic of C_3 or C_4 plants. The dry matter computation is performed on the basis of the following equation:

$$P = P_{max} \; min \; (\; f_T, \; f_G, \; f_A, \; f_F, \; f_N) \; - \; R_T \qquad (3)$$

dove P_{max} represents the maximum quantity of dry matter assimilated by the stand and f_i function variyng between 0 and 1 of temperature, global radiation ,water availability, phenophase and nutrient while R_T is the respiratory loss.

The relations illustrated in the previous section were utilized for radiation, temperature and water availability. As to the phenophase the function to compute f_F is the following:

$$f_F = \frac{W_o \; e^{kt}k}{W_{max}} \qquad (4)$$

where W_o = the stand weight at t = 0
W_{max} = the stand weight at LAI = 5
k = relative growth rate
t = time from emergence

The function for computing the nutrient effect was the following:

$$f_N = 1 - e^{-0.000067 \; N} \qquad (5)$$

where N = Kg nitrogen.

5. Results

Computation was performed at this stage for the meteorological station of Tuscany. In tab. II , III the results were summarized concerning the beginning of winter growing period and summer growing period together with the lenght of winter and summer growing period. The distribution order in tables follows the belonging of the stations to different watershed.

As aspected the most relevant factors influencing the distribution are the distance to the coast, the height and the belonging to the same watershed. The last conditions seems to be extremely relevant so in fig. 4 - 5 was put in evidence the degree of similarity, between 1 and 0.6, among the stations by the application of cluster analysis to the beginning and to the lenght of the season. The groupment results to be more stressed for the winter season than for the summer. In the first case temperature was in most cases the limiting factors with an high degree of variability between station and station while for the summer season the beginning of dry period was limiting and was widespread all over the region. Only on the higher stations was delayed but at the same time also the date of beginning was delayed.

Once the calculations of beginning and lenght were performed, we compute for every day the productivity index. Computing was done for a period relative to the degree day summation of winter wheat and maize for summer.

Table II - Lenght of the winter season

Climatological station	h m.	Average date	S.D.	Day average number	S.D.
Viareggio	3	23.10	23.9	271	31.6
Livorno	3	3.11	25.6	226	35.7
Portoferraio	25	22.11	35.5	183	43.2
Massa	65	14.10	19.7	251	24.7
Grosseto	8	2.11	30.0	212	36.4
Lucca	20	14.10	16.3	268	29.5
Pistoia	88	9.10	21.4	278	27.5
Pescia	81	21.10	28.5	278	37.2
S.Miniato	137	26.10	31.2	253	33.3
Pisa	11	24.10	30.3	247	45.3
Firenze	51	16.10	21.7	255	32.7
Prato	74	14.10	23.9	256	28.7
Empoli	27	25.10	29.8	235	41.7
Ponteginori	66	19.10	21.5	234	29.1
Montescudaio	241	28.10	28.6	229	33.8
Campagnatico	140	31.10	29.6	221	34.0
Scansano	500	28.10	30.7	255	37.5
San Sepolcro	330	15.10	19.4	243	44.8
Arezzo	277	23.10	29.6	229	38.8
Siena	348	4.11	29.4	237	32.4
Montalcino	564	23.10	18.8	248	20.2
Chianciano	459	28.10	29.7	228	43.5
Pienza	499	10.11	31.9	216	34.7
Firenzuola	223	5.10	20.3	223	40.5
Vallombrosa	955	3.10	21.6	175	28.2
Camaldoli	1111	29.9	21.3	170	20.4
S.Marcello	625	30.9	14.5	267	38.5

In fig. 6 - 7 two examples are reported for two stations at different height 50 m and about 800 m. As we can see for the winter crop for the second station there is a long period of winter rest and the values of the indices differ one from another due to the effect of morphology and temperature. The last one controlling the crop growth rate. For the summer season nevertheless the dry period isn't limiting, the productivity is greatest in the lower station due to the temperature.

We can conclude that despite of necessary refinement to do in the future, the method of agroclimatic classification performed seems to be sufficiently reliable to account for the most important processes involved in the agricultural productivity.

Table III - Lenght of the summer season

Climatological station	h m.	Average date	S.D.	Day average number	S.D.
Viareggio	3	31.3	20.7	87	36.1
Livorno	3	16.3	50.0	50	38.5
Portoferraio	25	25.1	21.9	73	41.3
Massa	65	14.4	17.3	47	25.6
Grosseto	8	16.4	51.4	38	30.8
Lucca	20	1.4	30.0	69	25.5
Pistoia	88	15.4	14.0	80	37.8
Pescia	81	11.4	16.0	90	54.9
S.Miniato	137	13.4	14.9	69	30.1
Pisa	11	15.4	16.4	53	30.7
Firenze	51	10.4	16.0	58	32.5
Prato	74	8.4	15.0	64	26.8
Empoli	27	17.4	12.6	53	32.3
Ponteginori	66	17.4	14.9	42	23.2
Montescudaio	241	21.3	44.3	53	35.9
Campagnatico	140	24.4	38.0	35	24.9
Scansano	500	26.4	14.8	46	26.3
San Sepolcro	330	21.5	50.0	47	21.7
Arezzo	277	24.5	39.8	44	27.3
Siena	348	30.4	17.8	49	20.1
Montalcino	564	15.3	34.6	72	28.9
Chianciano	459	13.5	42.6	44	18.9
Pienza	499	2.5	15.3	36	20.7
Firenzuola	223	14.5	11.8	49	29.0
Vallombrosa	955	10.6	11.9	45	30.4
Camaldoli	1111	12.6	8.3	59	48.0
S.Marcello	625	20.5	8.4	55	22.7

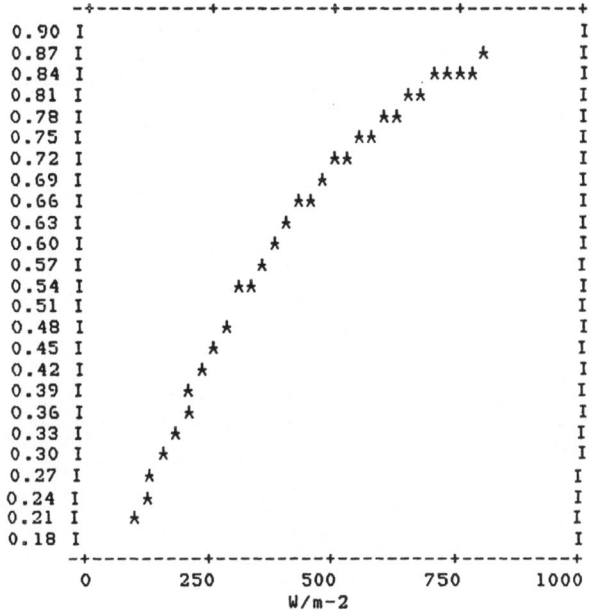

Fig.1 - Relative effect of global radiation on dry matter production.

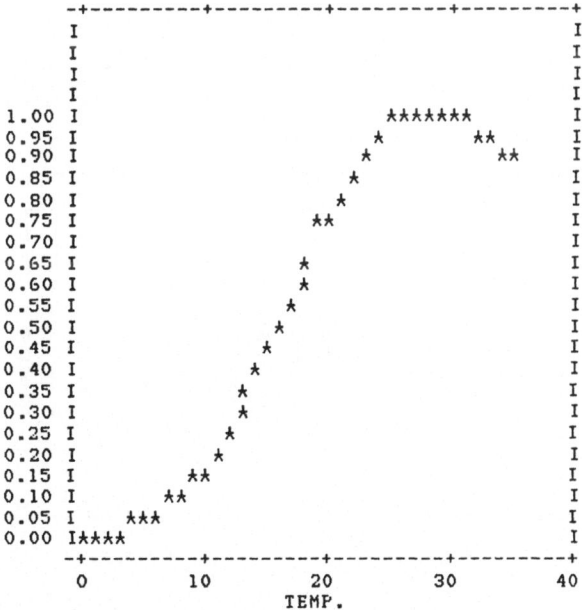

Fig.2 - Relative effect of temperature on dry matter production.

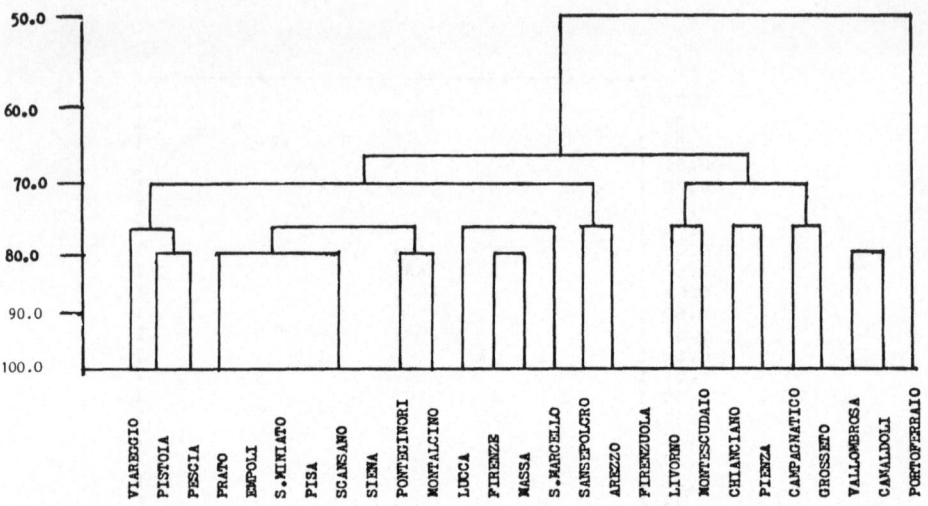

Fig.4 – Cluster classification of stations for winter season lenght and season beginning date.

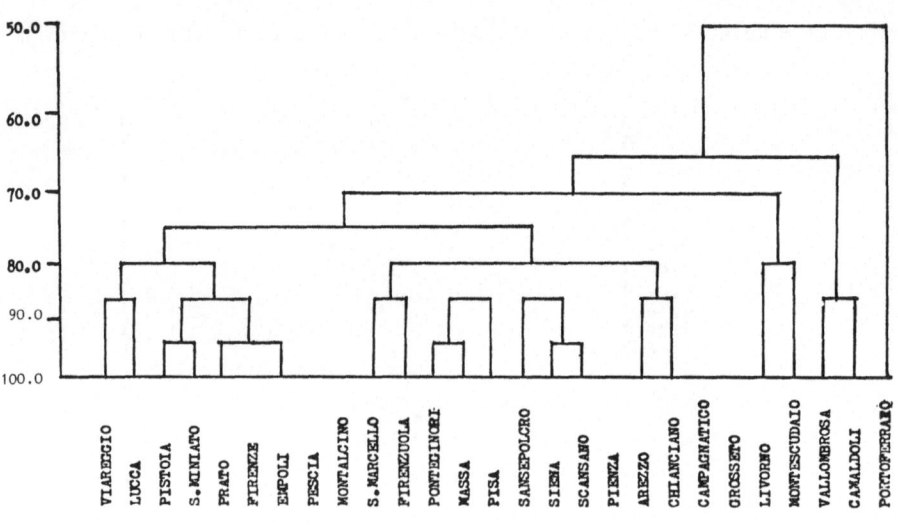

Fig.5 – Cluster classification of stations for summer season lenght and season beginning date.

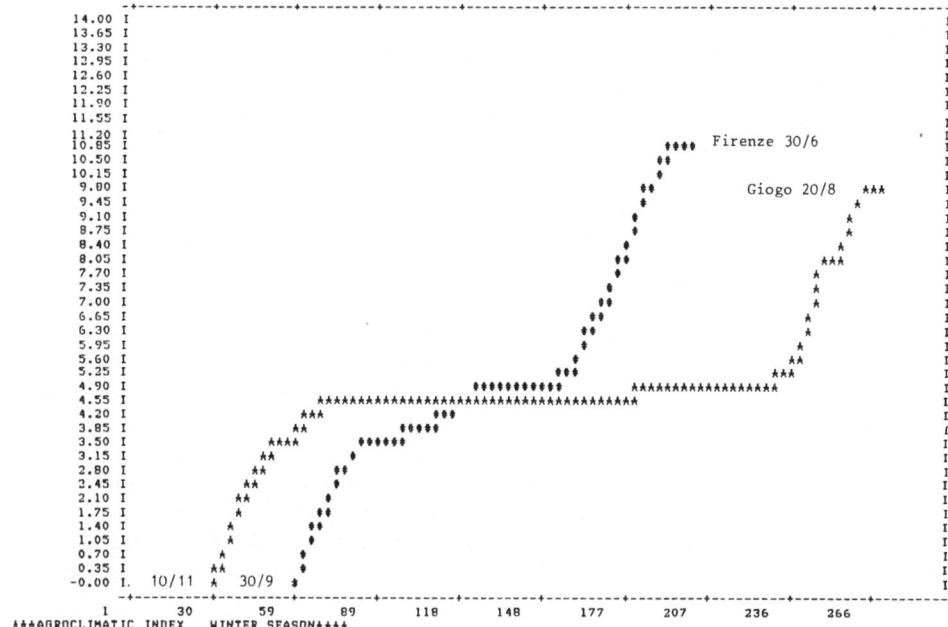

Fig.6 – Summation of winter index of productivity.

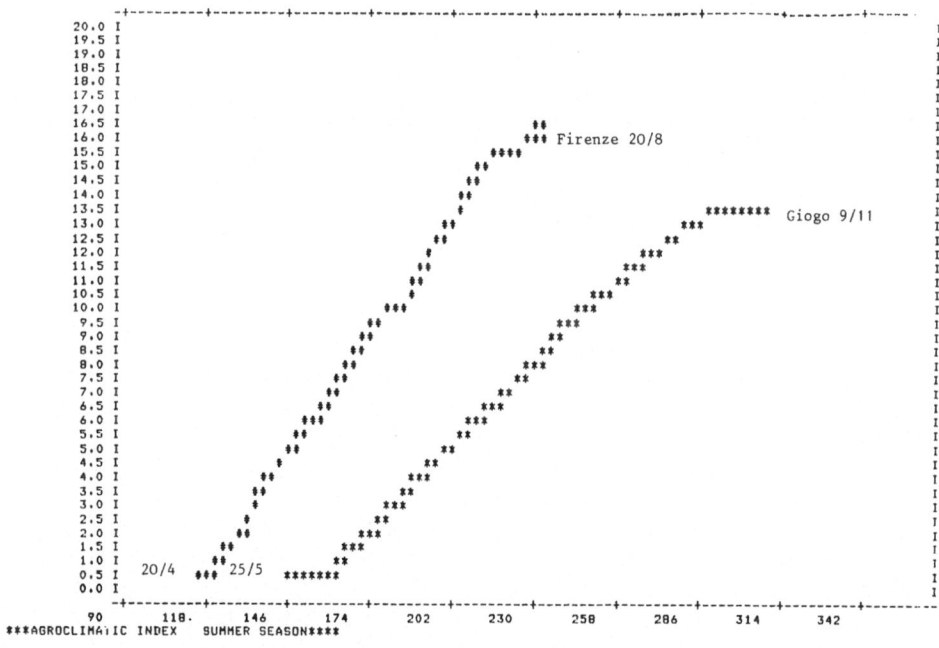

Fig.7 – Summmation of summer index of productivity.

REFERENCES

1. AUDERBERG, (1973). Cluster analysis for application. Academic Press
    New York
2. DOOREMBOS, J., KASSAM, A.H. (1979). Yield response to water. FAO irri-
    gation and drainage paper.
3. EVANS, G.C. (1982). The quantitative analysis of plant growth. Black-
    well Scientific Publications
4. KIRKBY, M.J. (1969). Infiltration through flow and overland flow. In
    Chorley J., Phisical Hydrology, Methuen e Co.
5. LIETH, H. (1975). Primary productivity of the biosphere springer. Ver-
    lage.
6. MARACCHI, G., and al. (1978). Simulazione di una coltura di mais. Rivi-
    sta di Agronomia, Anno XII, n.1-2, 103-198
7. MARACCHI,G., CONESE, C., MIGLIETTA, F. (1982). Agroecological classifi-
    cation. In Proceeding EARSEL Workshop
8. MARACCHI, G., VENEZIAN, M.E., LOSAVIO, N., MASTRORILLI, M., PALOSCIA,
    S., RASCHI, A., ZIPOLI, G. (1982). Evaluation of ET by stomatal resi-
    stance, leaf water potential and crop temperature. ICID Bulletin.
9. MARACCHI, G., BACCI, L., RASCHI, A., VAZZANA, C. (1984). Evaluation of
    crop growth by ecophysiological parameters. In Proceeding First Inter-
    national Symposium on water relations in fruit crop.
10. RIPLEY, E.A., REDMANN, R.E. (1976). Grassland in J.L.Monteith Vegeta-
    tion and the atmosphere. Academic Press.
11. SAUER, H.R. (1978). Assimilation model for Grassland Primary Producer
    Phenology and Biomass Dynamics. In G.S.Innis, Grassland simulation mo-
    del. Springer - Verlag.
12. THORNLEY, J.H.M. (1976). Mathematical models in Plant Phisiology. Aca-
    demic Press.
13. VENEZIAN, M.E. (1984). Modelling sorghum productivity under water stress
    Proceeding ICID Congress Fort Collins, 22 May.

C H A P T E R   V   :   G E N E R A L   T O P I C S

- World climate research programme, plan and implementation

- CO_2 programme of US department of energy

- Satellite climatology - Study of the earth radiation budget from in flight ultrasensitive accelerometer techniques

- Measurement of the radiation pressures by accelerometry: a new way to the determination of the earth radiation budget - Result obtained with CASTOR/CACTUS experiment

WORLD CLIMATE RESEARCH PROGRAMME, PLAN AND IMPLEMENTATION

V.P. MELESHKO
Joint Planning Staff for WCRP
WMO/ICSU, Geneva, Switzerland

1.   INTRODUCTION

Over the last 15 years, non-governmental organizations and governmental bodies in various countries have shown an increasing interest in understanding of the causes of the observed short-term climate variations, and have expressed concern about the growing influence of human activity on the climate on regional and global scales. As a response to this concern, the Eighth Congress of the World Meteorological Organization (WMO) decided to establish the World Climate Research Programme (WCRP) in 1979. In the same year, WMO and the International Council of Scientific Unions (ICSU) signed an agreement on the joint planning and implementation of this programme. The two organizations established an international body, the Joint Scientific Committee (JSC) with the main task of providing scientific guidance in climate studies requiring international efforts.

2.   OBJECTIVES OF THE PROGRAMME

The main objectives of the WCRP are to determine to what extent climate can be predicted and the extent of man's influence on climate. The primary approach for achieving these objectives is based on the use of physical-mathematical models capable of simulating and, eventually, predicting climate changes over a wide range of space and time scales.

The Joint Scientific Committee, together with the Committee on Climate Changes and the Ocean (CCCO), the WMO Commission for Atmospheric Sciences (CAS) and other international organizations, prepared a Preliminary Plan for the WCRP which was published at the beginning of 1981 (1). The Plan identifies priority problems and focusses attention on the study of climatologically signficant physical processes occurring in the climate system, diagnostic studies and formulation of a comprehensive list of data requirements for various observational programmes.

In adopting this Plan three years ago, the JSC recognized that it contained a few shortcomings, in particular concerning the oceanographic component of the programme and therefore the JSC proposed to consider the Plan as provisional until a number of projects could be adequately defined.

A new scientific Plan was recently prepared and reviewed by the Joint Scientific Committee during its fifth session in March 1984. The Plan is now available in the form of a draft document which is expected to be published and widely distributed before the end of this year (2). The goals of the WCRP are formulated in terms of three specific objectives or streams of climate research, each corresponding to different time scales.

The first stream aims at establishing the physical basis for the prediction of weather anomalies on time scales of one to two months. This goal requires observing the initial value of the ocean surface temperature and sea ice fields, and making progress in the ability to predict the relatively rapid changes of the land surface boundary conditions, e.g. the amount of stored soil water.

The second stream aims at predicting the variations of the global climate over periods up to several years. The largest contribution to the variations of the global atmosphere which may be predictable on interannual time scales is now seen to be the influence of the oceans and especially, the tropical oceans in which large-scale circulation and temperature anomalies can be forced by remote atmospheric events. Other physical factors, such as interannual variations of sea ice in the Arctic and Antarctic Oceans need also to be taken into consideration.

Finally, the third stream aims at characterizing variations of climate over periods of several decades and assessing the potential response of climate to either natural or man-made influences, such as the increase in the atmospheric concentration of carbon dioxide. This stream will build upon the results of the first and second stream activities, and will include a major oceanographic effort to observe and model the global oceanic circulation.

3.   RESEARCH AND OBSERVATIONAL PROJECTS FOR THE WCRP

The JSC has adopted a number of projects which constitute the essential part of the Plan and internationally co-ordinated activity. The status of the projects are different: some are at the state of formulation of objectives and feasibility studies, others are under implementation. Let us consider the major ones.

3.1   International Satellite Cloud Climatology Project

The available information on the three-dimensional cloud distribution over the globe is insufficient for validation of climate models and development of improved methods for parameterization of the various types of clouds in these models. The International Satellite Cloud Climatolgy Project (ISCCP) was approved as the first project of the WCRP (3) and began its operational phase in July 1983. Its objectives are:

-     To obtain a global data set on reflected solar and long wave outgoing radiation which is used to derive cloud data.

-     To stimulate and co-ordinate research on methods for determining the physical properties of clouds using meso-scale radiation measurements and to develop optimal algorithms for determining the spatial distribution of clouds.

ISCCP has two components, operational and research. The operational part takes advantage of the global coverage provided by the current and planned international array of geostationary and polar-orbiting meteorological satellites during the 1980's to produce a five-year global satellite radiance and cloud data set. The research component of ISCCP will co-ordinate studies to validate the cloud climatology, to improve simulation of cloud effects in climate models, and to investigate the role of clouds in the radiation budget of the atmosphere. Validation will involve comparative measurements at a number of test areas selected as representative of major cloud types and meteorological conditions. Complimentary efforts within the framework of WCRP will promote the use of the resulting ISCCP data sets in climate research.

The ISCCP concept envisions global and diurnal coverage of the earth by the one to two polar orbiting and five geostationary meteorological satellites now in operation leading to the creation of a five year radiance and cloudiness data set. Thus far, commitments have been obtained from all satellite operators with the exception of the Indian Meteorological Department in New Delhi.

Data products from all meteorological satellites, with the exception of INSAT, are being routinely collected by the appropriate Sector Processing Centres (SPCs) and delivered to the Global Processing Centre (GPC) in accordance with project requirements. All SPCs are now fully operational.

A cloud algorithm pilot study was conducted over the past two years leading to the selection of an interim operational algorithm for retrieving cloud physical properties from the remotely sensed radiances. This study involved an intercomparison of the performance of nine candidate techniques when applied to a common FGGE satellite data set. The details of this study have been published as WCP-73 report (4).

3.2   Programme on Land Surface Processes and Climate

Simulation experiments with atmospheric general circulation models have indicated the sensitivity of weather patterns and short-term climate anomalies to changes of soil moisture, thus pointing to the possibility that the storage of water in the ground provides a form of memory to the climate system which might contribute to climate predictability. Recent studies show that a larger scale anomaly of soil moisture can persist over several months and effect the air-surface temperature and circulation of the atmosphere.

The current GCMs incorporate land surface and hydrological processes in rather simplified forms which do not require detailed information about soil properties. The variety of vegetation types is usually ignored and the concept of "soil moisture" for a grid-size area is not properly defined. Climate modellers are facing the problem of developing parameterization schemes which would make it possible to evaluate the bulk surface fluxes over an inhomogeneous grid-size area including different types of vegetation. To validate such parameterization schemes, comparison with an empirical data base including separate determinations of energy and water fluxes is required under typical atmospheric and land surface conditions. For this reason the JSC felt that a pilot project over an area of 100 x 100 km^2 or slightly less was needed to provide the best possible conventional meteorological and hydrological measurements, together with independent determinations of energy and water vapour fluxes near the surface.

The meeting of experts held in Geneva at the end of last year approved the basic principles regarding the design of a Pilot Hydrological-Atmospheric Experiment and formulated the guidelines for selecting the experimental site and conducting the observational study (5).

Two sites were tentatively identified as suitable for the pilot experiment. These are the Washita River Basin in the southern Great Plains (USA) and the south-western region of France. Both sites have well developed networks of hydrological and meteorological observations and satisfy the climatological and surface topography requirements.

It is expected that the planning and implementation of the first pilot experiment, as well as subsequent similar field experiments, will be undertaken at the initiative and under the responsibility of a national agency in the host country, with the co-operation of other institutions, as appropriate.

Active co-operation of the international scientific community is also needed to proceed with the tasks of organizing process studies in this field, in comparing experimental results and translating these findings into improved formulations of land surface and atmosphere exchange processes in GCMs. JSC-V (March 1984) endorsed the proposal to establish a research programme on Land Surface Processes and Climate for the purpose of maintaining an effective dialogue between the hydrological community and atmospheric scientists and offering scientific guidance for the conduct of observational projects in the field.

3.3   Tropical Ocean and Global Atmosphere (TOGA)

Recent theoretical and empirical research has shown a close relationship between SST anomalies in the tropical oceans and large-scale variations of the surface pressure in the southern hemisphere. It has also been found that these anomalies influence the intensity of the Indian Monsoon, the formation of droughts in China and northeast Brazil and persistent anomalous atmospheric circulation in the middle latitudes of the northern hemisphere. It is now recognized that these atmospheric anomalies are the most significant signal of short-period variations in the global atmosphere and are therefore of interest to the WCRP.

At the beginning of 1982 the JSC and CCCO discussed the problem of interannual variability of the tropical oceans and the global atmosphere and decided to organize the TOGA Programme with the major objectives (6):

(i)   To determine to what extent the time dependent behaviour of the tropical oceans and global atmosphere system is predictable on time scales of months to several years, and to understand the mechanisms of this behaviour.

(ii)   To study the feasibility of modelling the coupled ocean-atmosphere system for the purpose of predicting its variations on time scales of months to years.

A number of expert meetings were arranged jointly by the JSC and CCCO since 1982 with the task of defining the scientific and technical activities that must be organized internationally to achieve the objectives of the TOGA Programme and to provide scientific guidance for its conduct.

A plan prepared by the TOGA Steering Group includes the following major components:

(i)   An oceanographic observational component consisting of (a) a measurement programme to provide, for a ten-year period, a description of the month-to-month variability of the temperature field and currents in the upper layer of the tropical oceans in the latitude band 20°N to 20°S.

(ii)   A limited programme of atmospheric observations and data processing activities, supplementary to those of the World Weather Watch, designed to provide a detailed description of the month-to-month variations of the global atmospheric circulation, thermodynamics and hydrological cycle for a ten-year period.

(iii)   An air-sea flux measurement programme aimed at providing a 10-year data set of monthly values of the fluxes of momentum, heat and moisture across the air-sea interface.

(iv)   The development of tropical ocean models, and of appropriate techniques for the assimilation of data as required to specify the initial values for numerical prediction.

(v)   An atmospheric modelling activity to establish the sensitivity of the atmospheric regime to various possible forcings on time scales of several months to several years and to develop atmospheric models in a form suitable for coupling with the ocean.

Several intensive observational projects are planned and will be implemented as national or multi-national initiatives and will contribute significantly to knowledge of the important processes in the tropical oceans ["EPOCS", "TROPIC HEAT" (USA),, "ERFEN" (Chile, Columbia, Equador and Peru), "Section" (USSR): "FOCAL-SEQUAL" (USA-France)]. However, the TOGA Programme as a whole requires an internationally co-ordinated measurement programme to provide a consistent ten-year record of the basic geophysical variables, needed to describe the variability of the coupled tropical ocean and global atmosphere system, against which the models can be validated.

The duration of TOGA is determined basically by the time scale of the Southern Oscillation which is the dominant mode of interannual variability over the globe as a whole. Recent strong El Niño/Southern Oscillation events occurred in 1972-73, 1976-77 and 1982-83. The historical record shows that such events can occur as frequently as two years apart, but on occasions there have been more than ten years between events. It is hoped that the ten-year period chosen for TOGA will encompass at least one, and possibly two, strong events. Accordingly, it is planned that the TOGA programme should begin on 1 January 1985 and continue for a ten-year period.

TOGA is considered as a major research activity within the stream 2. A Scientific Plan for the TOGA Programme was presented for review at the Conference on the TOGA Scientific Programme which was held in Paris in September. The Conference, which was attended by about 200 scientists from all over the world, clearly showed the great interest of the scientific community in this Programme.

3.4   World Ocean Circulation

A World Ocean Circulation Experiment (WOCE) can now be realistically planned because new techniques for observing the oceans, especially measurements from satellites, make it possible to collect a global ocean data set.

The overall objectives of this Experiment are (7):

(a)   To collect the data necessary to develop and test ocean models useful for predicting climate change.

(b)   To determine the representativeness of the specific WOCE data sets for the long term behaviour of the ocean, and to find methods for determining long term changes in the ocean circulation.

WOCE will be the main experimental activity within stream 3 of WCRP. The timetable for WOCE is determined, to a large extent, by the launchings of oceanographic satellites. Recognizing the uncertainty inherent in the development schedule of new satellite systems, the planning of WOCE must proceed on the basis of best schedule estimates. Several other observational components, while generally demonstrated in principle, also require substantial technological developments in order to be operational for large-scale use during the intensive field phase. It is planned therefore that WOCE will continue through the next decade and will include an intensive field observation phase during the five-year period 1989-1994.

The main internationally organized activities in relation to this experiment are being carried out under the guidance of the WOCE Steering Group established jointly by the JSC and CCCO.

3.5   Sea-Ice

The important role of sea ice in the climate system has been well recognized in a variety of empirical and theoretical studies. It is known that the extension of sea ice undergoes significant variations during the annual cycle, ranging up to 90% of its maximum value in the Antarctic and 60% in the Arctic. In addition low frequency variations are superimposed on the mean annual cycle producing significant interannual changes.

A large variety of sea ice models with different degrees of sophistication has been developed, some of which include an explicit treatment of ice dynamics. Such models are capable of reproducing realistically the annual cycle of the extent and thickness of sea ice when driven by prescribed forcing from the ocean and the observed atmospheric fields. However, the problem of coupling sea ice models with atmospheric general circulation models, to constitute a single interactive system, has not yet been solved successfully. The validation of such coupled models is hampered by the lack of adequate observations which should include the surface stress and energy fluxes at the ocean-ice and ice-atmosphere interfaces.

A meeting of experts on sea ice and climate modelling, which was organized in Geneva in December 1983, proposed that an internationally coordinated project aimed at determining the atmospheric and oceanic forcings on sea ice and improving the performance of the GCMs in the polar regions would be an efficient means to promote the required scientific developments (8).

In the Arctic, where a larger data base exists and where knowledge of the physical processes is more advanced, the activities envisaged would encompass primarily the use of satellite remote sensing data, information collected by automatic data buoys, and theoretical and modelling studies.

In the Antarctic, where much less is known about ice drift, ice characteristics, and the physical processes in the ocean, ice and atmosphere, the
project should encompass a full range of international experimental and
theoretical investigations.

The JSC endorsed, in principle, the proposal of establishing an international programme on monitoring and modelling of sea ice in the context of
climate research, as an integral part of the WCRP. The Committee requested
that the scope and specific objectives of this programme be further studied
and defined.

3.6   Assessment of the climate response to increase of atmospheric CO_2

Three international organizations, UNEP, WMO and ICSU have agreed to
conduct jointly an international assessment of the impact of CO_2 on climate
which is expected to be completed in 1985. These organizations have given the
responsibility to the International Meteorological Institute in Stockholm
(IMI) for the co-ordination of the necessary preparatory activities in co-
operation with international bodies and national research institutions engaged
in this task. They also agreed that sound scientific grounds for this assessment should be based on the consideration of the major aspects of the CO_2
problem.

The JSC has taken responsibility for the assessment of the climate
response to an increased atmospheric concentration of CO_2, and the early
detection of CO_2 effects on climate.

Several assessment of the potential climate change, induced by atmospheric CO_2 have been made over the last few years. The studies suggest that
a doubling of the CO_2 content of the atmosphere would produce an increase in
global mean air surface temperature of the order of $3 \pm 1.5°C$. There is a
feeling that the mean temperature may be realistic, but the ranges of uncertainties can hardly be considered acceptable.

At its fifth session in March 1984 the JSC expressed the view that the
current climate models could only simulate the global and zonally averaged
feature of climate with some confidence and are not so advanced as to be able
to predict regional climate changes. This is mainly due to shortcomings in
parameterization of major physical processes and the inadequate spatial resolution of models. The recent numerical experiments with increased atmospheric CO_2 showed a substantial degree of model dependence in relation to
the predicted regional scale features. However, the JSC thought that useful
qualitative inferences could be made if an appropriate methodology based on
diagnostic studies were developed to relate regional climate changes   to
zonally averaged climate quantities.

The JSC rather cautiously expressed its view regarding the detection
of climate changes. It stated that the existing observational evidence does
not allow concluding with confidence that the observed variations are caused
specifically by the increasing concentration of atmospheric CO_2. On the
other hand, the surface warming indicated by the record of the global mean
temperature is not inconsistent with the computed effect of the observed increase of atmospheric CO_2 since the beginning of the industrial revolution.

Regarding the schedule for preparation of the IMI report and the date of the Conference on CO_2 evaluation, it is expected that the draft version will be available to the reviewers at the beginning of next year and the Conference is planned for October 1985.

3.7 Programme of numerical experimentation

A special effort is being directed towards developing climate models in which the various elements of the climate system are coupled together. The JSC Working Group on Numerical Experimentation is currently undertaking an intercomparison project for the evaluation of the parameterization schemes used to determine cloud amounts and types for the purpose of radiation calculations.

The results obtained so far show that the simulation of clouds associated with synoptic scale systems in middle latitudes were generally realistic but that some types of cloud, for instance at the top of the boundary layer, were substantailly different from one model prediction to another, none of which indicated high skill. There were also considerable differences in simulations of convective clouds in the tropics where a correct prediction is especially important.

During the last two years, WGNE devoted particular attention to studies of the impact of tropical sea surface temperature (SST) anomalies on the global atmosphere. Experiments with a typical El Niño SST anomaly have demonstrated a clear response in the tropics; models have also been able to simulate the Southern Oscillation. The response of the atmosphere at higher latitudes is not so clear and seems to be dependent upon the flow configuration and eddy activity. A new series of experiments, based on the observed 1982/83 anomaly was suggested and preliminary results were reported during the JSC/CCCO Symposium on coupled Ocean/Atmosphere Models in Lèige in May 1984.

3.8 Other WCRP activities

Other activities organized by the JSC in collaboration with CCCO, WMO CAS and IAMAP Radiation Commission include:

- assessment of the impact of aerosols on climate

- sensitivity of climate to other radiatively important trace gases

- intercomparison study of radiation codes used in GCMs.

4. DATA FOR CLIMATE RESEARCH

Research into the climate system, whether by means of numerical experiments using climate models or by diagnostic studies based on observations of the present and past climates, requires the support of an adequate and reliable data base. Unlike the First GARP Global Experiment (FGGE), which was limited in time to one year with two short observing periods of intensive observations, the WCRP requires the collection of a fairly complete global data set covering a relatively long period of time to describe the variability of the different components of the climate system and their interaction. The WCRP therefore relies heavily on operational programmes providing observations in the atmosphere (World Weather Watch) and in the ocean (Integrated Global Ocean Services System) as well as new observing systems that could be capable of providing global data sets describing the climate during a long period of time (for instance, measurements from experimental oceanographic satellites).

The WCRP gives much attention not only to comprehensive monitoring of the climate but also monitoring of those factors (or parameters) which may be responsible for long term climate variations (e.g. concentrations of CO_2, trace gases, aerosols, etc.).

It is noted in the Plan for the WCRP that in order to evaluate the performance of climate models and to provide adequate monitoring of climate variability, the following types of data are required:

- Basic atmospheric parameters (temperature, wind, humidity and surface pressure);

- Characteristics of the radiation balance and cloudiness (net fluxes of solar and long-wave radiation at the boundaries of the atmosphere, albedo at the earth's surface, etc.);

- Oceanographic parameters (SST, wind stress, sea surface level, heat content of the mixed layer and characteristics of the surface current);

- Parameters of the atmospheric water balance (precipitation, run off);

- Parameters of the cryosphere (extent of snow cover and sea ice ice thickness, etc.);

- Composition of the atmosphere (CO_2, aerosols, ozone and radiatively active gases).

The climatic data sets which have been complied up to the present time are far from complete. Relatively long time series of observations are available only for certain parameters observed at the earth's surface and they are extremely limited in spatial coverage. Space-borne measurements have opened up a unique opportunity for measuring some climatic parameters such as cloudiness, radiation budget at the top of the atmosphere; albedo of the earth surface, sea surface temperature, sea ice, etc. One can hope that sufficiently reliable data will soon be accumulated on these parameters. For example, the planned observing programmes covering 5-10 years (International Satellite Cloud Climatology Project, TOGA, WOCE, etc.) will provide new extremely useful data which will be needed not only for studying individual physical processes in the climate system, but also for testing the model performance.

Although the Plan for WCRP does not explicitly address the development of paleoclimatological research it is recognized that proxy records provide unique information about the variability of climate, various causal factors and possible mechanisms involved in climatic changes. It is assumed that ongoing international co-operation in paleoclimatology will proceed under the guidance of specialized commissions and scientific unions within ICSU.

5.   IMPLEMENTATION OF WCRP

The scientific Plan for the WCRP was reviewed by the WMO Executive

Council at its 36th session in June 1984. The Council expressed satisfaction about the adequacy of the specific research activities, laid out in the Plan, with the general objectives of the WCRP. With regard to follow-up activities for implementing the Programme, the EC considered that consultations at the intergovernmental level should take place in the form of informal planning meetings bringing together the representatives of all agencies which would be prepared to participate in the implementation of WCRP.

REFERENCES

1.   Preliminary Plan for the World Climate Research Programme. Geneva, January 1981, WCP-2

2.   Plan for the WCRP. Draft version, 1984

3.   International Satellite Cloud Climatology Project (ISCCP). Preliminary Implementation Plan, November 1982, WCP-35

4.   The Interntational Satellite Cloud Climatology Project (ISCCP) - Cloud Analysis Intercomparison, March 1984, WCP-73

5.   Report of the Meeting of Experts on the Design of a Pilot Atmospheric Hydrological Experiment for the WCRP, December 1983, WCP-76

6.   Plan for the TOGA Scientific Programme, Working Draft, July 1984

7.   Report of the First Session of the JSC/CCCO WOCE Scientific Steering Group, Woods Hole, USA,3-5 August 1983, WCRP-69

8.   Report of the Meeting of Experts on Sea-Ice and Climate Modelling, Geneva, 12-16 December 1983, WCP-77

CO$_2$ PROGRAMME OF US DEPARTMENT OF ENERGY

F.A. Koomanoff, Director
Carbon Dioxide Research Division
Office of Basic Energy Sciences
U.S. DOE

During the past three years excellent progress has been made in developing an understanding of carbon dioxide and its effects on climate and vegetation. In addition, the indirect effects or consequences of changes in vegetation and climate have been studied so as to define, identify, and characterize the needed data and methodologies to understand potential effects. These indirect effects or consequences include fisheries, forests, hydrology, agriculture and human health. We have kept the question of sea level as a special project due to its importance. The U.S. Department of Energy (DOE) working with the American Association for the Advancement of Science (AAAS) are in the process of completing the series of five State of the Art (SOA) reports. These reports include carbon cycle, climate, first detection of a CO$_2$ climate change, vegetative response, and indirect effects. A companion volume which will serve as an index to all volumes is also being prepared by 87 scientists. The reports, which are going through extensive peer review by the AAAS, utilize more than 300 reviewers with more than 20% being non-U.S. scientists. The SOA's will be published in 1985. This will be followed by a Statement of Findings report which, besides synthesizing what is known and unknown and the uncertainties of a scientific knowledge base of CO$_2$, will be the foundation for future research directions for the next decade. We expect to have detailed discussions with the international science community on these directions and attempt to develop joint collaborative scientific research bonds with the international community.

In the following, a few figures are presented to highlight the CO$_2$-Research.

THE CO$_2$ PROGRAM

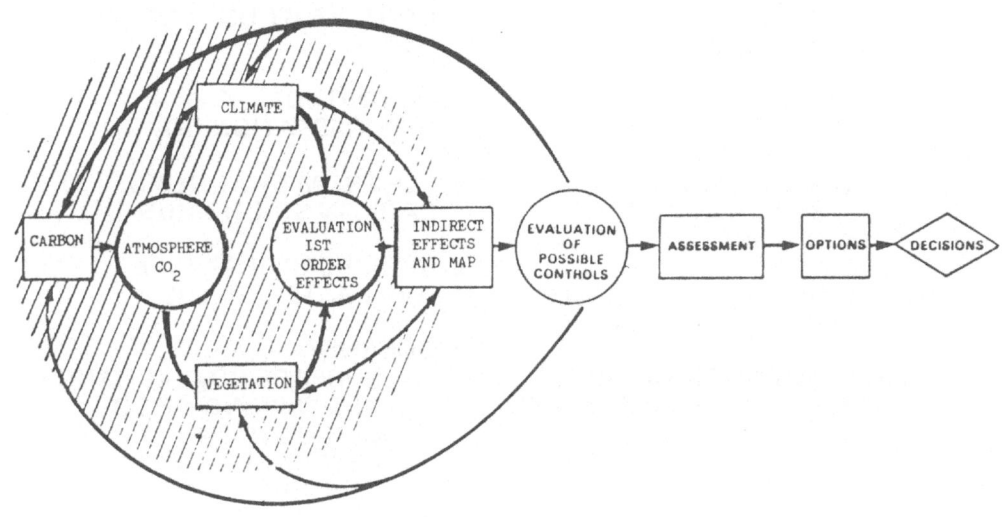

CLIMATE–CO$_2$

KNOWNS

- MODELS ESTIMATE A RANGE
  $\bar{T} = +1.5°$ TO $4.5°C$

- DATA INDICATE A RANGE
  $\bar{T} = +1.5°$ TO $3.0°C$

UNCERTAINTIES

- DIFFERENCES BETWEEN MODELS

- OCEAN & CLOUD DYNAMICS

- PREINDUSTRIAL CO$_2$ LEVELS
  (250 - 290 ppm)

- TEMPERATURE & OTHER DATA TIME SERIES RECORDS

- DIFFERENCES BETWEEN MODELS AND DATA

CARBON CYCLE

KNOWNS

- ATMOSPHERIC CO_2 INCREASING (3 GT/YR)

- HISTORICAL FOSSIL EMISSION DOCUMENTED

- 55 TO 60% FOSSIL FUEL RELEASE REMAINS IN ATMOSPHERE

- 30 TO 60% FOSSIL FUEL CO_2 ABSORBED BY OCEANS

- DEFORESTATION CO_2 IS 10 TO 40% OF FOSSIL FUEL RELEASE

UNCERTAINTIES

- UP TO 2.5 GT UNACCOUNTED FOR

- MAGNITUDE OF DEFORESTATION CO_2

- CO_2 FROM CHANGES OF CARBON SOIL-HUMUS

- BIOSPHERE RESPONSE TO CO_2 & CLIMATE CHANGE

- OCEANS
  - SURFACE CO_2 EXCHANGE
  - DEEP MIXING OF CO_2/ CIRCULATION
  - RESPONSE TO CLIMATE CHANGE

TECHNICAL ISSUE

o MODELS DO NOT AGREE WITH OTHER MODELS, NOR DO MODELS COMPARE WELL TO DATA.

o STARTED COMPARISON OF MODELS.

- INTERNATIONAL RADIATIVE MODEL COMPARISON THE RESULTS OF WHICH WERE DISCUSSED

  IN ITALY THIS AUGUST (WMO, DOE SPONSORED)

- GCM INTERCOMPARISON

o WHY? --

- SIMPLE MODEL WITH 15 YEAR OCEAN LAG

- TEMPERATURE FOR A DOUBLING (I.E. MODEL SENSITIVITY) VS. TEMPERATURE CHANGE

  1850 - (1961-1980 AVERAGE)

- DIFFERENCE IS 0.3 - 0.75°C

- IF PRE-INDUSTRIAL CO_2

  240-280 PPM, THE HIGH SENSITIVITY MODELS MUST BE QUESTIONED

- THIS, OF COURSE, DOESN'T MEAN LOWER SENSITIVITY MODELS ARE CORRECT -

  E.G. THESE DO NOT AGREE ON REGIONAL DETAIL

FOSSIL CO$_2$ EMISSION: HISTORICAL, 1860–PRESENT

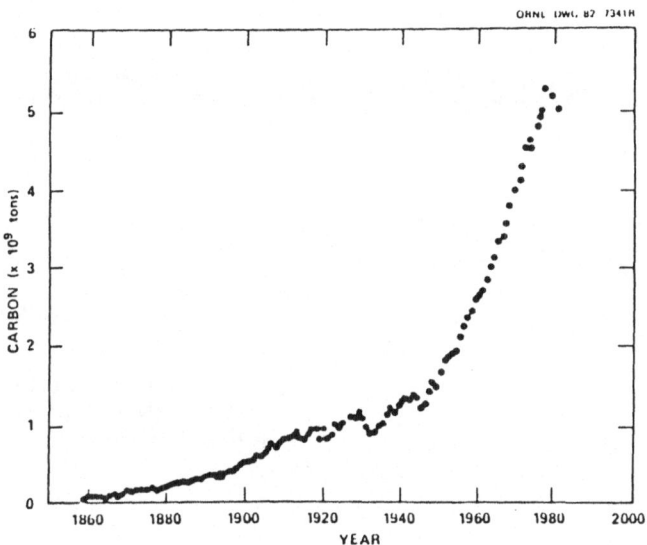

ORNL DWG. 82 7341R

FOSSIL CO$_2$ EMISSIONS BY TYPE OF FUEL

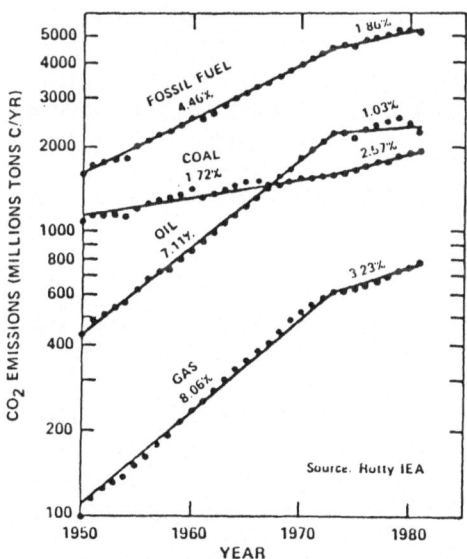

TRACE GASES AND AEROSOLS

UNCERTAINTY

- MODELING SUGGESTS ADDITIONAL 0.5° TO 2°C OVER $\bar{T}$ CO_2 INDUCED CHANGE

- TRENDS FOR INDIVIDUAL SPECIES

- SOURCES/SINKS (ATMOSPHERIC CHEMISTRY) RELATIONSHIPS

- COMBINED RADIATIVE EFFECTS (NOT A SIMPLE SUMMATION)

HISTORICAL CARBON DIOXIDE BUDGET
(10^6)

	FY78	FY79	FY80	FY81	FY82	FY83	FY84	TOTAL
DOE	1.5	2.9	6.0	9.7	11.9	9.1	12.5	53.6
OTHER AGENCIES	~5.0	~5.0	6.5	9.7	9.5	10.4	12.4	58.5
TOTAL	6.5	7.9	12.5	19.4	21.4	19.5	24.9	112.1

SATELLITE   CLIMATOLOGY

STUDY OF THE EARTH RADIATION BUDGET FROM IN FLIGHT
ULTRASENSITIVE ACCELEROMETER TECHNIQUES

J.P. ROZELOT
C.E.R.G.A.
Avenue Copernic
06130 GRASSE
France

It is well known now that the use of triaxial microaccelerometer on board
a satellite of CASTOR type (1) permits with a very good accuracy the measurement of all the accelerations due to surface forces acting on the satellite and more especially those due to Sun and Earth radiation pressures (2).
As an example, relative variations of 5.10^{-10} ms^{-2} can be detected and this
allows a direct measurement of the Earth's radiation balance (net flux). It
is also possible to put in evidence slight differences of Earth infrared
radiation over oceans and continents or over tropical and equatorial zones.

Even if one cannot be certain that variations in solar radiation flux are
the key to our climatic variations, the energy absorbed by the Earth's atmospheric system is clearly the motor of atmospheric circulation and thus,
a great constraint of our climate. So it is very important to get accurate
data over long period of time with a very sure method. In flight ultrasensitive accelerometer techniques are until now the only method of direct measure of Earth radiation budget because this measure is a differential one.

The results which are listed in the paper hereafter presented by Boudon and
the previous studies made under an ESA contract (3) proved the feasability
of such measurements with an accelerometer on board a spherical satellite
coated with a skin whose thermo-optical properties are not wavelength dependent, the accelerometer satellite system having a behavior of a broad-field
(4π steradians) radiometer and a broad spectral range (0.2 to 60 μm).

Other simulations have clearly shown that the solar irradiance can be computed within an accuracy of 3W/m² on a daily measurement (0.2 %). Moreover,
a statistical analysis enables to reach a level of 0.4 % on the absolute
value of the solar irradiance and 0.06 % - 0.1 % on the relative one.

These values and the fact that the experiment is not a very expansive one thus appear to be sufficiently promising for meriting consideration of its use in climatology.

REFERENCES

(1) BOUDON, Y. ; 1984, thèse de Doctorat d'Etat, "Détermination de la pression de radiation par microaccélérométrie dans le cadre de l'expérience CACTUS", Université Paris VI.

(2) MAINGUY, A.M. : 1984, thèse de Doctorat d'Etat, "Proposition d'un principe de mesure absolue du bilan radiatif basé sur l'accélérométrie spatiale", Université Paris VI.

(3) MAINGUY, A.M., BERNARD, A., DUHAMEL, T., GIRARD, A., GREGOIRE, G., JUILLERAT, R., MARCHAL, P., MONTIGNY, F., PASTRE C., "BIRAMIS, évaluation numérique de la précision globale du système" , rapport technique n°7/3373/SY, contrat ESA n°3454/77, mars 1980

MEASUREMENT OF THE RADIATION PRESSURES BY ACCELEROMETRY: A NEW WAY
TO THE DETERMINATION OF THE EARTH RADIATION BUDGET.
RESULT OBTAINED WITH CASTOR/CACTUS EXPERIMENT

Y. BOUDON

Centre d'Etudes et de Recherches Géodynamiques et Astronomiques

Avenue Copernic

06130 GRASSE,   France

ABSTRACT

In May 1975 an ultra sensitive accelerometer CACTUS have been placed
on board the satellite CASTOR to carried out several aeronomical studies.
The threshold of the apparatus was so good that statistical analysis has
permitted to reach the study of solar radiation pressure and infrared
radiation pressure of the Earth. From these results we obtained by this
new method the first evaluation of the Earth radiation budget during one
year (July 1975 - July 1976) in the intertropical zone. The evolution
obtained shows a good agreement with the radiometric results derived from
a composite of data spanning 14 years (1964 - 1977). These results show that
accelerometric method, (such as BIRAMIS system) can constitue a new way in
the field of climatologist research.

INTRODUCTION

The purpose of this paper is mainly concerned with the study of atmospheric thermal balance by measurement of radiation pressure. More specifically, we shall emphasize how the Earth radiation budget can be deduced
with accurate precision from ultrasensitive accelerometers on board satellites.

In the 60s, a new generation of satellites were developed in order to
measure very low forces which act on passive satellites when they move in
a very low atmospheric density. Such surface force measurements are only
possible with ultrasensitive microaccelerometers, which are able to detect
very low accelerations. A lot of spatial experiments have been made in this
way, among them can be quoted : SAN MARCO 1 and 2 (1964 and 1967), OV 15

and 16 (1968) from the SPADES experiment, TRIAD (1972) with a triaxial accelerometer and ATMOSPHERIC EXPLORER (1973).

The CACTUS accelerometer developed at ONERA and making up the payload of the satellite CASTOR (CNES) permitted the measurement of all the accelerations due to surface forces acting on the satellite. This measurement were in fact obtained with a better sensitivity than that expected before launch. So that the great quantity of data acquired and a well adapted processing technique permitted to reach the study of radiation pressures and it has been possible to get information about the radiation fields around the Earth in tropical zone.

These results have led the European Space Agency (ESA) to study the utilization of such a system for the Earth Radiation Budget measurement.

THE CACTUS MICROACCELEROMETER - PRINCIPLE   (1)

The apparatus consists of a small ball - proof mass - (mass m = 580 g and 4 cm diameter) inside a spherical cavity. The ball is held is suspension within the cavity by 3 electrostatic forces produced by electrodes disposed along three orthogonal axes (fig. 1).

At the equilibrium, the acceleration of the cavity-satellite system is equal to the acceleration of the ball, and if $\vec{F_E}$ represents the non-gravitational forces due to the atmospheric drag and the radiation pressure

$$\frac{\vec{F_E}}{M} = \frac{\vec{F_B}}{m}$$

where   M = total mass of the satellite (with the ball)

          m = mass of the ball

and $\vec{F_B}$ = resultant of forces linking ball and cavity

$$\vec{F_B} = \vec{F_C} + \vec{F_P}$$

Denoting F_P as that part of internal perturbations which can be modelled and F_C the command force developed by the servo-action transmitted by telemetry to the ground, the measured acceleration is defined by :

$$\vec{\gamma} = \frac{\vec{F}_E}{M} - \frac{\vec{F}_P}{m}$$

The CACTUS measurement range extends from 10^{-9} m/s² to 10^{-4} m/s².

THE CASTOR SATELLITE

The satellite almost spherical (diameter 0.8 m) and weightiness 76 kg was launched (May 1975) into an orbit with the following parameters : apogee 1270 km, perigee 270 km and inclinaison of orbit to equator 30°. At the time of the launching, the scientific aim was twofold : at apogee test the sensitivity of the device and measure the atmospheric density up to an altitude of about 600 km.

The value of the accelerometer threshold, expected to be 10^{-8} m/s² was in fact 10^{-9} m/s² and through a statistical process of in-flight measurements, it has been possible to reach 10^{-10} m/s². Relative variations of 5.10 m/s² have even been detected. The time life of the CACTUS was scheduled for only 6 months but the good quality of the measurements, the conception of the satellite itself and the results obtained led keeping the telemetry system in operation up to the descent of the satellite (Feb. 1979). And it is one of the reasons for which a statistical analysis was made possible and from it, to put in evidence slight differences of Earth infrared radiation over oceans and continents, and even on tropical and equatorial zones, for example (2) (3) (4) (5).

RESULTS FOR THE EARTH INFRARED RADIATION FIELD   (6)

The choice of selected measurements along short arcs of orbit higher than 800 km to reduce the drag and taken during night led to a large set of measurements (from July 1975 to July 1976) of acceleration due to the IR radiation pressure of the Earth, after modelling drag and perturbating forces. Figure 2 shows the latitudinal variations obtained with two maxima at the tropics and the N-S asymmetry. In the same figure one can see the seasonal displacement of the minimum close to the equator, according to the displacement of the intertropical convergence zone. Cloudiness and precipitations are typical of this zone (called equatorial trough) which is colder than tropics.

Radiometric values computed by Stephens et al (15) are plotted on the same figure and the agreement is quite good between the two curves.

This result shows the capability of the accelerometer to detect the cloudiness of the sub-satellite surface. So a specific study has been made using pictures provided by geostationary satellite GEOS, and giving the cloud cover over the 230°E 340°E longitude. The correlation between the radial acceleration due to the IR radiation and the cloud cover at the same epoch appears very well as shown in fig. 3.

SOLAR RADIATION (7)

For the determination of the direct solar radiation pressure over the satellite, use is made of small parts of orbits at high altitude on both sides of the night-day and day-night transitions : when the satellite moves along such an arc, the solar radiation pressure is the only quickly varying effect. A two-side extrapolation technique enables to separate this effect, and gives the acceleration due to the direct solar radiation pressure.

Monthly means, computed from 100 to 150 transitions, are plotted in fig.4. The upper curve shows the raw values, for a 3-year period; the values reduced to a 1 A.U. Earth-Sun distance are plotted below, together with an estimate of the standard deviation. Gaps in data are due to bad orbit configurations.

THE RADIATION BUDGET

The Earth's radiation budget can be expressed as

$$\mathcal{B} = S - R - E$$

where S = incident solar energy, 98% of which is contained between wave-
 lenghts of 0,3 and 4 μm
 R = reflected solar energy
 E = energy emitted by the planet (4-50 μm)

From classic spatial measurements of radiometry, $\mathcal{B}$ can be reached by a determination of each of the components S, R and E. All of them are measured by different devices, with a field more or less open and operating in different spectral fields. Moreover, all the measurements are based upon two basic physical processes which are the transformation of the radiative

- 344 -

energy received by the sensors on board satellite, either in thermal energy, or in electric energy. The link between the different scales of measurements is not obvious and it is necessarily afterward to integrate in order to cover all the spectrum and all the space.

The originality of utilizing a spaceborne micro-accelerometer for the determination of the Earth's radiation balance is that it provides an integrated measurement of the radiation budget. The micro-accelerometer can be considered as a wide field-of-view sensor, the spacecraft skin being the sensitive area through which the various radiation fluxes are transformed into forces and accelerations.

These two techniques (radiometric and "accelerometric") should not be opposed but rather considered as complementary ones since various aspects should be taken into account when comparing the two methods such as quality of the scientific informations obtained (spectral distribution, accuracy on individual radiation, flux components...), space and time resolution (global scale, synoptic scale...).

An analysis of the transformation "flux into acceleration" taking into account the different wavelenght ranges of the various fluxes, a particular shape (sphere) of the satellite embarking the accelerometer and various possible materials for the spacecraft skin has permitted to define requirements for the payload and the satellite system called BIRAMIS project (8) (9) (10) (11) (12) (13).

ZONAL EARTH RADIATION BUDGET MEASURED BY CASTOR/CACTUS

In spite of the system was not well fitted to the determination of this budget an attempt of this assessement was made using the CASTOR/CACTUS data.

The mean value of the acceleration due to the solar radiation pressure and reduced to 1 A.U. is found equal to $3.019 \ 10^{-8}$ m/s². This value is associated to the 1376 W/m² one given by HICKEY et al. (14) and one can then compute the mean visible radiation pressure coefficient A. The IR coefficient is unknown but is set, in a first attempt, equal to the visible one.

In this condition the local instantaneous budget is given by

$$\xi_n = \frac{Mc}{\pi r^2 A} \gamma_n \qquad W/m^2$$

where M is the satellite weight

       r is the satellite radius

      c is the speed of light

and γ_n is the radial acceleration due to resultant flux (directed, reflected and emitted)

and the budget at an altitude h, integrated over period T over the surface Σ

$$\mathcal{B}_{h\ T} = \frac{1}{24} \int_H \frac{1}{\Sigma} \iint_\Sigma \xi_n \ d\sigma \ dH$$

A selection of short arcs of orbit for which the conditions for a good modelisation of secondary effects are fulfilled, was made for the period July 75 - July 76. For each of them, a mean component of the radial acceleration was computed, corrected for all the perturbating effects, flagged with its local hour and accumulated in a time-space bin. A simple numerical scheme was then to take care of the diurnal variation, and monthly means of the radiation budget were obtained for 6 latitudinal zones of 10° width, from 30°S to 30°N. (17)

Figure 5 shows the variation of the annual mean with latitude and figure 6 presents the time variation of the radiation budget for each of the 6 zones, compared with the results computed from the data gathered by STEPHENS et al (15) over a longer period.

The qualitative agreement is good, whereas a systematic difference of about 20 W/m² appears, especially in the nothern hemisphere. Part of this discrepancy might be due to a wrong value assigned to the IR coefficient.

CONCLUSION

We can now conclude that the CACTUS experiment has make a positive contribution to our knowledge in analysis of the Earth's radiation balance. In addition to that, these results show clearly that it is possible, with a good choice of the skin cover of the satellite, an adequately defined orbit and a good knowledge of the attitude parameter, to gather a very large amount of data and to monitor the Earth radiation budget with a valuable accuracy, at least on a zonal basis. This is the objet of the BIRAMIS project, studied by ONERA and ESA (16).

ACKNOWLEDGEMENTS

This research has been supported by G.R.G.S.* in CERGA Observatory in collaboration with ONERA and CNES. The study of BIRAMIS project was made by ONERA through contract with ESA. We would like to express our achnowledgements to J.P. ROZELOT who took an important part in the definition of the manuscript and allowed us the presentation of this paper to the E C climatology symposium.

* Groupe de Recherches de Géodésie Spatiale

BIBLIOGRAPHY

(1) DELATTRE M., BEAUSSIER J., BERNARD A., BOULAY J.L., BOUTTES J., FAVE J.,
     GAY M., GUIBERT J.P., JUILLERAT R., LAROCHE P. et MAINGUY A.M. Les essais
     en orbite de l'accéléromètre CACTUS. Publication ONERA, n° 1976-5.

(2) BOUTTES J., DELATTRE M. and JUILLERAT R. Qualification in orbital flight
     of the CACTUS high sensitive accelerometer. XIX Cospar Meeting Philadelphia
     June 1976.

(3) BARLIER F., BOUDON Y., FALIN J.L., FUTAULLY R., VILLAIN J.P., WALCH J.J.,
     MAINGUY A.M. and BORDET J.P. Preliminary results obtained from the low-g
     accelerometer CACTUS. XIX Cospar Meeting Philadelphia, June 1976.

(4) BEAUSSIER J., MAINGUY A.M., OLIVERO A. and ROLLAND R. In orbit performance
     of the CACTUS accelerometer. XXVIIè Congress International Astronautical
     Federation Anaheim, California, october 1976.

(5) BERNARD A., GAY M., MAINGUY A.M., JUILLERAT R., WALCH J.J.; BOUDON Y.,
     BARLIER F. and LALA P., Radiation pressure determination with the CACTUS
     accelerometer. XX Cospar Meeting, Tel Aviv, 1977.

(6) BOUDON Y., BARLIER F., BERNARD A., JUILLERAT R., MAINGUY A.M., and WALCH J.J
     Synthesis of flight results of the CACTUS acceleration below 10^{-9}g; Acta
     Astronautica, vol.6, n°9, sept 79, 29è congres I.A.F. Dubrovnik Yougoslavie
     1-8 October 1978.

(7) BOUDON Y., MAINGUY A.M., Application des mesures de pression de radiation
     à la surveillance de la constante solaire, colloque du CNES, à Toulouse
     en septembre 1980 : Soleil et Climat.

(8) CROMMLYNCK D., Utilisation d'un accéléromètre super Cactus monté sur un
     satellite pour l'observation directe du bilan radiatif. ESA "Song" Elman
     1978.

(9) BARLIER F., PASTRE C., MAINGUY A.M., JUILLERAT M., PAILLOUS A., ROMERO M.,
     WALCH J.J., BOUDON Y. and CAZENAVE A. Study of atmosphere thermal balance
     by measurement of radiation pressure. ESA Journal, vol. 2, pp.28-36, 1978.

(10) MAINGUY A.M., BOUTTES J., JUILLERAT R., et BARLIER F. Mesure du bilan
     radiatif de la Terre à l'aide d'un accéléromètre ultrasensible Projet
     Biramis. Colloque International sur l'évolution des atmosphères plané-
     taires et de la climatologie de la Terre, Nice, octobre 1978.

(11) MAINGUY A.M., BERNARD A., ROMERO M., BOUTTES J., and BARLIER F. Biramis.
     Mesure du bilan radiatif de la Terre par microaccélérométrie spatiale.
     Congrès de la Société Française de Physique, Toulouse, juin 1979.

(12) MAINGUY A.M. - Projet BIRAMIS- Mesure du bilan radiatif de l'atmosphère terrestre par satellite accélérométrique. ONERA - Note technique 1981-7.

(13) MAINGUY A.M., BERNARD A., DUHAMEL T., GIRARD A., GREGOIRE G., JUILLERAT R., MARCHAL P., MONTIGNY F., PASTRE C. BIRAMIS - Evaluation numérique de la précision globale du système. Rapport technique n°7/3373/SY. Contrat ESA. n° 3454/77, mars 1980.

(14) HICKEY J., STOWE L., JACOBOWITZ H., PELEGRINO P., MASCHOFF R., ARKING A., HOUSE J., INGERSOLL A., and VONDER HAAR T.H. Initial Solar irradiance determination from NIMBUS 7 cavity radiometer measurements. SCIENCES 208, 281-283, 1980.

(15) STEPHENS G.L., CAMPBELL G.G. and VONDER HAAR T.H. Earth radiation budgets J.G.R. 86, n° C10 9739, 9760, October 1981.

(16) MAINGUY A.M., Proposition d'un principe de mesure absolue du bilan radiatif basé sur l'accélérométrie spatiale. Thèse d'Etat. Université Pierre et Marie Curie, Paris VI, novembre 1984.

(17) BOUDON Y., Détermination de la pression de radiation par microaccélérométrie dans le cadre de l'expérience spatiale CASTOR/CACTUS. Contribution à la mesure du bilan radiatif dans la zone intertropicale. Thèse d'Etat. Université Pierre et Marie Curie, Paris VI, novembre 1984.

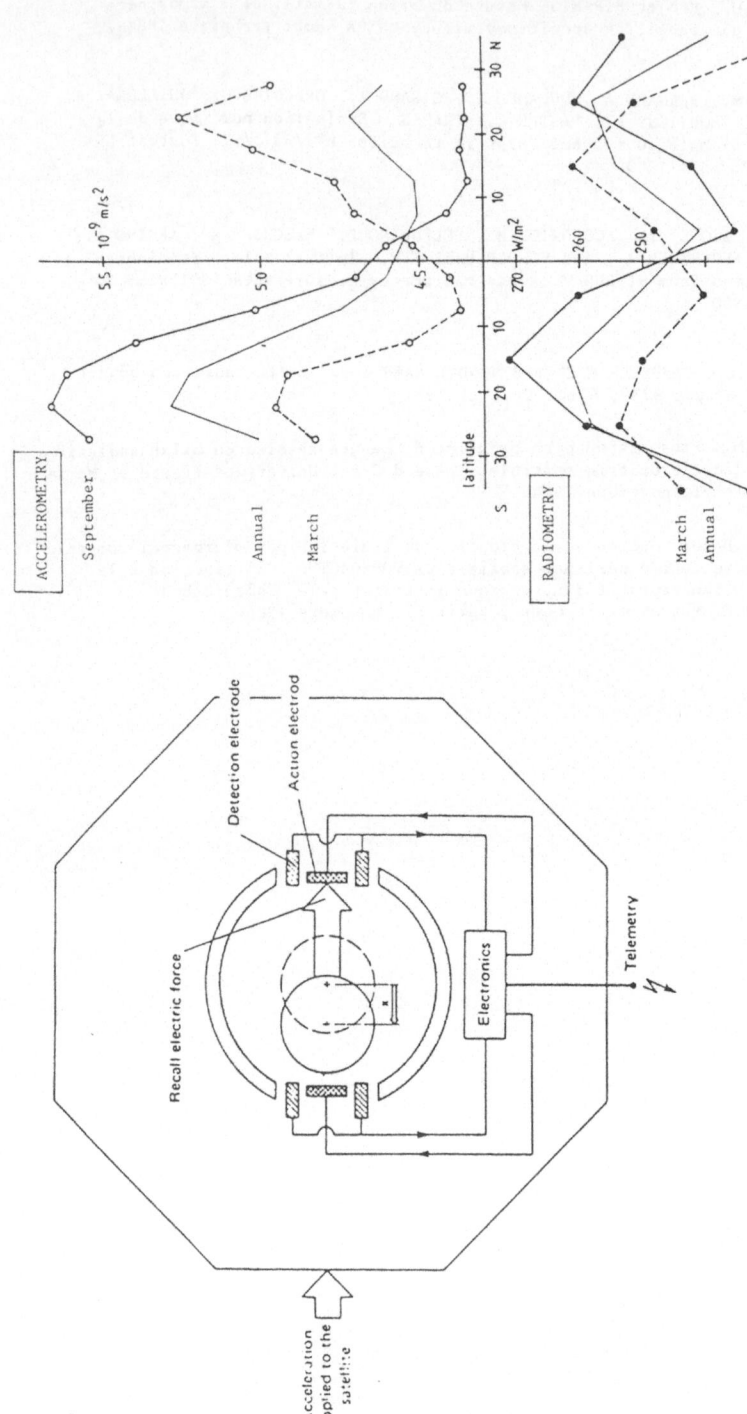

Fig. 2. Annual and Seasonal variations in latitude of the IR radiation of the Earth measured top : by accelerometer (July 1975–July 1976)

below : by radiometry.

Mean result over several years (Stephens et al.)

FIG. 1. PRINCIPLE OF ACCELEROMETER

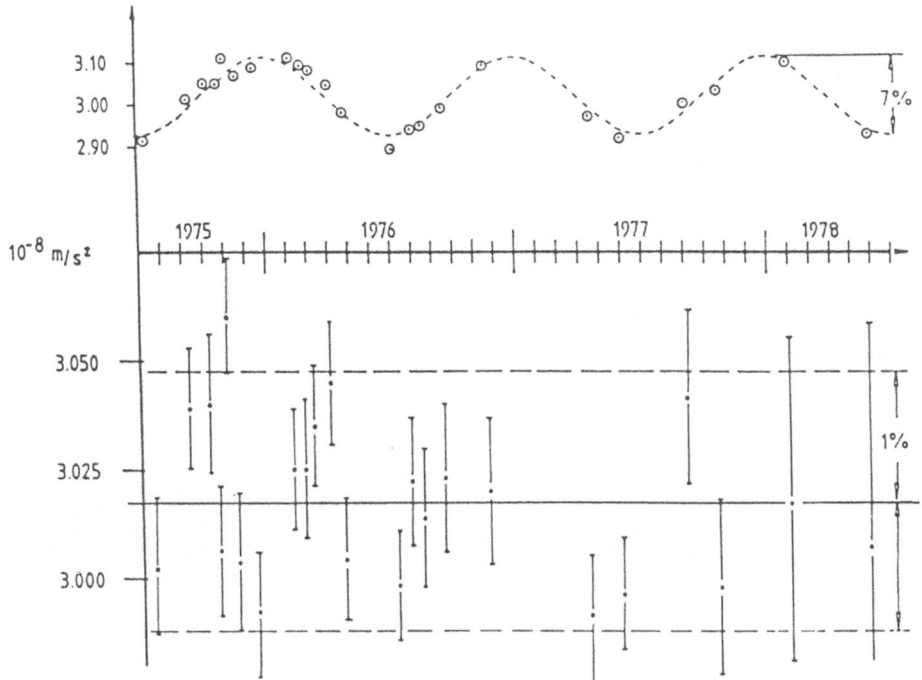

<u>Fig. 4</u> . Accelerations due to the direct radiation pressure from the Sun durir 3 years (monthly averages or on about 100 transitions) Values reduced to a 1 A.U. Earth-Sun distance are plotted below.

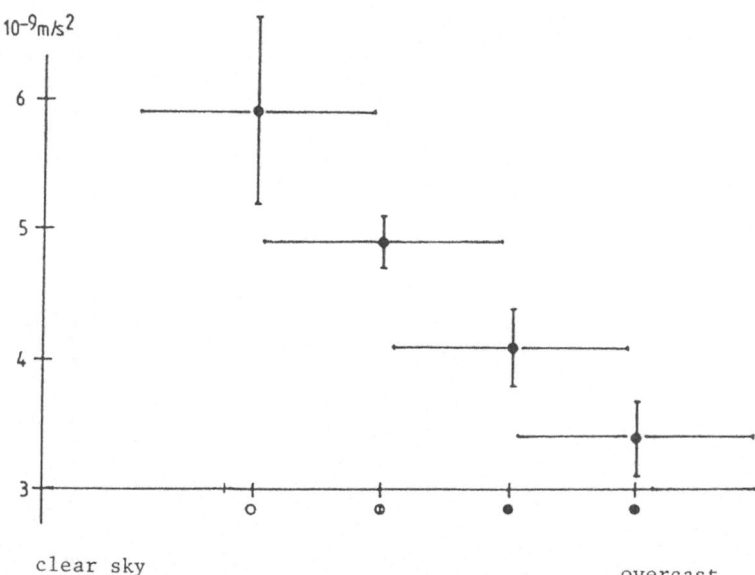

<u>Fig. 3</u>. Correlation between the cloudiness of the sub-satellite surface and the IR acceleration measured.

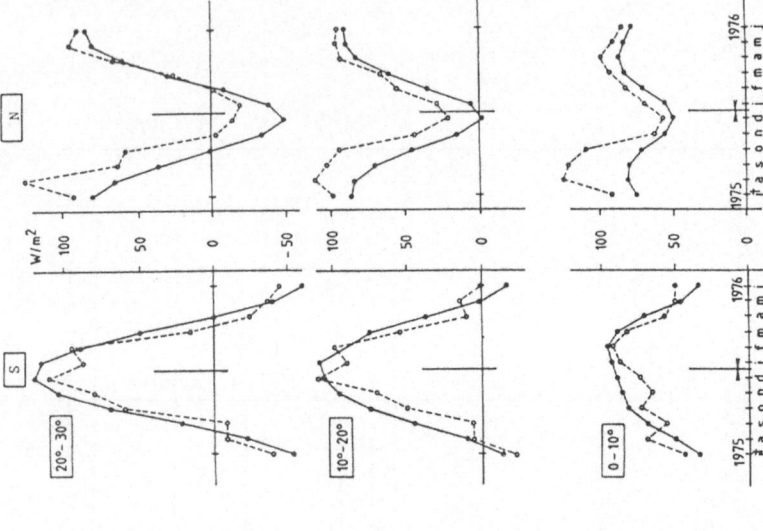

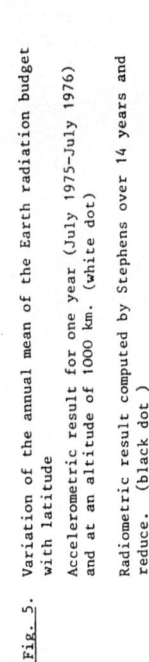

Fig. 6. Comparaison of the time variations of the Earth radiation
budget for different latitude zone between :

- Accelerometric CASTOR/CACTUS results (white dot)
- Radiometric results obtained by Stephens et al. over 14 years.
(black dot)

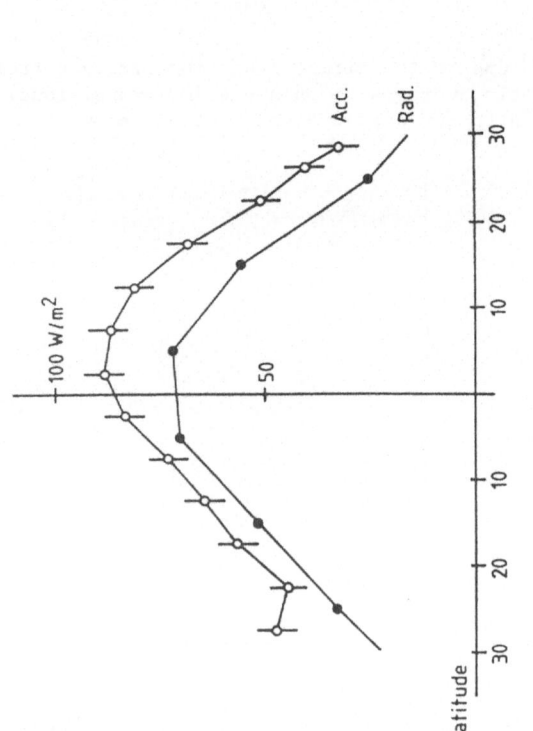

Fig. 5. Variation of the annual mean of the Earth radiation budget
with latitude

Accelerometric result for one year (July 1975-July 1976)
and at an altitude of 1000 km. (white dot)

Radiometric result computed by Stephens over 14 years and
reduce. (black dot)

LIST OF PARTICIPANTS

AUBRY, M.
Department E.O.E. MST
Ministère Recherche et Technologie
1, rue Descartes
F - 75 500 PARIS

BENZI, R.
Centro Scientifico IBM
Via Giorgione 129
I - ROMA

BERGER, A.L.
Université Catholique
B - LOUVAIN LA NEUVE

BONNEFILLE, R.
Lab. de Géologie du Quat.
CNRS - Université de Luminy
F - 13 288 MARSEILLE Cédex 9

BOURDEAU, P.
D-G Science, Research and Develop.
200, rue de la Loi
B - 1040 BRUXELLES

BRUEHL, Ch.
Max Planck Institut für Chemie
Saarstrasse, 21 (POB 3060)
D - 6500 MAINZ·

CANTU, V.
Via Trevignano, 8
I - 00062 BRACCIANO

DANSGAARD, W.
Haraldsgade, 6
DK - 2200 COPENHAGEN

DAY, W.
Rothhamsted Experimental Station
UK - HARPENDEN, Herts

DUPLESSY, J.C.
CFR / CNRS
F - 91 190 GIF-SUR-YVETTE

FANTECHI, R
D-G Science, Research and Develop.
200, rue de la Loi
B - 1040 BRUXELLES

FISCHBACH, U.
Inst. f. Physikal u. Theoret. Chemie
Universität Frankfurt / Main
D - 6000 FRANKFURT

FLOHN, H.
Meteorological Inst. of the Univers.
D - 53 BONN 1

FRENZEL, B.
Botanisches Institut
D - 7000 STUTTGART 70

GASCARD, J.C.
CNRS / Museum National
43, rue Cuvier
F - PARIS

GAUDRY, A.
CNRS / CFR
F - 91 190 GIF-SUR-YVETTE

GEURTS, H.
KNMI
NL - DE BILT

GHAZI, A.
D-G Science, Research and Development
200, rue de la Loi
B - 1040 BRUXELLES

GILCHRIST, A.
Met. Office
UK - BRACKNELL

GORMANN, V.O.
Nat. Board for Science & Technol.
IR - DUBLIN

HASSELMANN, K.
Max Planck Inst. f. Meteorologie
D - 2000 HAMBURG

HEKSTRA, G.P.
Ministerie VROM
Postbus 439
NL - 2260 LEIDSCHENDAM

HINZPETER, H.
Max Planck Inst. f. Meteorologie
D - 2000 HAMBURG

JONES, M.B.
Botany Department
Trinity College
IR - DUBLIN 2

KATSOULIS, B.
Met. Institute
National Observatory of Athens
GR - ATHENS

KILLWORTH, P.D.
Dept. of Applied Mathematics
Silver Street
GB - CAMBRIDGE CB3 9EW

KOK, C.J.
KNMI,
NL - DE BILT

KOOMANOFF, F.
Director Carbon Dioxide Research
Division DOE - ER12
USA - 20 545 WASHINGTON DC

LALAS, D.P.
National Observatory of Athens
(8 Wayne State University, USA)

LAVAL, K.
LMD E.N.S.
24, rue Lhomond
F - PARIS

LIETH, H.
FB5 Biologie Univ. Osnabrück
D - 4500 OSNABRUECK

MALCHER, J.
Met. Inst. Frankfurt University
D - FRANKFURT

MARACCHI, G.
Inst. Agronomia
Università di Firenze
I - FLORENCE

MERLIVAT, L.
LGI - DPC - CEN SACLAY
F - 91 190 GIF-SUR-YVETTE

MITCHELL, J.
Meteorological Office
UK - BRACHNELL, Berks

NICOLIS, C.
Inst. d'Aéronomie Spatiale de Belg.
3, av. Circulaire
B - 1180 BRUXELLES

NIHOUL, J.
Université de Liège
Inst. de Physique B5
B - LIEGE

O'KANE, J.P.
Dept. of Civil Engineering
University College, Earlsfort Terrace
IR - DUBLIN

PAEPE, R.
Kwartairgeologie
Vrije Universiteit Brussel
Pleinlaan, 2
B - 1050 BRUSSELS

PALMER, T.N.
Met. Office
UK - BRACKNELL

PONS, A.
Université A-4 Marseille
Faculté des Sciences St. Jérôme
F - MARSEILLE

PRODI, F.
Fisbat - CNR
Via de' Castagnoli 1
I - 40 126 BOLOGNA

PUGH, D.
Nerc
UK - SWINDON

RAYNAUD, D.
LGGE / CNRS B.P. 96
F - 38 402 ST. MARTIN D'HERES   Cédex

RENNER, V.
Deutscher Wetterdienst
Frankfurter Str. 135
D - 6050 OFFENBACH

REYNAUD, L.
Lab. Glaciologie / CNRS
F - GRENOBLE

ROTONDO' P.P.
D-G Information, Market and Innovation
L - 2920 LUXEMBOURG

ROZELOT, J.P.
CERGA
Avenue Copernic
F - 06 130 GRASSE

SCHNELL, Ch.
Dornier System
D - FRIEDRICHSHAFEN

SCHOENWIESE, C.D.
Met. Inst. Frankfurt University
D - FRANKFURT

SCHUURMANS, C.G.E.
KNMI
NL - DE BILT

SLINGO, A.
U.K. Met. Office
UK - BRACKNELL

SPERANZA, A.
Fisbat - CNR
I - 40 126 BOLOGNA

SPITZY, A.
Geol. Inst. Univ. Hamburg
Bundesstrasse 55
D - 2000 HAMBURG 13

TANNHÄUSER, I.
Bundesministerium f. Forschung
und Technologie
D - 53 BONN

THOMPSON, N.
Meteorological Office - METO SA
London Road
UK - BRACKNELL Berks RG12 2SZ

TRICOT, C.
Université Catholique de Louvain
B - 1348 LOUVAIN LA NEUVE

TURON, J.L.
Dept. Géologie et Océanographie
Avenue des Facultés
F - 33 405 TALENCE Cédex

VAN CAMPO, M.
Université des Sc. et Techn. du
Languedoc
F - MONTPELLIER

VAN DER MERSCH, I.
Université Catholique de Louvain
Chemin du Cyclotron
B - LOUVAIN LA NEUVE

VAN ENGELEN, A.
KNMI
Kon. Wilhelminalaan, 10
NL - DE BILT

WATTS, W.A.
Trinity College
IR - DUBLIN 2

WIGLEY, T.
Climatic Res. Unit,
University of East Anglia,
UK - NORWICH

LIST OF AUTHORS

ASCENCIO-PARVY, J.M. ; (247)

BARNOLA, J.M. ; (240)
BAUER, H. ; (219)
BERGER, H. ; (16, 77)
BOUDON, Y. ; (341)
BOURDEAU, Ph. ; (V)

CLAUSEN, H.B. ; (45)
CLERC, J. ; (28)
COUTEAUX, M. ; (28)

DAANSGAARD, W. ; (45)
DAHL-JENSEN, D. ; (45)
DAY, W. ; (287)
DE BEAULIEU, J.L. ; (28)
DUPLESSY, J. ; (8, 28)
DUPRAT, J. ; (28)

FARMER, G. ; (97)
FLOHN, H. ; (61)

GASCARD, J.A. ; (207)
GAUDRY, A. ; (247)
GILCHRIST, A. ; (10)
GUNDESTRUP, N. ; (45)

HAMMER, C.U. ; (45)
HASSELMANN, K.H. ; (10, 172)
HATZIOTIS, M.E. ; (113)
HEISE, E. ; (219)
HINZPETER, H. ; (258)
HUS, J. ; (113)

JALUT, G. ; (28)
JINGXING, L. ; (113)
JONES, M.B. ; (294)

KILLWORTH, P.D. ; (195)
KOOMANOFF, F.A. ; (334)

LABEYRIE, L. ; (28)
LAMBERT, G. ; (247)
LAVAL, K. ; (151)
LIVADITIS, G. ; (113)

MARACCHI, G. ; (309)
MARIOLAKOS, I. ; (113)
MELESHKO, V.P. ; (324)
METTOS, A. ; (113)
MIGLIETTA, F. ; (309)
MITCHELL, J.F.B. ; (228)
MOYES, J. ; (28)

NIHOUL, J.C.J. ; (163)

OGILVIE, A.E.J. ; (97)

PAEPE, R. ; (113)
PESTIAUX, P. ; (77)
PFAENDTNER, J. ; (219)
PONS, A. ; (28)
PUJOL, C. ; (28)

RAYNAUD, D. ; (240)
REILLE, M. ; (28)
RENNER, V. ; (219)
ROZELOT, J.P. ; (3, 339)

SABATIER, R. ; (28)
SABOT, V. ; (113)
SCHNELL, Ch. ; (270)
SCHOENWIESE, C.D. ; (16)
SCHUURMANS, C. ; (23)
SLINGO, A. ; (139)
SPERANZA, A. ; (132)

THOREZ, J. ; (113)
TURON, J.L. ; (28)

VAN CAMPO, M. ; (28)
VAN OVERLOOP, E. ; (113)
VANHOORNE, R. ; (113)

WATTS, W.A. ; (8, 101)
WEBER, K.-H. ; (61)
WIGLEY, T.M. ; (97)
WILSON, H. ; (139)